T0304015

Data Science for Engineers

With tremendous improvement in computational power and availability of rich data, almost all engineering disciplines use data science at some level. This textbook presents material on data science comprehensively, and in a structured manner. It provides conceptual understanding of the fields of data science, machine learning, and artificial intelligence, with enough level of mathematical details necessary for the readers. This will help readers understand major thematic ideas in data science, machine learning and artificial intelligence, and implement first-level data science solutions to practical engineering problems.

This book:

- Provides a systematic approach for understanding data science techniques
- Explains why machine learning techniques are able to cross-cut several disciplines.
- Covers topics including statistics, linear algebra and optimization from a data science perspective.
- Provides multiple examples to explain the underlying ideas in machine learning algorithms
- Describes several contemporary machine learning algorithms

This textbook is primarily written for undergraduate and senior undergraduate students in different engineering disciplines including chemical engineering, mechanical engineering, electrical engineering, electronics and communications engineering for courses on data science, machine learning and artificial intelligence.

Data Science for Engineers

Raghunathan Rengaswamy

Resmi Suresh

CRC Press
Taylor & Francis Group
Boca Raton London New York

CRC Press is an imprint of the
Taylor & Francis Group, an **informa** business

First edition published 2023
by CRC Press
6000 Broken Sound Parkway NW, Suite 300, Boca Raton, FL 33487-2742

and by CRC Press
4 Park Square, Milton Park, Abingdon, Oxon, OX14 4RN

CRC Press is an imprint of Taylor & Francis Group, LLC

© 2023 [Raghunathan Rengaswamy and Resmi Suresh]

ISBN: 978-0-367-75426-6 (hbk)
ISBN: 978-1-032-40549-0 (pbk)
ISBN: 978-1-003-35358-4 (ebk)

DOI: 10.1201/b23276

Typeset in CMR10 font
by KnowledgeWorks Global Ltd.

Publisher's note: This book has been prepared from camera-ready copy provided by the authors.

This book is an outgrowth of the course "Mathematical Foundations of Data Science" that I have taught several times to an inter-disciplinary group of students at IIT Madras and the several other courses that I have delivered through our IIT Madras incubated start-up GITAA Pvt. Ltd. A course called "Data Science for Engineers" has been offered in the IIT Madras NPTEL platform several times, and some of the material that is presented in this book was prepared toward delivery of this course. I thank all the students who participated in these courses and who through their questions helped me learn and appreciate the subject better. I have had the good fortune of guiding several excellent graduate students throughout my career. My research groups at IIT Bombay, Clarkson University, Texas Tech, and IIT Madras have been critical to my career progression and learning post my PhD; I thank all of them. I certainly do have the best set of faculty colleagues, collaborators, and friends at IIT Madras. I want to particularly thank Shankar Narasimhan, who has been an intellectual sparring partner for many years. I am grateful to all the students who checked the problems for errors and suggested other corrections to the book. Finally, I thank my wife Suchitra and children Abhishek and Aadarsh and my extended family for being there for me at all times. My mom and dad, who are not there anymore, will always be there in my thoughts.

Raghu

Being a beginner in the field of data science, working on this book was a great learning experience. Special thanks to Prof. Raghu, my advisor and co-author of this book, for always believing in me and motivating me from the time I started working with him. His confidence in my abilities has always been my strength. Developing problems and solutions that convey the concepts in the simplest way possible was very challenging. Feedback from various lectures that I have taught for GITAA Pvt. Ltd. and for my course at IITG on data science has really helped me to revise the book. I would like to thank students from my research group at IITG and students from the SENAI group at IITM for their help in reviewing this book and for their valuable suggestions. Special thanks to Sukanya and Srimathi for all the help and for putting up with my deadlines. Finally, I would like to thank my dearest family and friends for their affection, encouragement, and care throughout.

Resmi

Contents

Preface

With tremendous improvement in computational power and availability of rich data, almost all engineering disciplines are caught in the data science (DS) tsunami. Technical developments in the area of DS are happening at a break-neck speed. In the current scenario, there is a cacophony of information on DS, machine learning (ML), and artificial intelligence (AI). There are several claims and counter-claims that are made regarding the effectiveness of these approaches. A beginner learner would find all of this bewildering and difficult to comprehend. Further, the mathematical background that is needed to understand this field seems daunting. In the midst of all this, the reality is that it is imperative that all engineering students have a good understanding of this field, much like how almost all engineering students are expected to have a reasonable knowledge and grounding in mathematics. While there are multiple resources that are available for learning this field, they are generally not organized to suit a first-level course in an undergraduate engineering curriculum. Most of these resources are targeted at professionals to enable them to reskill and quickly transition into this domain.

The aim of this book is to present material on DS comprehensively and in a structured manner such that this book can be used in an undergraduate engineering classroom. The intent is to introduce students to basic concepts in DS, ML, and AI. The orientation is toward providing a conceptual understanding of the field with a sufficient level of mathematical detail required for beginners. At the end of the course, students will be expected to understand major thematic ideas in DS, ML, and AI and be able to implement first-level DS solutions to practical engineering problems of interest.

One of the important aspects of DS is that it cuts across several disciplines. This would necessarily mean that there are core topics that are largely domain agnostic (much like mathematics, statistics, and so on) and these ideas can be isolated from domain specific details. As a result, the material is generalized and couched in a mathematical framework.

The first two chapters in this book provide an introduction to the field, discuss key ideas, and explain the differences between DS, ML, and AI. Further, reasons why ML algorithms are so effective are also described in a conceptual fashion. There are some critical ideas that allow ML algorithms to be applicable across several disciplines. One of these is that the type of problems that ML algorithms solve can be largely bucketed into just two categories: function approximation and classification. Additionally, through the notion of feature engineering, problems in different domains can be connected to these two types of problems. These two important concepts are explained through several examples in the second chapter. Other ways in which ML algorithms are classified, such as, supervised, unsupervised, parametric, and non-parametric are also detailed in this chapter.

At this point it will become clear that three topics—linear algebra, optimization, and statistics—are of critical importance for a thorough understanding of ML and AI techniques. Three chapters are devoted to these in this book; one for each of these topics. Material covered from these vast topics is chosen based on their direct relevance to the understanding of the fundamental underpinnings of DS, ML, and AI. In keeping with this approach, material presented will also have a decidedly DS and ML flavor.

This is followed by two chapters where several ML algorithms are described in a simple manner using all the concepts introduced in the previous chapters. In summary, this book is targeted at a beginner learner. This book will demystify the areas of DS, ML, and AI and will allow the reader to think of these areas as a cohesive whole and not as fragmented algorithms. Further, the most important mathematical concepts—linear algebra, optimization, and statistics—that underlie these fields are presented in a simple, understandable manner, mainly from the viewpoint of relevance to the field. There are two ways in which the material presented in this book can be used in an undergraduate classroom. One approach is to use this book in two courses in separate semesters where the first part of this book leading up to optimization is taught in the first course as a purely mathematical foundations course, and a second course follows that up with all of the algorithms and case studies. In this case, the basic mathematical techniques can be taught in great detail and all the algorithms can be covered comprehensively. If the whole material has to be covered in a single semester, then the instructor has to shorten the mathematical foundational aspects and also pick a few of the algorithms to illustrate key ideas in ML and AI.

A quick note about the writing style adopted in this book. The material is presented in a largely conversational style that we believe is suited to the current generation of readers. DS, ML, and AI are very rich fields and the mathematical topics that underlie these areas are vast and quite sophisticated. This book could also have been written very formally in a theorem-proof format. Since this book is targeted at a first-level undergraduate course and the material to be covered is considerable, detailed proofs and background material have not been included. However, mathematical rigor is maintained and all the required background material are organically introduced in each of the chapters. As a result, there are no appendices in the book.

Authors Bio

Raghunathan Rengaswamy is the Marti Mannariah Gurunath Institute Chair Professor, Dean of Global Engagement, and a core member of the Robert Bosch Center for Data Science and AI (RBC-DSAI) at IIT Madras. He is a co-founder and Director of three IITM incubated companies. Raghu's work is in systems engineering, data science, ML, and AI techniques. His work in these areas has resulted in more than 140 international journal papers, one textbook, two US patents, several conference papers, and presentations. His work has been well cited and scores of students have gone through his MOOC courses: "Data Science for Engineers" and "Python for Data Science." He has received awards for his research: Young Engineer Award for the year 2000 awarded by INAE, and the Graham faculty research award at Clarkson University in 2006. He has also received teaching awards: Omega Chi Epsilon professor of the year award at Clarkson in 2003, and Dr. Y.B.G. Varma award for teaching excellence at IIT Madras in 2018. He was elected a fellow of Indian National Academy of Engineering in 2017.

Resmi Suresh is an Assistant Professor in the Department of Chemical Engineering at IIT Guwahati. She earned her BTech from NIT Calicut and her PhD from IIT Madras. Before joining IITG in 2019, she did her post-doctoral research at Columbia University. She has a background in process systems engineering. Her recent work has focused on the application of data science and systems engineering in various engineering systems, particularly in the energy sector.

1 Introduction to DS, ML, and AI

Tremendous improvements in computational power and improved access to data of all variety has led to considerable excitement and, at the same time, pressure to derive value out of the existing data infrastructure. This trend is not specific to any particular discipline; rather this is trans-disciplinary. From an engineering viewpoint, the emergence of internet of things (IOT) is an important contributor to improvements in the quantity and quality of data that can be collected. Data storage technology has become affordable and hence, enormous amount of data are being stored. Further, big data frameworks are enabling computationally effective means of utilizing stored data. Additionally, popular programming platforms, such as Python, are open-source, easy to use, and rich in the availability of algorithms. Furthermore, unprecedented growth of new-age technologies are likely to further accelerate the adoption of Machine Learning (ML) and Artificial Intelligence (AI) techniques. Consequently, it has become imperative that all engineers be trained in the burgeoning fields of Data Science (DS), ML, and AI, not only at a graduate level, but also at an undergraduate level. An understanding of DS algorithms and an ability to apply these algorithms is now a requirement for every practicing engineer. In this chapter, basic definitions and concepts in the fields of DS, ML, and AI are explored to enable the reader to generate an initial understanding of these topics.

1.1 DEFINITIONS OF DS AND ML

We start by defining the term data science. The most appropriate definition is "Data science is the science of deriving insights from data". An important component of this definition is the word "insights". Insights imply some derived knowledge that goes beyond mere presentation of data. Supposing one were presented with yields of a certain product from two reactors (A and B) for different batches. Without any further analysis, this is mere data. However, if one were to calculate the averages of these yields and verify that average yield of reactor A is better than reactor B, then one would have derived the insight that reactor A is operating better than reactor B. The notion of average, a statistical concept, is used to derive this insight. One could further drill down and identify the reasons for this yield improvement and ascribe a reason such as better reactor design or better catalyst has been used and so on. Such additional insights would require assessing domain specific assumptions and in some cases, generation of further data. Two key points that emerge

DOI: 10.1201/b23276-1

out of this discussion are that mathematical and statistical tools are required to derive insights, and assumptions underlie insights that are derived. The notion that assumptions are the bedrock for not only generating insights but also in the development of ML and AI techniques is an important one and we will revisit this idea several times throughout this book. In the example that was described, statistics was used as the tool for DS; however, ML, AI algorithms or any other tool such as first principles modeling can also be used to derive insights. If one were to use such an overarching definition for DS, then it becomes clear that engineers have been working in the field of DS from time immemorial.

ML is a field that has captured the imagination of all engineers. There are a number of interpretations and definitions for ML in literature. It is not uncommon for a single concept to have multiple definitions in a growing field. For the purposes of this book, we will define ML as the "Science of mimicking human learning". Interestingly, there are definitions in the literature that describe ML as learning without explicit programming and so on. We will eschew such romantic view of ML and interpret this field as simply an attempt at mimicking human learning. This definition will serve us well, particularly when we want to discuss the field from the viewpoint of engineering applications. In general, we include several aspects as belonging to human learning. For example, we learn to: distinguish between different situations, predict how long it might take to reach a destination, play games, become proficient at languages, identify our dislikes, acquire knowledge, recognize competencies, and so on. Comprehensively defining all aspects that have to be included in human learning itself is a daunting task. At the same time, one cannot work with abstract or all-encompassing notions of learning if the aim is a systematic understanding of the field given our current limited understanding. To this end, in Figure 1.1, we decompose human learning into three components that can each be understood towards an understanding of ML. We will describe each of these components presently.

This view is particularly useful for us to later categorize and understand ML techniques, acquire insights into why these approaches work so well, and why they are, in some sense, universally applicable. One could also view this decomposition as a framework to formalize various aspects of learning. As a caveat, we do not yet have a clear idea of a comprehensive set of modules needed to mimic all aspects of human learning. Hence, we do not claim our definition of ML to be complete in any sense.

1.2 WHAT IS LEARNT IN ML ALGORITHMS?

ML algorithms learn models that allow them to reason and make decisions. Models can be used in function approximation or classification tasks. In a function approximation problem, given data, which are categorized as inputs and outputs, the aim is to represent the outputs as functions of inputs. There are several examples of such learning in the engineering domain. For example, one

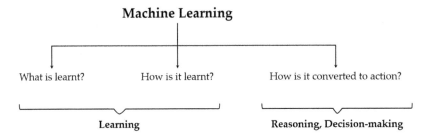

FIGURE 1.1 Components in machine learning.

could learn a function that captures how yields change as a function of various reactor designs and operating parameters; learn how efficiency changes with respect to inputs and operating conditions in an energy conversion equipment and so on. Mathematically, the aim is to write, $y = f(x_1, \ldots, x_n, p_1, \ldots, p_m)$. In this equation, f is a function that is identified and it is parameterized by $p_1, \ldots, p_m$. If there are multiple outputs, there will be one function for each of the outputs. There are several important clarifications that need to be made here. First, one does not know, *a priori*, the function that will model the data adequately. Further, the function f may be realized in multiple ways. The realization could be parametric, non-parametric, a tree-like structure, as rules, as a graph, and so on. Popular models can be classified as either parametric or non-parametric.

The other type of problem is the classification problem. In a classification problem, given data partitioned into different classes, the aim is to derive a decision function that can help assign classes to new datapoints. Let us consider a simple binary classification problem, where data has to be classified into one of two pre-defined classes. In this case, ML algorithms learn a decision function as below:

$$h(x_1, ..., x_n, p_1, ..., p_m) \geq 0 \text{ if data belongs to class 1}$$

$$h(x_1, ..., x_n, p_1, ..., p_m) < 0 \text{ if data belongs to class 2}$$

There are several problems related to classification in engineering domain. One could classify if a motor is working normally or not, a plant is operating normally or not and so on. Engineers are able to look at data and make these assessments intuitively. An implicit mental model that is developed allows one to perform this activity. The expectation is that ML algorithms will capture this mental model in a digital form.

1.3 HOW DOES LEARNING HAPPEN IN ML?

The second component in Figure 1.1 relates to how these models are learnt, which will include both an appropriate function and the parameters in the function. As an example, a linear function is $f = ax + b$, with parameters a and b. If ML has to be generally applicable then ML algorithms should be able to learn any function (shown as the universal set of functions in Figure 1.2) from which data has been generated. Modeling data that could be generated from any function in a universal set of functions is not feasible; hence, ML algorithms model the data using candidate functions that are chosen from a bag of functions as shown in Figure 1.2. This choice could be made in a trial-and-error fashion; once a function is chosen the parameters that are best suited to represent the data are identified such that the model (function and parameters together) provides an adequate representation of the data. The quality of approximation depends on both the actual function and the candidate function that is used. One could evaluate a large number of candidate functions for this approximation. Another way of realizing this approximation is to restrict the number of candidate functions to a set of functions called the basis functions and build a structure using these functions to approximate the true function. In this case, for the best approximation, one would search for the best structure.

At every step of the iteration, for a candidate function, the parameter values that provide the best fit for the data are identified. This would require an optimization problem to be solved. Usually optimization problems are solved using multiple iterations where the solution is gradually improved. This process is called learning in ML literature and the equation that modifies the parameter values so that the model fit improves is called the learning rule. As an example, backpropagation is one popular learning rule that is used in training neural networks.

Further, there are two approaches to learning that are possible: supervised and unsupervised learning. In supervised learning, values for the target that is being learnt or class labels are provided as a part of the learning process. In unsupervised learning, target values or class labels are not available while training. Supervised learning is also called dissimilarity-based learning and unsupervised learning is called similarity-based learning. At this point, it might be difficult for readers unfamiliar with this topic to fathom how one can learn without being told what they are learning. Nonetheless, this will be further explained and clarified in Chapter 2.

1.4   DECISION-MAKING IN ML

The third component in Figure 1.1 is how learning is converted into actions. Many of the popular ML models by themselves do not address this question; however, this is an important component in our understanding of human learning. When we say that someone learns to bicycle, we do not mean that they just have a mental model of how to ride a bicycle but they can actually use a bicycle for transporting themselves. In general, once a model is identified, it has to be utilized to take some decisions. This would require definition of an objective for the action. In the example of cycling, the objectives might be to

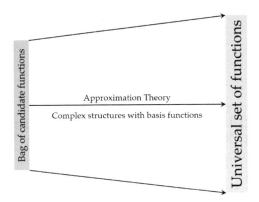

FIGURE 1.2 Approximating a true function.

go fast, not fall down, and so on. The decisions are body position with respect to the cycle, rate of leg movement, application of break, etc. Decisions that optimize the chosen objective need to be identified. When we attempt to mimic human learning, this is also solved as an optimization problem and becomes a part of the ML solution. In contrast, there are also ML techniques where it might be hard to separate these aspects (learning a model and decision making) as distinct components.

In some cases, the optimization problem will turn out to be combinatorial and tree searches might be employed to make decisions. An excellent example of this is how some traditional ML techniques play games such as chess. In chess, players make moves so that their board position in the game improves. This automatically means that there must be some mental model that converts a board position to a notional value. This is a function approximation problem, where the input to the function is the board position. Let us assume that this functional form has been learnt. In order to use this functional form to make a decision, an ML algorithm might search a tree as shown in Figure 1.3. This figure shows a tree which has been unraveled to depict one move of the player and a move by the opponent in response to the move by the player. It is easy to see that each node in the tree represents a board position and hence a value could be associated to each node. At this point, rules or algorithms that take several factors into account could be used to identify the move that a player should make based on the unraveled board positions and their corresponding values. While the picture depicts only two levels of the tree, powerful computers could search multiple levels down this tree. While this can also be viewed as solving an optimization problem, *i.e.,* finding a move that maximizes board value, this is a slightly different approach to optimization than a standard gradient-based approach (discussed in Chapter 4). With these components (what, how, reasoning, and decision-making), most of the ML techniques can be well-explained, particularly for engineering problems.

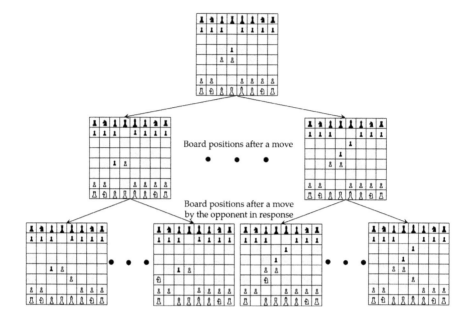

FIGURE 1.3 Tree search for chess.

1.5 DISCUSSION ON AI

We now come to a discussion on AI. If one were to look at all the literature (archival, online, social media, and so on), there would be multiple viewpoints and interpretations, many of them blurring the lines between ML and AI. This is not surprising as there are several gray areas that can be labeled as either ML or AI. We make a slight modification to our definition of ML to distinguish ML and AI. AI is the science of mimicking human intelligence. We will use a slightly broader notion of human intelligence, which of course also includes human learning abilities. From a philosophical viewpoint, human intelligence could include several aspects such as leaps of insight, meta-cognition, notion of consciousness, and so on. These terms are also discussed when one talks about artificial general intelligence (AGI). There are other fields such as emergence that might have a strong bearing on how AGI might be achieved. However, it is important to note that this expansive view of AI is still developing and there is no consensus on how to achieve AGI, leave alone established methods for achieving AGI, or when (and if) it will be achieved.

From an engineering viewpoint, the traditional notion of AI and ML are quite different. The focus of traditional AI (ideas from a couple of decades ago) was on knowledge representation, reasoning, and decision-making. The notion of data driving AI systems was not prevalent. Multiple ideas were explored for knowledge representation such as: expert rules, qualitative differential equations, ontologies, petri-nets, bond graphs, and so on. Similarly, reasoning and search strategies are also well-studied. Some of these are breadth or depth first search, and A-star algorithm. The first successful AI systems were the expert systems, where knowledge was represented through rules, and reasoning and decision-making activities were achieved through search of these rules, and conflict resolution when multiple rules become applicable to a given situation. These were followed by AI systems that used different types of models as described earlier for a variety of applications. Qualitative simulation, for example, simulates and predicts the behavior of the system through qualitative differential equations built using different ontological assumptions. Qualitative process theory is another such approach in AI for modeling physical systems. While currently not in the mainstream of AI research, one would expect some of these ideas to be hybridized with purely data learning approaches, when the limitations of purely data driven AI/ML become an impediment to further progress.

The technology that blurs the difference between ML and AI the most is neural networks. Neural networks can be quite easily viewed as ML algorithms in the sense that they perform function approximation and classification tasks and require a large amount of data for their successful application. The most recent work on neural networks, the deep networks, in popular notion straddles ML and AI. If one were to look at current literature, deep neural networks and AI are being used synonymously. There are multiple reasons for this. First, problems that traditional AI techniques have been attempting to solve in fields such as image processing and natural language processing have been solved in a spectacular fashion by deep neural networks. Since it is believed that AI concepts are needed to solve this problem and deep networks have achieved remarkable success, deep networks are thought of as mainstream AI techniques. Further, as described before, considerable focus of traditional AI techniques was on representation schemes. While originally neural networks were treated as black-boxes (and to a large extent they are treated as such even now), successful deep network architectures such as convolution neural networks have an in-built structure that can lead to interpretations vis a vis knowledge representation. Convolution neural networks have convolution layers, followed by what are called pooling layers, followed by full-connected networks and so on. If one were looking at image recognition problems, then spatial convolution can be thought of as representing local features. Further layers can be thought of as a making decisions (such as classifying pictures and so on). In such a view of neural networks, the two important components of AI (knowledge representation and reasoning) are both embedded in the structure leading to the idea that deep neural networks are AI systems.

In summary, we have provided definitions for DS, ML, and AI that highlight both the commonalities and differences between these ideas. Considerable success has been achieved in all of these fields in the last few years. In particular, great strides have been made in natural language processing and image processing systems using neural network systems and architectures. ML techniques have become standard and are being used in several engineering applications. A framework to understand these areas was described in this chapter. In Chapter 2 of the book, we will expand on these ideas through the prism of the framework described till now.

1.6   WHY ARE ML AND AI TECHNIQUES EFFECTIVE?

In this section, we will explain the reason for this sudden and tremendous renewal in interest in the fields of ML and AI. Clearly, the potential of these technologies has been understood for a long time but the results were not commensurate with the expectations. The manner in which we mimic human learning has a large computational cost associated with it and computational power a couple of decades back was not at a scale that allowed meaningful progress to be made. Due to the dramatic improvements in computational capability, largely driven by Moore's law, ML algorithms are able to solve realistic engineering problems now. In some cases, these techniques perform much better than earlier simply because of this computational power. Assume that we are given data for two variables and are asked to identify a relationship between these variables. In this case, one could simply visualize the data in a 2D plot and identify if there is an underlying relationship between the variables. This could be done analytically also. If the number of datapoints is small and if one were to evaluate only a small number of candidate functions, then this is an easy problem to solve. One might set-up a least squares objective and identify the best fit parameters for the candidate functions, and choose the function that best fits the data. However, consider a scenario where there are hundreds of variables, millions of sample points, and the objective is to identify if there are relationships between these variables. At this scale it becomes simply difficult to make progress using traditional techniques.

In other words, when the number of variables and quantity of data scale-up, uncovering underlying relationships between the variables becomes an incredibly complex problem. Additionally, since the underlying functions that capture the relationships between the variables are unknown, the number of candidate functions that need to be evaluated will also scale-up for a reasonable solution to be identified. All of these are simply beyond human capabilities. ML algorithms, on the other hand, with increased computational power at their disposal and solid mathematical ideas assisting them, can surmount the challenges of exploding data, variables, and options. This is depicted in Figure 1.4.

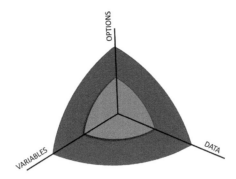

FIGURE 1.4 Effectiveness of ML techniques.

It is not hard to imagine that when a ML algorithm uncovers a complex relationship between multiple variables and uses that to make decisions, it might feel like pure magic to us. Nonetheless, it is important to note that, while impressive, the ability of ML algorithms are largely a result of solid mathematical foundational ideas, most notably from linear algebra, statistics, and optimization.

2 DS and ML – Fundamental Concepts

In this chapter, we expand upon the fundamental ideas in ML that were introduced in the previous chapter. A typology of ML techniques is depicted in Figure 2.1. ML techniques can be viewed from the perspective of the problems that they solve, the type of model that is built, and also the training philosophy. Further, utilizing the notion of feature engineering, we show how to convert a conceptual problem in any domain into a concrete problem solvable using ML algorithms. This allows one to appreciate the generality of ML algorithms. A variety of problems that ML algorithms solve across domains are then described.

Subsequently, ML algorithms will be conceptualized as tools to visualize multi-dimensional data. This viewpoint allows one to understand ML algorithms through the prism of the assumptions that underlie these algorithms; this is useful for a deep understanding of ML algorithms. While the problems solved using ML are bucketed into two categories, it is important to understand that DS problems, as posed originally, might not directly fit into this categorization. However, in many cases, the original problem can be broken down into sub-problems that are identified as function approximation or classification problems. This idea leads to the distinction between an algorithm and a solution and is summarized through a conceptual framework for developing DS solutions.

2.1 CLASSIFICATION AND FUNCTION APPROXIMATION

Classification and function approximation are two types of problems that ML algorithms predominantly solve. In classification problems, a set of labeled data are provided and the aim is to develop a decision function that can identify the correct label, given new data. The labels provide the class information. The simplest problem type is the binary classification problem, where all the data are tagged as belonging to either Class-1 or Class-2. Multi-class and multi-dimensional problems are the other types. Multi-class classification problems are characterized by a single label for a datapoint with the label drawn from a set: Class-1, Class-2, ..., Class-k. Multi-dimensional classification problems have more than one label for each datapoint with the labels drawn from different sets. Multi-class classification problems can be viewed as a special case of multi-dimensional problems. An example of a binary classification problem is one where an image is being classified as a cat or a dog. A multi-class classification problem is one where process data is being classified

DOI: 10.1201/b23276-2

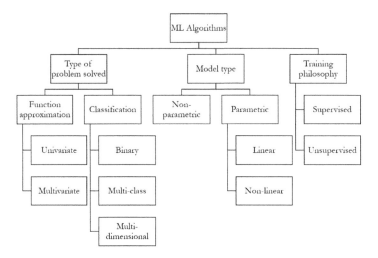

FIGURE 2.1 Typology of ML algorithms.

as being *normal*, *fault*$_1$, ..., *fault*$_k$, and a multi-dimensional classification problem is one where each of the fault classes are further categorized into several levels of severity.

Figures 2.2a and 2.2b pictorially depict a binary classification problem in two variables. In these figures, data are distributed in a two-dimensional plane. Each datapoint is a vector with two components $[x_1, \ x_2]^T$. There are two classes to which the data can belong, identified using different shapes in the figure. The aim of an ML algorithm is to arrive at a mathematical function (decision function) of the variables and parameters that will allow one to decide if a datapoint belongs to a particular class or not. A linear decision function will suffice for the example shown in Figure 2.2a. On the other hand, for the case shown in Figure 2.2b, a nonlinear decision function is required.

Figures 2.2c and 2.2d illustrate a function approximation problem in two variables. The data are again distributed in a two-dimensional plane with each datapoint being characterized by two variables $[y \ u]^T$. In the classification case, each datapoint was assigned a label and the variables themselves were assumed to be independent. In contrast, in the function approximation case, one assumes that some of the variables are output or dependent variables ($\mathbf{y}$) and the other variables are input or independent variables ($\mathbf{u}$). The aim here is to relate the output variable(s) as a mathematical function of the input variable(s). The problem shown in Figure 2.2c is an univariate regression problem, and in this case, a linear function will capture the relationship between y and u. However, for the data shown in Figure 2.2d, a nonlinear function will be required.

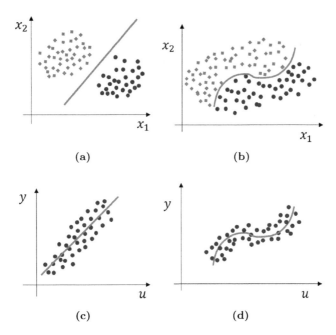

FIGURE 2.2 Linear and nonlinear classification and function approximation.

There are other interesting geometric aspects that can be noticed from Figure 2.2. In classification problems, a decision function from which the data
belonging to different classes are as far away as possible is identified. In contrast, in a function approximation problem, the data are spread close to the
function approximator. In classification, variables need not be characterized
as inputs and outputs, whereas in many function approximation problems this
distinction is usually drawn. It is also possible to solve function approximation
problems without this distinction, which is described later in the book.

While it is easy to decide if a linear or a nonlinear function is needed for
classification and function approximation for data in two dimensions, it is
not possible to make this decision, visually, for higher dimensional problems,
with a large number of classes or outputs. ML algorithms learn the required
functional form automatically.

2.2 MODEL FORMS

In Chapter 1, ML techniques were discussed from the viewpoint of what is
learnt and how it is learnt. What is learnt are functional forms that can be
used to predict or classify. These functional forms are also interchangeably
described as models. Any model form would have parameters that have to be

identified. Learning, to a large extent, refers to identifying appropriate values for these parameters. At a higher level, a functional form that is appropriate for the particular problem at hand could also be learnt. In this section, a brief overview of the different model forms are provided.

Models can be classified as parametric or non-parametric. Parametric models are functions characterized by parameters, for example, $f(x_1, x_2, \ldots, x_n, p_1 \ldots, p_m)$, where the model is parameterized by the parameters $p_1, \ldots, p_m$. A multivariate linear function $a_1 x_1 + \ldots + a_n x_n + b$ is an example of this form with parameters $a_1, \ldots, a_n, b$. Non-parametric models, on the other hand, are not explicit functions but predict or classify using all (or part) of the data. To illustrate this, consider a datapoint $(x_1, \ldots, x_n)$, for which an output (y) value needs to be predicted. In a parametric approach, the values of the **x** variables will be substituted in a model to predict y. There are multiple approaches for prediction using non-parametric models. One possible non-parametric approach is to identify the k neighbors of **x** using an Euclidean distance measure and predict y as the average of all the neighbors' outputs. This approach can be used to predict the output of any new data. This is called the k-NN approach as we will see later.

There are some obvious differences between these two approaches. In the parametric approach, a model form needs to be specified, while explicit analytical functional forms are not required in non-parametric models. Further, in the parametric approach, the extracted parameters are sufficient for future predictions and in the non-parametric approaches, data is required for future predictions.

Another viewpoint is that the models are either global or local representations of data. This is an interesting viewpoint from an ML perspective. This idea can be explained by a simple one dimensional example. Assume that we have data for one independent (x) and one dependent (y) variable and a model of the form $y = f(x)$ needs to be built. Let us also assume that the range is $-10 \leq x \leq 10$. If a model $f(x) = a + bx + cx^2$ is built with the parameter values identified from all the data, this is a global model for the data. If for the same problem, the data is split into four regions based on x values and a model is built for each region, then this would lead to localized representation of data. Each of the individual models are valid only in their local regions; however, finer features in localized regions can be captured by these models at the cost of not providing global insights. Global models, on the other hand, might provide more general insights about the problem, and result in a compact representation while compromising on the identification of finer features.

2.3 TRAINING PHILOSOPHY

Training of ML models can be realized through supervised or unsupervised training paradigms. Supervised learning is also called similarity-based learning and unsupervised learning is called dissimilarity-based learning. It is easy to explain supervised and unsupervised learning from the viewpoint of classification problems. Consider the problem of identifying fraudulent credit card transactions. If prior analysis has been done, the transaction data can be provided with a label with two possibilities, "authentic" or "fraudulent". When we build a classifier using this data we call that supervised learning. The term "supervised" refers to the fact that already identified labels are used while learning. Unsupervised learning is where the same problem is solved without the knowledge of labels for the data. The algorithm has to label and learn the authenticity of the transactions without any prior analysis, hence the name unsupervised learning. Approaches for unsupervised learning will be described in detail in Chapters 6 and 7. The notion of unsupervised learning is valid for both classification and function approximation problems. Function approximation algorithms that are unsupervised in nature will also be described later.

2.4   GENERALITY OF DATA SCIENCE/ML SOLUTIONS

An interesting aspect of DS/ML is that these ideas cross-cut several disciplines. Unlike traditional disciplines which have very well defined boundaries, DS/ML techniques are being applied in all domains such as engineering, social sciences, natural sciences, and so on. In fact, the algorithm development is largely domain agnostic. It is worthwhile to understand how this separation of the algorithm development and domain of application is possible, while the universality of the algorithms is still retained. Two important ideas are critical for the universality of ML algorithms. First, a large number of problems in various domains can be classified into just two categories of problems: function approximation and classification. It is not to be misconstrued that every problem is immediately recognizable as one of these two categories (it is another matter that many of these can be), but if one were to break-down a problem statement through a series of logical decompositions, the sub-problems that result will largely be in these two categories. Solving these sub-problems and consolidating these solutions within the decomposition framework will solve the original problem. The second important concept is that problems in various domains can be converted into function approximation and classification problems using the notion of feature engineering. Put together, these two ideas make ML algorithms universally applicable. In this section, we will discuss these two ideas. Table 2.1 depicts problems in several domains and categorizes them as either a classification or a function approximation problem.

2.4.1   Examples of Function Approximation Problems

Function approximation problems have already been introduced in this chapter. In this section, we will describe function approximation problems described in Table 2.1 in a little more detail.

TABLE 2.1

Problems in Different Domains and Categories

Problem	Type
Predicting materials property for different chemicals	Function Approximation
Predicting scores in a cricket match	Function Approximation
Predicting mechanical properties of a part	Function Approximation
Predicting the value of a board position in a game of chess	Function Approximation
Fraud detection in credit card transactions	Classification
Distinguishing objects - Self-driving cars	Classification
Detecting failures in equipment	Classification
Classifying emails	Classification

2.4.1.1 Predicting Materials Property for Different Chemicals

Consider the problem of predicting properties of interest, such as glass transition temperature for polymers. There might be considerable data available for glass transition temperatures of different polymers. ML algorithms can be trained using this data to predict glass transition temperature of a new polymer structure for which prior data does not exist.

2.4.1.2 Predicting Scores in a Game of Cricket

Another interesting example is from a sport that is passionately followed in India, cricket. In a game of cricket, at the end of the game, the team that scores more runs than the other is the winner. The total number of runs that a team will score can be predicted at any point in the game.

2.4.1.3 Predicting Mechanical Properties of a Part

Another interesting problem in the field of mechanical engineering is the prediction of properties of an assembled part. An example is prediction of the strength (breaking point) of a part when a force is applied. Strength data available for different existing designs can be used to predict part strengths for new designs.

2.4.1.4 Predicting Value of a Board Position in Chess

Tremendous improvements in AI technologies have resulted from analyzing and replicating how humans play games. A prime example of this is the interest in machines playing chess. Progress made over the years has resulted in chess playing programs that humans can't beat anymore. An AI program playing chess will be required to numerically evaluate the value of any board position that it encounters.

2.4.2   Examples of Classification Problems

Similar to the previous function approximation examples, many problems in disparate disciplines can be solved as classification problems. In this section, a few of these (Table 2.1) are described.

2.4.2.1   Fraud Detection in Credit Card Transactions

Millions of credit card transactions are made every day. While a large majority of them are authentic, a small minority of them are fraudulent transactions. ML algorithms can be trained to classify credit card transactions as either "authentic" or "fraudulent".

2.4.2.2   Distinguishing Objects—"Self-driving Cars"

One of the problems in "self-driving" cars is the identification of different objects that are encountered. This is a classification problem with a large number of classes (as opposed to the binary problem described for credit card transactions).

2.4.2.3   Detecting Failures in Built Systems/Equipment

Classification problems abound in the engineering domain. Failure detection and diagnosis problems are of relevance in all engineering domains. Civil engineers aim to build bridges that do not collapse. Similarly, chemical engineers endeavor to build reactors that will never explode. This requires that engineered systems are continuously monitored so that tell-tale signatures that predict potential problems are quickly recognized and addressed.

2.4.2.4   Classifying Emails

Our email boxes are populated with mostly relevant emails but interspersed with spam emails. Algorithms that are able to recognize spam emails and remove them automatically from the inbox would be of considerable value. This is a classification problem in the natural language processing domain.

2.4.3   Feature Engineering as a Connector Between Domain and Data Science/ML

While many problems can be categorized as either function approximation or classification problems, this in itself does not make ML algorithms universally applicable in a domain agnostic fashion. There has to be some universal connector that maps problems in different domains into ML recognizable problems. Feature engineering is the connector that makes this transformation possible. Feature engineering converts the conceptual problem into a data matrix as described below that is utilized by ML algorithms. This idea will be demonstrated using two examples from the previous discussion.

Consider the function approximation (FA) problem for polymer property prediction. Property data for multiple polymers is assumed to be available as shown in Table 2.2(a), where the polymer is identified by a chemical formula. The chemical formula is usually a string consisting of several characters (such as CH_4). While this is clearly a FA problem, the input available for the problem (the string) is unlikely to be informative enough for a model to be constructed; more explanatory features need to be generated as inputs.

The features chosen should characterize the polymer in multiple ways, providing information for predicting the property of interest. Some example features are: 1D descriptors such as structural fragments and chemical groups, 2D descriptors such as angles and graph properties, 3D descriptors such as volume, steric properties and finally, derived descriptors from first principles calculations. The features can be categorical or quantitative (real numbers). Structural fragments are categorical and angles are real numbers. Table 2.2(b) depicts this conversion from the original problem to the feature engineered problem with n datapoints.

TABLE 2.2

Function Approximation Example — Property Prediction

    (a) Original FA problem   (b) Feature engineered FA problem

Formula (F)	Property (P)
M_1	P_1
$\vdots$	$\vdots$
M_n	P_n

F	1DD	2DD	3DD	DD	P
M_1	$1D_1$	$2D_1$	$3D_1$	d_1	P_1
$\vdots$	$\vdots$	$\vdots$	$\vdots$	$\vdots$	$\vdots$
M_n	$1D_n$	$2D_n$	$3D_n$	d_n	P_n

Table 2.3 describes the original problem and the feature engineered problem for credit card transaction classification. Similar to the property prediction case, mere transaction ID as shown in Table 2.3 (a) would not be enough to perform classification. More meaningful information about each of these transactions is required. Table 2.3 (b) describes possible informative inputs such as the card holder gender, type of purchase, cost of purchase, time of purchase, and place of purchase. These features collectively build a digital purchase profile for customers, deviations from which can be used to detect potentially fraudulent transactions.

In summary, domain problems are converted to ML problems through the development of an input data matrix (X) using feature engineering. The X

TABLE 2.3

Classification Example—Credit Card Transactions

<table>
<tr><td colspan="2" align="center">(a) Original
classification problem</td><td colspan="7" align="center">(b) Feature engineered
classification problem</td></tr>
</table>

Transaction (T)	Authentic (A)
T_1	Y/N
$\vdots$	$\vdots$
T_n	Y/N

T	Gender	Type	Time	Cost	Place	A
T_1	M/F	Ty_1	Ti_1	C_1	P_1	Y/N
$\vdots$	$\vdots$	$\vdots$	$\vdots$	$\vdots$	$\vdots$	$\vdots$
T_n	M/F	Ty_n	Ti_n	C_n	P_n	Y/N

matrix will have as many rows as samples and as many columns as features that are generated. In a supervised classification problem, each row is annotated with a label. In function approximation, some of the features are identified as outputs and others as inputs and models relating the inputs to outputs are identified. The features are of various data types depending on the domain and the information that the feature represents.

2.5 DATA CLASSIFICATION

Data can be categorized into several types and each datatype allows only certain mathematical operations. It is important to understand operations that are appropriate for different datatypes. The first type is the nominal data. Nominal data is predominantly used for labeling, searching, sorting, and so on. Examples are: an ID for each row of data in a matrix (row1, row2, ...), jersey number in sports (23 in basketball and 10 in soccer have special significance) and so on. Mathematical operations are not performed on these variables; however, searches could be based on such variables. Using jersey numbers in basketball, a statement such as 23 > 10 does not make sense since jersey numbers represent players. However, useful information can still be gleaned from these variables based on a domain understanding. In soccer, we know that the best player in the team typically wears jersey number 10.

The second type is the ordinal data. Ordinal data provides a notion of order; hence these variables can be ordered in an ascending or descending order. For an ordinal temperature variable that takes values "hot", "hotter" and "hottest", "hot < hotter < hottest" is a valid statement. Rank of a student in a class is another ordinal variable. While ordering is possible with ordinal variables, subtraction is not appropriate. The difference between ranks 1 and 2 is not equivalent to difference between ranks 2 and 3. Variables that allow ordering and interval comparison are called interval variables. Continuous temperature measurement is an interval variable where ordering and subtraction are appropriate. Statements such as today is 2 degrees hotter

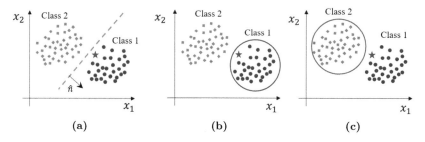

FIGURE 2.3 ML as a tool to understand multi-dimensional data: Classification problems.

than yesterday are meaningful. However, division is not appropriate for such variables as origin is not defined. Consequently, statements such as today is two times hotter than the other day do not make sense. Interval variables for which an origin is defined are called ratio variables. The salary that a person draws is an example of a ratio variable. It would make sense to say that someone draws x times more or less salary than another person.

While the datatypes described above have been used in management literature quite extensively, from an ML perspective, data can be simply viewed as either categorical or numerical. Categorical data can be viewed as nominal or ordinal and this characterization will inform the algorithms of how these variables are to be treated. Similarly, numerical data is classified as discrete or continuous. Discrete variables are ones which take values from a countably finite set of values, whereas continuous variables can take values from an infinite set of numbers.

2.6 VIEWING ML ALGORITHMS AS TOOLS TO UNDERSTAND MULTI-DIMENSIONAL DATA

Another viewpoint that is extremely useful in understanding ML algorithms is to conceptualize them as tools to visualize multi-dimensional data. This notion also helps one appreciate the importance of understanding the assumptions that underlie the ML algorithms. Consider a situation where data for two classes in two dimensions is provided. The data can be visualized in a 2D plot, marking all the data corresponding to class 1 as circles and all the data corresponding to class 2 as squares as shown in Figure 2.3.

A new point shown as a "star" will be classified as belonging to class 1 in all three cases shown in Figures 2.3a, 2.3b, and 2.3c. A line separating the two classes (linear), and circles as shown in Figures 2.3b and 2.3c are some of the possible decision boundaries. Mathematically, if for Class-1, $h(\mathbf{x}) \geq 0$, then $h(\mathbf{x})$ for the three cases are of the form $a_1 x_1 + a_2 x_2 + b_1$, $r_1^2 - (x_1 - c_1)^2 - (x_2 - c_2)^2$, $(x_1 - c_3)^2 + (x_2 - c_4)^2 - r_2^2$ respectively (with appropriate values for $a_1, a_2, b_1, r_1, r_2, c_1, c_2, c_3, c_4$).

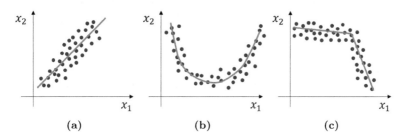

FIGURE 2.4 ML as a tool to understand multi-dimensional data: Function approximation problems.

Figure 2.4 shows data for three different cases in a function approximation problem where x_2 and x_1 are dependent and independent variables, respectively. In the first case, it is clear that a linear model would suffice but in the other two cases, a linear model would be insufficient. It can be seen that a quadratic model would work well in the second case and two piecewise linear models might be an option for the third case. This kind of detailed analysis is possible with 2 or 3 dimensional data; however, such analysis is not possible for larger dimensional data. Generating insights through multiple 2-D visualizations is akin to many blind men characterizing an elephant based on the region of the elephant that they are exploring.

Since multi-dimensional data cannot be effectively visualized, an assumption-validation-refinement cycle can be explored to understand the data. The process is the following. First, certain assumptions about the multi-dimensional data are made, and a corresponding ML technique that has been developed based on similar data assumptions is chosen. The data is then analyzed using the chosen ML technique and the model performance is verified. Acceptable performance of the chosen ML model is an indication that the assumptions that underlie the ML algorithm match the hypothesized assumptions about the data. Unacceptable performance leads to the conclusion that the assumptions driving the ML algorithm do not characterize the data well. This will lead to a revision of assumptions and another choice of an ML technique. Convergence of this process leads to characterization of multi-dimensional data, with insights analogous to actual visualization being generated. This viewpoint allows one to approach ML algorithms as an overarching framework for characterizing and deriving insights from multi-dimensional data.

In summary, ML algorithms where the underlying assumptions resonate with data characteristics will result in excellent characterization of data by the model. This highlights the idea that the superiority of one ML algorithm over another cannot be established purely based on performance in application studies. The foregoing discussion does not imply that ML algorithms cannot be compared. There are several appropriate metrics for comparison of ML algorithms such as generality (how many different data characterizations does an ML structure address), accuracy (statistical characterization of error rates), robustness, and computational tractability.

2.7 A FRAMEWORK FOR SOLVING DATA SCIENCE PROBLEMS

Until now, the main ideas that allow ML algorithms to solve problems in different domains were described. Data science solutions using ML algorithms have several other components to them. It is important to appreciate this difference between data science solutions and ML algorithms. Most data science problems start as a concept or an idea. For example, while there might be data available, some of the data could be missing and a data science solution that utilizes this imperfect data might have to be developed for a specific application. Off-the-shelf ML algorithms will usually not exist to solve such problems. A data science solution to this problem will require several sub-problems to be solved; some of these sub-problems will be solvable directly using ML algorithms. Another example from sports is where one is tasked with identifying whether luck played a role in the outcome of a game. Again, this problem cannot be directly solved using an off-the-shelf ML algorithm. Rather, a structured data science solution to this problem will have elements that can be solved using ML. Hence, a solution and an algorithm have to be viewed differently. A solution will consist of several inter-connected elements, each of which will incorporate requisite algorithms as needed. A rational framework for developing data science solutions is shown in Figure 2.5. The various components in the framework will be explained through a simple example. The remaining chapters of the book will describe the various components that are alluded to in this section. Readers may revisit this section after completing five chapters of this book.

2.7.1 Data Imputation

The data imputation problem that we describe here is very common for data that comes from engineering domain. This problem will be used to explain the data science problem solving framework depicted in Figure 2.5.

2.7.1.1 Start: Problem Arrival

A data scientist has been told that there is considerable data for several variables but the dataset might be incomplete. The engineering team asks the data scientist to explore if a completely curated dataset could be returned back to them for further analysis.

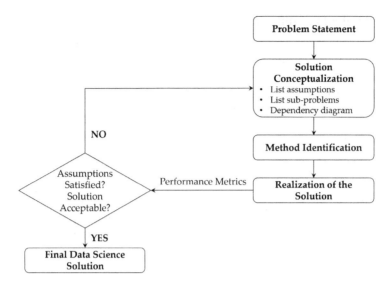

FIGURE 2.5 Framework for data science problems.

2.7.1.2 Problem Statement

The data scientist inspects that data and wants to formalize the problem into a mathematical framework. Consider a problem where data (m samples) for n variables of interest are provided. The data can be assembled into a data matrix A with m rows and n columns. Now the mathematical problem is, given a matrix A, which is not completely filled, how does one fill or impute entries that are missing? This crisp problem statement converts the original problem statement into a mathematical statement.

2.7.1.3 Solution Conceptualization

In this step, various elements of the solution are conceptualized. Imputation of the unknown values will require assumptions about the underlying data. Two possible assumptions are: variables are related through a model, or each variable is a random variable characterized by a statistical distribution. In the first case, relationships need to be identified and in the second case, a distribution needs to be assumed. The assumptions can be further refined by specifying linear relationships in the first case and normal distribution (refer Chapter 5) in the second case. This completes the solution conceptualization. Methods consistent with the assumptions to impute the missing data are identified next.

2.7.1.4   Method Identification

Method identification depends on the assumptions made. Under the assumption of linear relationships and no error, the rank (r) of the data matrix can be computed. There are $n - r$ relationships that can be identified (Chapter 3). As long as the number of missing datapoints in each row is less than $n - r$, the identified relationships can be used to impute data. If $r = n$, this method is not feasible as there are no relationships that can be used to impute data. If one assumes that there are linear relationships and there is also noise in the data, then a principal component analysis (described in Chapter 6) can be used to impute values for the missing data.

   If all the variables are assumed to come from the same distribution, then all the data can be combined together for the computation of statistics. An average value for all the variables can be computed and every missing value can be imputed with this data. Another possibility is to impute with the median value identified from the dataset.

2.7.1.5   Solution Realization

Solution realization refers to developing computer codes to implement a solution algorithm. Any programming platform such as Python, R or Matlab could be used. The data science solution to the problem will comprise of several modules of code. A logical flow for the data imputation problem is provided below.

(a) Import data
(b) Characterize missing data (% of missing data, rows where there are no missing data and so on)
(c) Partition original data into complete and incomplete sets
(d) Build model on a sub-selection of complete set
(e) Impute values for remaining data of the complete set and validate
(f) Impute data for the missing fields for the incomplete set

2.7.1.6   Assess Assumptions

Assessing assumptions is a critical part of the data science framework. One approach to assessing assumptions is to split data into training and testing data, build models with the training data and assess the model predictions on test data. In this particular case, both the training and testing data should come from samples that are complete (with no missing values). Another approach to assessing the overall solution in this case is to use the model built from imputed data (imputed complete data) for a downstream application and compare the performance against a reference solution where a model built only from the originally complete dataset is used. If the model from imputed data does better, then the imputation solution has value.

2.7.1.7 Validate-Revise-Assess Cycle

In the framework, irrespective of the approach used to assess the solution, the process is complete if the assessment returns positive validation. If the assessment leads to the conclusion that the solution is not appropriate (or not of required quality), then the whole cycle is repeated starting with a new set of assumptions or modifications of the already made assumptions. A good data scientist will be able to quickly identify a reasonable set of assumptions for a solution of desired accuracy by learning from each iteration and tweaking assumptions appropriately.

2.8 CONCLUSIONS

The aim of this chapter was to describe several ideas in the fields of DS, ML, and AI, demystifying some of the terminology in the process. Various model forms and training strategies that are used in ML algorithms were described. The universality of ML algorithms was described as being due to two important ideas. One is that many problems or sub-problems that are encountered in engineering domains can be bucketed into two problem categories: classification and function approximation. Second, the generality as far as several domains is concerned comes from the notion of feature engineering. This is the process of identifying key inputs that provide enough information for the underlying problem to be solved. This feature engineering process converts problems in various domains to a unified format that can be solved using generic classification and function approximation algorithms. We explained this idea through two examples, one for classification and one for function approximation. Further, the notion of ML as a tool for visualizing multi-dimensional data was also introduced. This is an important viewpoint because this allows one to more closely inspect and understand the assumptions that underlie each of these algorithms. As a result, one could objectively evaluate these algorithms on an "apples-to-apples" basis.

A framework for developing data science solutions to problems in different domains was also presented. This is in general a somewhat unstructured activity and considerable creativity comes into play in terms of solution strategies. If two data scientists approach the same problem, it is very unlikely that they will both identify the same solution strategy. However, one would still need a framework to structure the data science solution process. One such framework is provided and discussed in detail using a data imputation example.

3 Linear Algebra for DS and ML

The aim of this chapter is to introduce the core ideas of linear algebra that are relevant for data science. Linear algebra is a vast field and cannot be covered in all its detail in a single chapter. Hence, the most important concepts from linear algebra that are useful in the field of data science have been identified and discussed, focusing on how they may be used for data science. Although linear algebra can be treated very theoretically and formally, in this chapter, key ideas are explained in a simple fashion without being too formal. However, relevant mathematical analysis is provided as and when appropriate. Data representation is an important aspect of data science and data is represented usually in the form of matrices. We will discuss this representation and various matrix concepts in Section 3.1. For data containing several variables, one would like to know how many of these variables are really important and if there are relationships between these variables. Further, if relationships exist between variables, how does one uncover them? Linear algebraic tools allow us to understand and answer these questions as discussed in sections 3.3 and 3.4.1. Finally, several ideas that are relevant for ML will be outlined.

3.1 MATRICES AS A CONCEPT FOR DATA ORGANIZATION

Typically, a matrix has a rectangular structure with rows and columns. In general, rows are used to represent samples and columns to represent the variables or attributes in the data. This is the convention that we will use in this book. Although there are many ways in which data can be organized, the row-column organization is a very intuitive and useful representation. Matrices can also be used to represent linear equations in a compact and simple fashion. If used to represent linear equations, components in the matrix are the coefficients of variables in these equations. For example, if there are two equations $\beta_{11}x_1 + \beta_{12}x_2 = y_1$ and $\beta_{21}x_1 + \beta_{22}x_2 = y_2$, they can be represented in the matrix form as $B\mathbf{x} = \mathbf{y}$ where $B = \begin{bmatrix} \beta_{11} & \beta_{12} \\ \beta_{21} & \beta_{22} \end{bmatrix}$, $\mathbf{y} = \begin{bmatrix} y_1 \\ y_2 \end{bmatrix}$ and $\mathbf{x} = \begin{bmatrix} x_1 \\ x_2 \end{bmatrix}$. Linear algebra provides tools to understand and manipulate matrices to derive useful knowledge from data.

Consider reactor operational data obtained from various sensors such as pressure, temperature, density, and viscosity sensors. Let us assume that sensors generated 1000 independent samples of Pressure (P, Pa), Temperature (T, K), Density $(\rho, kg/m^3)$, and Viscosity $(\eta, Pa\text{-}s)$. This data can be organized in a matrix as shown below.

DOI: 10.1201/b23276-3

$$
\begin{array}{c}
\\
1 \\
\vdots \\
1000
\end{array}
\begin{array}{cccc}
P & T & \rho & \eta \\
(Pa) & (K) & (kg/m^3) & (Pa\text{-}s) \\
\left[\begin{array}{cccc}
300 & 400 & 1000 & 200 \\
\vdots & \vdots & \vdots & \vdots \\
500 & 450 & 1200 & 350
\end{array}\right]
\end{array}
$$

Here, the first column corresponds to the values of pressure at different sample points. Similarly, the second and third columns correspond to the values of temperature and density at several sample points. In this matrix representation, each column represents a variable and each row represents a sample. For example, in the first sample, you will read that the value of pressure was 300 Pa, the value of temperature was 400 K, and the values of density and viscosity were 1000 kg/m^3 and 200 Pa-s, respectively.

While an engineering example was described above, matrix representation is very important for many applications. Consider how a computer might store data about pictures, such as the image in Figure 3.1. One way to represent this image in a computer is to represent it as a matrix of numbers. A small region in the picture can be broken down into multiple pixels and numbers from a range (black to white) assigned for each pixel. The matrix shown in Figure 3.1 represents a small region of the original picture. A larger matrix will represent the whole image. Storing a matrix with pixel values is equivalent to storing the image. Such a representation can be of use in image analysis applications. Although the rows and columns of the pixel matrix do not represent samples and variables as before, the rectangular structure remains the same. Each element in the matrix represents the pixel value at a corresponding location in the picture. To summarize, data matrices could be from various sources: from sensors, pixel values of images, or coefficients of variables in linear models.

FIGURE 3.1 Representing pictures as matrices in a computer.

3.2 MATRIX VIEW OF LINEAR ALGEBRA

Once a data matrix is assembled, some of the questions that can be posed from a data science perspective are:

a) Are all the attributes in the data matrix independent?
b) Are there linear relationships among variables—how does one identify these relationships?
c) If such relationships exist, how does one use this information in generating insights?
d) Can relationships between variables be leveraged for data compression?

The answers to these questions will be mapped to linear algebra concepts in this chapter.

Remark. *In general, when one talks about linear relations, both affine ($y = \beta x + c$) and strictly linear relations ($y = \beta x$) are considered. However, many of the linear algebra concepts are described using strict linearity. This is not really a restriction as any affine relation can be converted to strictly linear relation with the help of mean-centering. For example, let the relation between variables of interest be $y_1 = \beta x_1 + c$. Denote the mean value of y_1 by $\bar{y}_1$ and that of x_1 by $\bar{x}_1$. Consequently, $\bar{y}_1 = \beta \bar{x}_1 + c$. Subtracting $\bar{y}_1$ from y_1, we have*

$$y_1 - \bar{y}_1 = \beta(x_1 - \bar{x}_1) \tag{3.1}$$

If we denote $y_1 - \bar{y}_1$ by y_1' and $x_1 - \bar{x}_1$ by x_1', then the relationship is strictly linear, i.e., $y_1' = \beta x_1'$. In data-science applications, as a part of pre-processing, data is generally mean-centered first. Hence, all the concepts discussed in this chapter can directly be applied on the pre-processed data.

3.2.1 Rank of a Matrix

Rank of a matrix is defined as the number of linearly independent rows or columns of the matrix [42]. Consider the matrix $\begin{bmatrix} 1 & 2 & 1 \\ 2 & 4 & 0 \\ 3 & 6 & 0 \end{bmatrix}$ where each column represents a variable. It can be seen that the second column is twice the first column; the second variable is dependent on the first variable or vice versa. Thus, this matrix has two independent columns (or variables) and one dependent column (or variable). The rank of this matrix is 2. Rank of the data matrix is also the number of independent variables in the data.

Rank of a matrix that is less than the number of columns implies that some variables are dependent on others. This allows one to work with a reduced set of variables (also referred to as attributes). Assuming that there are many more samples than variables, rank of the matrix will identify the number of independent variables. Let the rank of the matrix with n variables and m samples be r. This automatically means that there are r independent variables and consequently, $n - r$ linear relations among variables. It is important to note that rank identifies only the number of independent variables, however, it does not identify which ones are independent. A method to identify the $n - r$ relationships is required.

Example 3.1. *Find the rank of the matrix,* $X = \begin{bmatrix} 2 & 3 \\ 4 & 5 \end{bmatrix}$.

Solution 3.1. *As the rows and columns are linearly independent (row 1 cannot be generated from row 2 using simple scalar multiplication or division or addition or subtraction), rank of the matrix X is 2.*

Example 3.2. *Find the rank of the matrix,* $X = \begin{bmatrix} 2 & 3 \\ 4 & 6 \end{bmatrix}$.

Solution 3.2. *As the second column is 1.5 times the first column, there is only one linearly independent column. Hence, the rank of the matrix X is 1.*

For matrices larger than dimension 2, it is difficult to identify the existence of linear relations by just looking at the numbers. A systematic approach to find the rank of the matrix is to convert it into a row echelon form. A matrix is in the row echelon form when the following conditions are satisfied.

a) The first non-zero entry (also called the pivot) in any row is to the right of the non-zero entry (pivot) in the previous row
b) All zero rows are below non-zero rows

Through a series of row operations with rearrangement of rows as needed, any matrix can be converted to its row echelon form. The number of non-zero rows in the row echelon form is the rank. The row operations can be formalized by a method called Gaussian elimination. This procedure is methodologically described below

1. Start with the first row.
2. If the row under consideration has all zeros, shift that row to the end and move to the next row. Continue this process until a non-zero row is identified. If there are no more rows terminate the procedure.
3. If the row under consideration is the last row terminate the procedure.
4. For any row under consideration, start with the diagonal entry.
5. If the entry is non-zero, using this pivot and row operations, convert all the elements in this column, in all the rows below, to zero. Continue from 8.
6. If the entry is zero, identify a non-zero entry in a row below (in the same column) and swap the rows. Perform operation 5 and continue from 8.
7. If the entry is zero and all the elements below are zero, consider the next element in the row. Continue from 5.
8. Proceed to the next row and continue from 2.

The example below illustrates the procedure for finding the rank of a matrix, by converting it to a row echelon form.

Example 3.3. *Find the rank of the matrix,* $X = \begin{bmatrix} 1 & 0 & 3 \\ 10 & 1 & 7 \\ -5 & 0 & -3 \end{bmatrix}$

Solution 3.3. *Elementary transformations to obtain the row echelon form for the given matrix are described below:*

a) $R_2 \leftarrow R_2 - 10R_1 \implies X_1 = \begin{bmatrix} 1 & 0 & 3 \\ 0 & 1 & -23 \\ -5 & 0 & -3 \end{bmatrix}$

b) $R_3 \leftarrow R_3 + 5R_1 \implies X_2 = \begin{bmatrix} 1 & 0 & 3 \\ 0 & 1 & -23 \\ 0 & 0 & 12 \end{bmatrix}$

It can be observed that matrix X_2 satisfies the first condition and is in the row echelon form. Since there are three non-zero rows, the rank of the matrix is 3.

The row echelon form that was used to understand the rank of the matrix can be used to answer the other data science questions that were posed. The number of non-zero rows represent the number of independent samples. This is because the samples corresponding to the zero rows can be written as a linear combination of the samples corresponding to the non-zero rows—this is precisely the reason why we could get the zero rows through row operations in the first place. In other words, if one were given the r samples from the X matrix that correspond to the non-zero rows in the modified matrix, all the other samples (or rows) in X can be written as a linear combination of the r rows. This is a useful result for data science because if one were given just the r rows and the constants to be used in the linear combinations then there is no need to store the remaining $m - r$ rows—they can always be reconstructed.

Example 3.4. *Find the rank of the matrix* $X = \begin{bmatrix} 1 & -1 & 3 \\ 10 & -10 & 18 \\ -5 & 1 & -3 \end{bmatrix}$

Solution 3.4. *Elementary transformations to obtain the row echelon form for the given matrix are described below:*

a) $R_2 \leftarrow R_2 - 10R_1 \implies X_1 = \begin{bmatrix} 1 & -1 & 3 \\ 0 & 0 & -12 \\ -5 & 1 & -3 \end{bmatrix}$

b) $R_3 \leftarrow R_3 + 5R_1 \implies X_2 = \begin{bmatrix} 1 & -1 & 3 \\ 0 & 0 & -12 \\ 0 & -4 & 12 \end{bmatrix}$

c) *Since the second diagonal element is zero, based on step 5 of the algorithm swap rows 3 and 2 to get* $X_3 = \begin{bmatrix} 1 & -1 & 3 \\ 0 & -4 & 12 \\ 0 & 0 & -21 \end{bmatrix}$

It can be observed that matrix X_3 satisfies the first condition and is in the row echelon form. Since there are three non-zero rows, the rank of the matrix is 3.

Example 3.5. *Find the rank of the matrix,* $X = \begin{bmatrix} 1 & -1 & 3 \\ 10 & -10 & 18 \\ -5 & 5 & -3 \end{bmatrix}$

Solution 3.5. *Elementary transformations to obtain the row echelon form for the given matrix are described below:*

1. $R_2 \leftarrow R_2 - 10R_1 \implies X_1 = \begin{bmatrix} 1 & -1 & 3 \\ 0 & 0 & -12 \\ -5 & 5 & -3 \end{bmatrix}$

2. $R_3 \leftarrow R_3 + 5R_1 \implies X_2 = \begin{bmatrix} 1 & -1 & 3 \\ 0 & 0 & -12 \\ 0 & 0 & 12 \end{bmatrix}$

3. *Since the second diagonal element is zero, based on steps 5 and 6 of the algorithm move to the next element in the second row.*

4. $R_3 \leftarrow R_3 + R_2 \implies X_3 = \begin{bmatrix} 1 & -1 & 3 \\ 0 & 0 & -12 \\ 0 & 0 & 0 \end{bmatrix}$

It can be observed that matrix X_3 satisfies the first condition and the zero rows are at the end and hence it is in the row echelon form. Since there is one zero row, the rank of the matrix is 2.

A much richer understanding of such relationships is possible when one views row and column operations as matrix multiplications. Matrix multiplication can be viewed as manipulation of columns and rows of matrices. For any matrix product equation $C = AB$, the columns of C are linear combinations of columns of A and rows of C are linear combinations of rows of B. This is easily seen through the following example.

Example 3.6. *Find the product of matrices* $\begin{bmatrix} 1 & 2 \\ 3 & 4 \end{bmatrix}$ *and* $\begin{bmatrix} 5 & 2 \\ 3 & 1 \end{bmatrix}$ *using element-wise multiplication, row operations, column operations, and sum of matrices with column-row products.*

Solution 3.6. *Traditional matrix multiplication*

$$\begin{bmatrix} 1 & 2 \\ 3 & 4 \end{bmatrix} \times \begin{bmatrix} 5 & 2 \\ 3 & 1 \end{bmatrix} = \begin{bmatrix} (1 \times 5) + (2 \times 3) & (1 \times 2) + (2 \times 1) \\ (3 \times 5) + (4 \times 3) & (3 \times 2) + (4 \times 1) \end{bmatrix} = \begin{bmatrix} 11 & 4 \\ 27 & 10 \end{bmatrix}$$

Linear combination of columns

$$\begin{bmatrix} 1 & 2 \\ 3 & 4 \end{bmatrix} \times \begin{bmatrix} 5 & 2 \\ 3 & 1 \end{bmatrix} = \begin{bmatrix} 5\begin{bmatrix}1\\3\end{bmatrix} + 3\begin{bmatrix}2\\4\end{bmatrix} & 2\begin{bmatrix}1\\3\end{bmatrix} + 1\begin{bmatrix}2\\4\end{bmatrix} \end{bmatrix} = \begin{bmatrix} 11 & 4 \\ 27 & 10 \end{bmatrix}$$

Linear combination of rows

$$\begin{bmatrix} 1 & 2 \\ 3 & 4 \end{bmatrix} \times \begin{bmatrix} 5 & 2 \\ 3 & 1 \end{bmatrix} = \begin{bmatrix} 1\begin{bmatrix} 5 & 2 \end{bmatrix} + 2\begin{bmatrix} 3 & 1 \end{bmatrix} \\ 3\begin{bmatrix} 5 & 2 \end{bmatrix} + 4\begin{bmatrix} 3 & 1 \end{bmatrix} \end{bmatrix} = \begin{bmatrix} 11 & 4 \\ 27 & 10 \end{bmatrix}$$

Sum of matrices with column-row products

$$\begin{bmatrix} 1 & 2 \\ 3 & 4 \end{bmatrix} \times \begin{bmatrix} 5 & 2 \\ 3 & 1 \end{bmatrix} = \begin{bmatrix} 1 \\ 3 \end{bmatrix}\begin{bmatrix} 5 & 2 \end{bmatrix} + \begin{bmatrix} 2 \\ 4 \end{bmatrix}\begin{bmatrix} 3 & 1 \end{bmatrix} = \begin{bmatrix} 11 & 4 \\ 27 & 10 \end{bmatrix}$$

Example 3.7. *For two matrices of sizes 3×3 and 4×5, find respective matrices that when pre or post multiplied with the given matrices, the product matrices will have the following characteristics*

1. *The third and first rows are swapped in the product matrix (row permutation matrix)*
2. *The second and first columns are swapped in the product matrix (column permutation matrix)*
3. *The third row is a sum of all the rows but the other rows are the same*
4. *The second column is a sum of the first and second columns and the third column is a sum of the first three columns*

Solution 3.7. *The following matrices will render the product matrices with the given characteristics*

1. *Pre-multiplication;* $\begin{bmatrix} 0 & 0 & 1 \\ 0 & 1 & 0 \\ 1 & 0 & 0 \end{bmatrix}$; $\begin{bmatrix} 0 & 0 & 1 & 0 \\ 0 & 1 & 0 & 0 \\ 1 & 0 & 0 & 0 \\ 0 & 0 & 0 & 1 \end{bmatrix}$

2. *Post-multiplication;* $\begin{bmatrix} 0 & 1 & 0 \\ 1 & 0 & 0 \\ 0 & 0 & 1 \end{bmatrix}$; $\begin{bmatrix} 0 & 1 & 0 & 0 & 0 \\ 1 & 0 & 0 & 0 & 0 \\ 0 & 0 & 1 & 0 & 0 \\ 0 & 0 & 0 & 1 & 0 \\ 0 & 0 & 0 & 0 & 1 \end{bmatrix}$

3. *Pre-multiplication;* $\begin{bmatrix} 1 & 0 & 0 \\ 0 & 1 & 0 \\ 1 & 1 & 1 \end{bmatrix}$; $\begin{bmatrix} 1 & 0 & 0 & 0 \\ 0 & 1 & 0 & 0 \\ 1 & 1 & 1 & 1 \\ 0 & 0 & 0 & 1 \end{bmatrix}$

4. *Post-multiplication;* $\begin{bmatrix} 1 & 1 & 1 \\ 0 & 1 & 1 \\ 0 & 0 & 1 \end{bmatrix}$; $\begin{bmatrix} 1 & 1 & 1 & 0 & 0 \\ 0 & 1 & 1 & 0 & 0 \\ 0 & 0 & 1 & 0 & 0 \\ 0 & 0 & 0 & 1 & 0 \\ 0 & 0 & 0 & 0 & 1 \end{bmatrix}$

3.2.2 LU Decomposition

There are two operations that form the basis for the algorithm that was used to convert a matrix into a row echelon form. These are the row additions and row swaps. Since each of the row operations use a row above to generate zeros in rows below, all of these are pre-multiplications by lower triangular matrices. In Example 3.3, the first row operation is a pre-multiplication by the

following matrix $\begin{bmatrix} 1 & 0 & 0 \\ -10 & 1 & 0 \\ 0 & 0 & 1 \end{bmatrix}$. All the matrix operations in the examples can
be viewed as pre-multiplication of the given matrix by a sequence of lower tri-
angular matrices interspersed with some permutation matrices, as and when
necessary. While the permutation matrices are identified as a part of the algorithm, if one knew the required permutations a-priori, a modified matrix
PX could be subjected to row operations without the need for permutations
during the process of conversion to row echelon form. At the end of the row operations, a row echelon matrix, which is an upper triangular matrix is derived.
Mathematically,

$$L_n \ldots L_1 PX = U \tag{3.2}$$

Using the result that the product of lower triangular matrices are lower triangular $(L_n \ldots L_1 = L')$ and the inverse of a lower triangular matrix is also a
lower triangular matrix $(L'^{-1} = L))$, equation 3.2 becomes

$$PX = LU \tag{3.3}$$

This result states that any matrix X (with appropriate row rearrangements)
can be decomposed as a product of a lower triangular (L) and an upper
triangular matrix (U) [39, 3].

Example 3.8. *Find the LU decomposition of the matrix,* $X = \begin{bmatrix} 1 & 3 \\ 2 & 6 \end{bmatrix}$.

Solution 3.8. *Performing the row operations on* $R_2 \leftarrow R_2 - 2R_1$, *we have*

$$\begin{bmatrix} 1 & 0 \\ -2 & 1 \end{bmatrix} X = \begin{bmatrix} 1 & 3 \\ 0 & 0 \end{bmatrix} \implies X = \begin{bmatrix} 1 & 0 \\ -2 & 1 \end{bmatrix}^{-1} \begin{bmatrix} 1 & 3 \\ 0 & 0 \end{bmatrix} = \underbrace{\begin{bmatrix} 1 & 0 \\ 2 & 1 \end{bmatrix}}_{L} \underbrace{\begin{bmatrix} 1 & 3 \\ 0 & 0 \end{bmatrix}}_{U}$$

Example 3.9. *Find the LU decomposition of the matrix,* $X = \begin{bmatrix} 1 & 1 & 0 \\ 2 & 3 & 1 \\ 4 & 5 & 1 \end{bmatrix}$.

Solution 3.9. *Consider the matrix* $X = \begin{bmatrix} 1 & 1 & 0 \\ 2 & 3 & 1 \\ 4 & 5 & 1 \end{bmatrix}$. *Through row operations*
$R_2 \leftarrow R_2 - 2R_1$ *and* $R_3 \leftarrow R_3 - 4R_1$,

$$\begin{bmatrix} 1 & 0 & 0 \\ -2 & 1 & 0 \\ -4 & 0 & 1 \end{bmatrix} \begin{bmatrix} 1 & 1 & 0 \\ 2 & 3 & 1 \\ 4 & 5 & 1 \end{bmatrix} = \begin{bmatrix} 1 & 1 & 0 \\ 0 & 1 & 1 \\ 0 & 1 & 1 \end{bmatrix}$$

Performing the row operation $R_3 \leftarrow R_3 - R_2$, *we get an upper triangular
matrix as shown below.*

$$\begin{bmatrix} 1 & 0 & 0 \\ 0 & 1 & 0 \\ 0 & -1 & 1 \end{bmatrix} \begin{bmatrix} 1 & 0 & 0 \\ -2 & 1 & 0 \\ -4 & 0 & 1 \end{bmatrix} \begin{bmatrix} 1 & 1 & 0 \\ 2 & 3 & 1 \\ 4 & 5 & 1 \end{bmatrix} = \begin{bmatrix} 1 & 1 & 0 \\ 0 & 1 & 1 \\ 0 & 0 & 0 \end{bmatrix}$$

Simplifying,

$$\underbrace{\begin{bmatrix} 1 & 0 & 0 \\ -2 & 1 & 0 \\ -2 & -1 & 1 \end{bmatrix}}_{L_1} \begin{bmatrix} 1 & 1 & 0 \\ 2 & 3 & 1 \\ 4 & 5 & 1 \end{bmatrix} = \underbrace{\begin{bmatrix} 1 & 1 & 0 \\ 0 & 1 & 1 \\ 0 & 0 & 0 \end{bmatrix}}_{U}$$

The resultant matrix (U) is upper triangular with one row completely zero. Thus, the rank of U is 2. Since rank of X = rank of U, rank of X=2. The LU decomposition of A can be written as

$$X = \begin{bmatrix} 1 & 1 & 0 \\ 2 & 3 & 1 \\ 4 & 5 & 1 \end{bmatrix} = \begin{bmatrix} 1 & 0 & 0 \\ -2 & 1 & 0 \\ -2 & -1 & 1 \end{bmatrix}^{-1} \begin{bmatrix} 1 & 1 & 0 \\ 0 & 1 & 1 \\ 0 & 0 & 0 \end{bmatrix}$$

$$\underset{X}{\begin{bmatrix} 1 & 1 & 0 \\ 2 & 3 & 1 \\ 4 & 5 & 1 \end{bmatrix}} = \underset{L}{\begin{bmatrix} 1 & 0 & 0 \\ 2 & 1 & 0 \\ 4 & 1 & 1 \end{bmatrix}} \underset{U}{\begin{bmatrix} 1 & 1 & 0 \\ 0 & 1 & 1 \\ 0 & 0 & 0 \end{bmatrix}}$$

Example 3.10. *Find the LU decomposition of the matrix,* $X = \begin{bmatrix} 1 & -1 & 3 \\ 10 & -10 & 18 \\ -5 & 5 & -3 \end{bmatrix}$.

Solution 3.10. *Consider the matrix* $X = \begin{bmatrix} 1 & -1 & 3 \\ 10 & -10 & 18 \\ -5 & 5 & -3 \end{bmatrix}$. *By performing the row operations* $R_2 \leftarrow R_2 - 10R_1$ *and* $R_3 \leftarrow R_3 + 5R_1$,

$$\begin{bmatrix} 1 & 0 & 0 \\ -10 & 1 & 0 \\ 5 & 0 & 1 \end{bmatrix} \begin{bmatrix} 1 & -1 & 3 \\ 10 & -10 & 18 \\ -5 & 5 & -3 \end{bmatrix} = \begin{bmatrix} 1 & -1 & 3 \\ 0 & 0 & -12 \\ 0 & 0 & 12 \end{bmatrix}$$

Performing the row operation $R_3 \leftarrow R_3 + R_2$, *a row echelon form as shown below results.*

$$\begin{bmatrix} 1 & 0 & 0 \\ 0 & 1 & 0 \\ 0 & 1 & 1 \end{bmatrix} \begin{bmatrix} 1 & 0 & 0 \\ -10 & 1 & 0 \\ 5 & 0 & 1 \end{bmatrix} \begin{bmatrix} 1 & -1 & 3 \\ 10 & -10 & 18 \\ -5 & 5 & -3 \end{bmatrix} = \begin{bmatrix} 1 & -1 & 3 \\ 0 & 0 & -12 \\ 0 & 0 & 0 \end{bmatrix}$$

Simplifying,

$$\underset{L_1}{\underbrace{\begin{bmatrix} 1 & 0 & 0 \\ -10 & 1 & 0 \\ -5 & 1 & 1 \end{bmatrix}}} \begin{bmatrix} 1 & -1 & 3 \\ 10 & -10 & 18 \\ -5 & 5 & -3 \end{bmatrix} = \underset{U}{\underbrace{\begin{bmatrix} 1 & -1 & 3 \\ 0 & 0 & -12 \\ 0 & 0 & 0 \end{bmatrix}}}$$

The resultant matrix (U) is upper triangular with one row completely zero. Thus, the rank of U is 2. Since rank of X = rank of U, rank of X=2. The LU decomposition of X can be written as

$$X = \begin{bmatrix} 1 & -1 & 3 \\ 10 & -10 & 18 \\ -5 & 5 & -3 \end{bmatrix} = \begin{bmatrix} 1 & 0 & 0 \\ -10 & 1 & 0 \\ -5 & 1 & 1 \end{bmatrix}^{-1} \begin{bmatrix} 1 & -1 & 3 \\ 0 & 0 & -12 \\ 0 & 0 & 0 \end{bmatrix}$$

$$\underset{X}{\begin{bmatrix} 1 & -1 & 3 \\ 10 & -10 & 18 \\ -5 & 5 & -3 \end{bmatrix}} = \underset{L}{\begin{bmatrix} 1 & 0 & 0 \\ 10 & 1 & 0 \\ -5 & -1 & 1 \end{bmatrix}} \underset{U}{\begin{bmatrix} 1 & -1 & 3 \\ 0 & 0 & -12 \\ 0 & 0 & 0 \end{bmatrix}}$$

Example 3.11. *Find the LU decomposition of the matrix* $X = \begin{bmatrix} 2 & 3 & 1 \\ 4 & 6 & 2 \\ 4 & 5 & 1 \end{bmatrix}$.

Solution 3.11. *When we perform a row operation* $R_2 \leftarrow R_2 - 2R_1$, *the second row becomes zero. Hence it is not possible to get a zero in the second column of the third row through standard row operations. However, if one were to swap rows 2 and 3 right at the beginning using the permutation matrix*
$P = \begin{bmatrix} 1 & 0 & 0 \\ 0 & 0 & 1 \\ 0 & 1 & 0 \end{bmatrix}$, *then PX can be multiplied by the matrix* $\begin{bmatrix} 1 & 0 & 0 \\ -2 & 1 & 0 \\ -2 & 0 & 1 \end{bmatrix}$ *to get a LU decomposition*

$$\underbrace{\begin{bmatrix} 2 & 3 & 1 \\ 4 & 5 & 1 \\ 4 & 6 & 2 \end{bmatrix}}_{PX} = \underbrace{\begin{bmatrix} 1 & 0 & 0 \\ 2 & 1 & 0 \\ 2 & 0 & 1 \end{bmatrix}}_{L} \underbrace{\begin{bmatrix} 2 & 3 & 1 \\ 0 & -1 & -1 \\ 0 & 0 & 0 \end{bmatrix}}_{U}$$

The row operations that we described are essentially linear combinations of rows to form new rows. This notion of linear combinations is of critical importance and can be generalized to both rows and columns, or more formally for vectors. Vectors are numbers stacked either horizontally (row vector) or vertically (column vector). A $m \times n$ matrix comprises of m rows $\in \mathbb{R}^n$ and n columns $\in \mathbb{R}^m$. Any $\mathbb{R}^k$ essentially means that there are k numbers that represent the vector; geometrically one can think of this vector as being in k dimensions. The rows vectors in a data matrix will typically represent one sample, where all the variable values are specified and column vectors represent the value that a variable takes across all samples.

Given a data matrix, it would be useful to know the number of independent samples and variables in the data. This would require that we formally define independence of vectors. In linear algebra, the notion of linear independence is used and k vectors ($\mathbf{x_1} \ldots \mathbf{x_k} \in \mathbb{R}^m$) are linearly independent (LI), if the only solution for $\beta_1\mathbf{x_1} + \ldots + \beta_k\mathbf{x_k} = 0$ is $\beta_1 = 0 \ldots \beta_k = 0$. To understand this consider three vectors. Let us assume that a non-zero solution exists. This implies $\mathbf{x_3} = -\frac{\beta_1}{\beta_3}\mathbf{x_1} - \frac{\beta_2}{\beta_3}\mathbf{x_2}$; $\mathbf{x_3}$ is a linear combination of $\mathbf{x_2}$ and $\mathbf{x_1}$, contradicting the fact that the vectors are LI. The LI idea can also be expressed in a matrix form. If we generate a matrix X with the vectors stacked in a column, then the columns of the matrix are LI if the only solution to the matrix equation $X_{(m \times k)}\boldsymbol{\beta}_{(k \times 1)} = 0$ is the trivial solution $\boldsymbol{\beta} = \mathbf{0}_{(k \times 1)}$.

We now arrive at the idea of subspaces. One generally defines a vector space before subspaces. Interested readers can refer to [42]. However, the material that we intend to cover can proceed without a detailed definition of vector spaces for now. For understanding the data science concepts in the book, a k-dimensional vector space can be considered as a collection of all possible vectors in $\mathbb{R}^k$. It is interesting to note that vector spaces can be defined for other objects such as matrices, polynomials, and so on. A subspace is an infinite collection of vectors in a vector space with two properties – additive and multiplicative closure. Let us denote this collection by $\mathbb{V}$. If two vectors are taken from the collection and added, the resultant vector will also be in the subspace. This is called additive closure. Mathematically

$$\forall \mathbf{x_i}, \mathbf{x_j} \in \mathbb{V}, \mathbf{x_i} + \mathbf{x_j} \in \mathbb{V}$$

If a vector from the subspace is multiplied by a scalar real number, the resultant vector will also be in the subspace. This is called multiplicative closure. Mathematically

$$\forall \mathbf{x_i} \in \mathbb{V}, \beta \mathbf{x_i} \in \mathbb{V}$$

Since multiplicative closure is true for any scalar, the zero vector is always a part of any subspace.

The notion of basis vectors allows one to characterize a subspace compactly. While there may be infinite vectors in a subspace, a smaller set of vectors that represent all the vectors in the subspace can be identified. All the vectors in the subspace can be represented as linear combinations of this subset of vectors. If this subset is also a minimal set, then these vectors are denoted as the basis vectors. In other words, the basis vectors span the subspace. Any minimal subset will suffice and this will not be an unique set, however, the number of vectors in equivalent basis sets will be the same. The number of vectors in the basis set is the dimension, Dim, of the subspace.

3.3 FUNDAMENTAL SUBSPACES

Subspaces arise naturally in data science problems. Four fundamental subspaces can be defined for any data matrix $X_{(m \times n)}$. These are: row space, column space, null space, and left-null space. We will first describe these subspaces and then understand the use of these definitions in data science.

3.3.1 Row and Column Spaces of a Matrix

Row space $(R(X))$ of a matrix X is the subspace comprising an infinite collection of all linearly independent rows of X. Since all linear combinations are allowed, additive and multiplicative closure are already guaranteed by construction and hence row space is a subspace. It is also apparent that the basis set should come from the rows of the matrix as every other vector in the subspace is a linear combination of the rows—that is the row vectors span the subspace. However, this might not be a minimal set as some rows could themselves be linear combinations of other rows. From the previous discussion on row operations, we know that dependent rows lead to zero rows in the U matrix. As a result, a basis for row space can be computed as follows.

a) Perform LU decomposition of PX ($P = I$, if no row swaps are required)
b) If there are r non-zeros in U, then a basis vector set for the row space is the first r rows of PX

The dimension of the row space $Dim(R(X))$ is equal to r. Interestingly, the first r rows of U could also be used as basis vectors for the row space of X.

Similarly, the column space of matrix X is the subspace comprising an infinite collection of all linearly independent columns of X. The same LU decomposition can be used to identify the basis vectors for the column space of X.

a) Perform a LU decomposition of PX ($P = I$, if no row swaps are required).
b) Identify the columns with pivots in U, the same columns in X will provide a set of basis vectors for the column space of X.

A note of caution here is that the columns of U with pivots are not a basis set for the column space of X, unlike in the case of the row space. However, since the number of columns with pivots is equal to the number of non-zero rows, the number of independent rows and the number of independent columns are the same. That is, the row rank is the same as the column rank. Alternately, the dimension of the column space is the same as the dimension of the row space.

3.3.2   Null and Left-Null Spaces of a Matrix

The two other subspaces are null space and left-null space. There is a subtle difference in the manner in which these subspaces are defined from the definition of row and column spaces. The null space of a matrix $(\mathcal{N}(X))$ is defined as a collection of all β such that $X\beta = 0$ [42, 25]. Notice that the null space is not defined as a linear combination of already known vectors but as solutions to vector equations. For this definition to describe a subspace, the vectors have to satisfy the additive and multiplicative closure properties. These are easy to verify. If β_1 and β_2 are from the null space, $\beta_1 + \beta_2$ should also be from the null space. That is $X(\beta_1 + \beta_2) = 0$. This implies $X\beta_1 + X\beta_2 = 0$, which is true because β_1 and β_2 are from the null space and hence $X\beta_1 = 0$ and $X\beta_2 = 0$. It can be easily verified that multiplicative closure also holds. The number of basis vectors for the null space is called the nullity of the matrix. A basis set for the null space can be identified through the following procedure:

a) Perform a LU decomposition of PX ($P = I$, if no row swaps are required).
b) Identify the columns with pivots, denote the variables corresponding to these columns as the basic variables.
c) Pick the equations corresponding to the non-zero rows (r of them will be there).
d) Identify $(n - r)$ solutions to these r equations keeping in turn one of the remaining variables (not basic) to be one and all other remaining variables to be zero.
e) Generate the $(n - r)$ consolidated vectors (basic and not basic) for each of the solutions. This will be a basis set for the null space.

From the construction itself it is clear that the rank of the matrix (r) determines the number of basis vectors in the null space ($n - r$). This is formalized as rank nullity theorem which states that the sum of rank and nullity of any matrix X is equal to the total number of variables of X or the number of columns of X (using the convention described earlier)[30, 3].

$$Dim(R(X)) + Dim(\mathcal{N}(X)) = n \qquad (3.4)$$

where $R(X)$ is the row space of X and $\mathcal{N}(X)$ is the null space of X. Nullity identifies the number of linear equations and rank identifies the number of independent variables.

The left null space is defined as a collection of all vectors β such that $\beta^T X = 0$. Similar to the null space, it can be easily verified that the left null space also satisfies additive and multiplicative closure properties. A basis set can be identified for the left null space by applying the procedure for null space generation outlined above to X^T since $\beta^T X = 0 \implies X^T \beta = 0$.

Example 3.12. *Find the dimensions of the four fundamental subspaces of the matrix,* $X = \begin{bmatrix} 1 & 0 & 3 \\ 10 & 1 & 7 \\ -5 & 0 & -3 \end{bmatrix}$.

Solution 3.12. *The key to finding the dimensions of the four fundamental subspaces of the matrix is to find the row rank of the matrix.*

Since row rank = column rank, dimension of column space = dimension of row space = rank. Rank nullity theorem provides the dimensions of the null and left null spaces.

From Example 3.3, the row rank of the given matrix is 3. Thus:

a) *dim (row space)* $= 3$
b) *dim(column space)* $=$ *dim (row space)* $= 3$
c) *dim (null space)* $= 3-$ *dim (row space)* $= 3 - 3 = 0$
d) *dim (left null space)* $= 3-$ *dim (column space)* $= 3 - 3 = 0$

Example 3.13. *Find the basis set for the row space and column space of the matrix* $X = \begin{bmatrix} 2 & 4 & 6 \\ 3 & 5 & 8 \\ 1 & 2 & 3 \end{bmatrix}$.

Solution 3.13. *To find the basis set, we first need to find the row echelon form. Using the row operations* $R_1 \leftarrow R_1/2$, $R_2 \leftarrow R_2 - 3R_1$, *and* $R_3 \leftarrow R_3 - R_1$,

$$\begin{bmatrix} 1 & 2 & 3 \\ 0 & -1 & -1 \\ 0 & 0 & 0 \end{bmatrix}$$

Since the last row is zero, the rank of the matrix is 2. The size of the row space and column space is 2. From the row echelon form, we have the basis set for the row-space as $\begin{bmatrix} 1 & 2 & 3 \end{bmatrix}^T$ *and* $\begin{bmatrix} 0 & -1 & -1 \end{bmatrix}^T$ *(non-zero rows of the row echelon form).*

Now, the pivots in the row echelon form are present in columns 1 and 2. Hence, the basis set for the column space as derived from the pivot columns consist of the vectors $\begin{bmatrix} 1 & 0 & 0 \end{bmatrix}^T$ *and* $\begin{bmatrix} 2 & -1 & 0 \end{bmatrix}^T$.

Example 3.14. *Find the nullity of the matrix* $X = \begin{bmatrix} 1 & 2 & 3 \\ 2 & 4 & 6 \\ 3 & 6 & 9 \end{bmatrix}$.

Solution 3.14. *Since the second row is twice the first row and the third row is thrice the first row, there is only one independent row. Thus, the rank of the matrix is 1. The nullity of the matrix is 2.*

Example 3.15. *Find the null space of the matrix* $X = \begin{bmatrix} 1 & 2 & 1 \\ 2 & 4 & 2 \\ 3 & 6 & 3 \end{bmatrix}$.

Solution 3.15. *LU decomposition of the matrix will lead to the upper triangular matrix,* $U = \begin{bmatrix} 1 & 2 & 1 \\ 0 & 0 & 0 \\ 0 & 0 & 0 \end{bmatrix}$. *To find null space,* $X\boldsymbol{\beta} = 0$ *where* $\boldsymbol{\beta} = \begin{bmatrix} \beta_1 & \beta_2 & \beta_3 \end{bmatrix}^T$. *The pivot is the first diagonal and hence* β_1 *is the basic variable.* β_2 *and* β_3 *are the free variables. The non-zero row equation is* $\beta_1 + 2\beta_2 + \beta_3 = 0$. *With* $\beta_2 = 1, \beta_3 = 0$, $\beta_1 = -2$, *this leads to one null space vector* $\begin{bmatrix} -2 & 1 & 0 \end{bmatrix}^T$. $\beta_2 = 0, \beta_3 = 1$, $\beta_1 = -1$ *leads to the second null space vector* $\begin{bmatrix} -1 & 0 & 1 \end{bmatrix}^T$.

Example 3.16. *Find the null and left null space of* $X = \begin{bmatrix} 1 & 2 & 1 \\ 2 & 4 & 3 \\ 3 & 6 & 4 \end{bmatrix}$.

Solution 3.16. *LU decomposition of the matrix will lead to the following* $U = \begin{bmatrix} 1 & 2 & 1 \\ 0 & 0 & 1 \\ 0 & 0 & 0 \end{bmatrix}$. *The pivots are in the first and third columns and hence* x_1 *and* x_3 *are the basic variables.* x_2 *is the other variable. The non-zero row equations are* $x_1 + 2x_2 + x_3 = 0$ *and* $x_3 = 0$. *With* $x_2 = 1$, *solving for the two equations leads to* $x_1 = -2$ *and* $x_3 = 0$, *this leads to the null space vector* $\begin{bmatrix} -2 & 1 & 0 \end{bmatrix}^T$. *Since the rank of the matrix is 2, the dimension of null space is 1 and hence there is only one basis vector. For left null space,* X^T *is considered. With no row permutation, we get the upper triangular matrix,* $U = \begin{bmatrix} 1 & 2 & 3 \\ 0 & 1 & 1 \\ 0 & 0 & 0 \end{bmatrix}$. *The non-zero row equations are* $\beta_1 + 2\beta_2 + 3\beta_3 = 0$ *and* $\beta_2 + \beta_3 = 0$. *The pivots are in the first and second columns and hence* β_1 *and* β_2 *are the basic variables.* β_3 *is the other variable. With* $\beta_3 = 1$, *solving for the two equations leads to* $\beta_1 = -1$ *and* $\beta_2 = -1$, *this leads to the left null space vector* $\begin{bmatrix} -1 & -1 & 1 \end{bmatrix}$.

3.4 DATA SCIENCE AND FUNDAMENTAL SUBSPACES

The four fundamental subspaces can be used to answer several data science questions. The rank of the data matrix directly provides the number of independent samples and variables. The row and column ranks being the same mean that there can be only as many independent samples as there are independent variables and vice versa. The basis vectors for the row space and the column space can be used to construct all the rows and columns of the data matrix respectively if the multiplicative constants in the linear combinations are also stored. This notion can be used in multiple applications related to denoising, data compression and so on, described later in this chapter. The null space and left null space provide relationships between the variables and samples respectively as discussed below.

3.4.1   Understanding Linear Relationships between Variables and Samples

The null space of a matrix X consists of all vectors $\boldsymbol{\beta}$ such that $X\boldsymbol{\beta} = 0$ and $\boldsymbol{\beta} \neq 0$. Nullity of a matrix is the number of vectors in the null space of the given matrix [30, 3]. Consider the matrix $X \in \mathbb{R}^{m \times n}$. If $m = n$ and all the vectors in X are linearly independent, then the only vector that satisfies $X\boldsymbol{\beta} = 0$ is the zero vector. However, even if there is one vector other than the zero vector in the null space, there will be infinite vectors just from scalar multiples of that vector. If two vectors $\boldsymbol{\beta}_1$ and $\boldsymbol{\beta}_2$ are in the null space of X, then $\alpha_1 \boldsymbol{\beta}_1 + \alpha_2 \boldsymbol{\beta}_2$ also lies in the null space of X where α_1 and α_2 are any scalars. Interestingly, the size of null space of the matrix provides us with the number of relationships that are among the variables. Every null space vector gives a relationship between the variables.

As an example, for a matrix of dimension 5, let the size of null space be 2. This it means that there are 2 relationships among these 5 variables. Consequently, of these 5 variables, only 3 are linearly independent and the remaining 2 variables can be calculated using the 2 relationships from the independent variables. The null space vectors ($\boldsymbol{\beta} \, \forall \, X\boldsymbol{\beta} = 0$) identify these linear relationships.

Let us suppose,

$$X = \begin{bmatrix} x_{11} & \cdots & x_{1n} \\ \vdots & \ddots & \vdots \\ x_{m1} & \cdots & x_{mn} \end{bmatrix} = \begin{bmatrix} \mathbf{x}_1 & \cdots \mathbf{x}_n \end{bmatrix} \qquad (3.5)$$

is a data matrix with n variables where x_{ij} represents the value of jth variable at the ith instant/sample. Assume that there is one vector in the null space of X and let it be $\boldsymbol{\beta} = \begin{bmatrix} \beta_1, \ldots, \beta_n \end{bmatrix}^T$. As per the definition, $\boldsymbol{\beta}$ satisfies all the equations given below:

$$x_{11}\beta_1 + x_{12}\beta_2 + \ldots + x_{1n}\beta_n = 0$$

$$\vdots$$

$$x_{m1}\beta_1 + x_{m2}\beta_2 + \ldots + x_{mn}\beta_n = 0$$

This means that, for every sample, the sum of the product of variables and $\boldsymbol{\beta}$ is zero. Since this relationship holds for every sample, we would assume that this is a true relationship or a model between all of these variables or attributes. In general, we can write

$$x_1\beta_1 + x_2\beta_2 + \ldots + x_n\beta_n = 0 \qquad (3.6)$$

where x_1 to x_n represent the variables. Thus, we have truly identified a linear relationship between the variables. Every null space vector corresponds to one such relationship and hence, the number of linear relations would be same as the nullity of the matrix.

For example, consider the matrix we discussed earlier $\left(X = \begin{bmatrix} 1 & 2 & 1 \\ 2 & 4 & 0 \\ 3 & 6 & 0 \end{bmatrix} \right)$. The rank of this matrix was identified as 2. Solving $X\beta = 0$, we get a single null vector $\beta = k \begin{bmatrix} 2 & -1 & 0 \end{bmatrix}^T$ where k is a constant. Note that any scalar multiple of β is also a null space vector. All of these vectors would give the same linear relationship, $2\mathbf{x}_1 - \mathbf{x}_2 = 0$, between the variables.

Example 3.17. *Find the linear relations that exist between variables (columns) in the data matrix* $X = \begin{bmatrix} 1 & 3 \\ 2 & 6 \end{bmatrix}$.

Solution 3.17. *The solution to* $X\beta = 0$ *gives a single null vector* $\beta = k \begin{bmatrix} -3 & 1 \end{bmatrix}^T$ *where* k *is a constant. Thus, the linear relationship between the two variables* x_1 *and* x_2 *in the data matrix is*

$$-3x_1 + x_2 = 0$$

Example 3.18. *Find the linear relations that exist between variables (columns) in the data matrix* $X = \begin{bmatrix} 1 & 2 & 1 \\ 2 & 4 & 2 \\ 3 & 6 & 3 \end{bmatrix}$.

Solution 3.18. *The solution to* $X\beta = 0$ *gives two null vectors* $\beta = k_1 \begin{bmatrix} -2 & 1 & 0 \end{bmatrix}^T$ *and* $\beta = k_2 \begin{bmatrix} 1 & 0 & -1 \end{bmatrix}^T$ *where* k_1 *and* k_2 *are constants. Thus, the linear relationships between the variables* x_1, x_2, *and* x_3 *in the data matrix are*

$$-2x_1 + x_2 = 0 \text{ and } x_1 - x_3 = 0$$

Similarly, left-null space vectors provide the relationships between samples. There will be as many relationships between samples as there are vectors in the left-null space. An interesting thing to note is that while the number of independent variables and the number of independent samples are the same, the number of relationships between samples and the number of relationships between variables are not the same.

3.5 SOLVING LINEAR EQUATIONS — MULTIPLE VIEWS

We will now discuss how to extract solutions for matrix equations. This is one of the most fundamental aspects of linear algebra with applications in data science.

Generalized linear equations can be represented in matrix form as $X\beta = \mathbf{y}$ where $X \in \mathbb{R}^{m \times n}$, $\beta \in \mathbb{R}^{n \times 1}$, and $\mathbf{y} \in \mathbb{R}^{m \times 1}$. m and n are the number of equations and variables respectively. In general, $\mathbf{y}$ is a known vector that is used on the right hand side.

There are two interpretations (row and column views) for the set of linear equations $X\beta = \mathbf{y}$[42].

a) Row view of $X\beta = \mathbf{y}$:

Consider the equation $X\beta = \mathbf{y}$ where $X \in \mathbb{R}^{m \times n}$ and $\beta \in \mathbb{R}^n$. Each row of the equation can be thought of as a hyperplane of dimension $n - 1$. A unique solution exists to this set of equations if there is a single point intersection of all the hyperplanes corresponding to the rows of the equations. Otherwise, there will be infinite number of solutions or if no intersection exists, there will be no solution to $X\beta = \mathbf{y}$.

b) Column view of $X\beta = \mathbf{y}$:

The columns of product AB are linear combinations of columns of A, *i.e.*, columns of AB are in the column space of A. So, if a solution exist to the equation set $X\beta = \mathbf{y}$, then the vector $\mathbf{y}$ can be written as a linear combination of columns of X or in other words, $\mathbf{y}$ is in the column space of X. If $\mathbf{y}$ is in the column space of X and the columns of X are linearly independent (X is full column rank), then $X\beta = \mathbf{y}$ will have a unique solution. If $\mathbf{y}$ is in the column space of X and the columns of X are linearly dependent, then $X\beta = \mathbf{y}$ will have infinite solutions. If $\mathbf{y}$ is not in the column space of X, then there will be no solution.

There are three cases that one needs to address:

a) $m = n$ (when the number of equations and variables are the same)
b) $m > n$ (when the number of equations is more than the number of variables)—Usually no solution
c) $m < n$ (when the number of equations is less than the number of variables)—Usually multiple solutions

We will discuss these cases independently and later, combine all the cases using the concept of pseudo inverse. One important result that will establish if a set of equations are solvable or not and valid for all the cases is the following. $X\beta = \mathbf{y}$ is solvable if and only if

$$Rank(X) = Rank(\begin{bmatrix} X & \mathbf{y} \end{bmatrix}) \tag{3.7}$$

The result basically states that if the rank of the combined matrix does not increase then $\mathbf{y}$ is not an independent column and hence, can be written as a linear combination of columns of X; this linear combination (β) is precisely what is being solved for in $X\beta = \mathbf{y}$ (column multiplication view).

3.5.1   Case 1: $m = n$ (Square Matrix)

If the matrix is full rank (determinant of X is non-zero), then it is both full row rank and full column rank, *i.e.*, the rank of the matrix $= m = n$. This means that all the equations are independent of each other or none of the equations can be written as the linear combinations of other equations. In this case, there is a unique solution to $X\beta = \mathbf{y}$, and that unique solution is

$\boldsymbol{\beta} = X^{-1}\mathbf{y}$, where X^{-1} is a $n \times n$ matrix such that

$$X^{-1}X = XX^{-1} = \begin{bmatrix} 1 & 0 & \cdots & 0 \\ 0 & 1 & \cdots & 0 \\ \vdots & \vdots & \vdots & \vdots \\ 0 & 0 & \cdots & 1 \end{bmatrix}_{n \times n} = I \text{ (Identity matrix)} \qquad (3.8)$$

Now, let us assume that the matrix is not full rank, i.e, the rank of the matrix is less than m (or n). This means that at least one row/equation can be written as a linear combination of other rows/equations. In this case, depending on $\mathbf{y}$, two situations might be encountered;

a) Infinite solution if the equations are consistent
b) No solution if the equations are inconsistent

If the LHS of the equations are linearly dependent, the equations are consistent when the same linear dependence is maintained on the right hand side (RHS). Alternatively, equations are consistent if $\mathbf{y}$ lies in the column space of X. If the equations are consistent, the linearly dependent equations can be ignored and only the linearly independent equations can be considered. Since the number of linearly independent equations is less than the number of variables, there will be infinite number of solutions.

To understand inconsistent equations, consider the example $\begin{bmatrix} 1 & 2 \\ 2 & 4 \end{bmatrix} \begin{bmatrix} \beta_1 \\ \beta_2 \end{bmatrix} =$ $\begin{bmatrix} 5 \\ 9 \end{bmatrix}$. Here, if we multiply the first row of X with 2, we get the second row of X. However, this is not true for the RHS. Such equations are referred as inconsistent equations. For this set of equations, we will never be able to find a solution that satisfies both the equations. The solution that satisfies $\beta_1 + 2\beta_2 = 5$ will always have $2\beta_1 + 4\beta_2 = 10$ and hence, will not satisfy the second equation: $2\beta_1 + 4\beta_2 = 9$. These arguments are valid for any inconsistent set of equations.

3.5.2   Case 2: $m > n$

In this case, we have more equations than variables. Let us first consider the case of a full rank matrix, that is $r = n$. Since the number of equations is greater than the number of variables, in general, not all equations can be satisfied. However, if $\mathbf{y}$ is such that all the equations are consistent (though number of equations is greater than the number of variables), then there will be an unique solution. This can be verified through equation 3.7. If this condition is not satisfied, then there will be no solution to the equations. However, one can identify an appropriate approximate solution by viewing this case from an optimization perspective.

Let us look at a solution to $X\boldsymbol{\beta} = \mathbf{y}$ or $X\boldsymbol{\beta} - \mathbf{y} = 0$ when the number of equations are more than the number of variables. Identifying a solution to

$X\boldsymbol{\beta} - \mathbf{y} = 0$ can be thought of as identifying an optimal $\boldsymbol{\beta}(= \boldsymbol{\beta}^*)$ such that the magnitude (defined later in Section 3.6.1) of the error $X\boldsymbol{\beta} - \mathbf{y}$ is minimum at $\boldsymbol{\beta} = \boldsymbol{\beta}^*$. If there is a perfect solution to the set of equations, then one will obtain an $\boldsymbol{\beta}^*$ such that $X\boldsymbol{\beta}^* - \mathbf{y} = 0$.

Let us denote $X\boldsymbol{\beta} - \mathbf{y} = \mathbf{e}$ (error). Notice that $X\boldsymbol{\beta} - \mathbf{y}$ is a vector and there will be as many error terms as the number of equations. Now, one could collectively minimize all the errors by minimizing the sum of absolute errors ($\sum_{i=1}^{m} |e_i|$, L_1-norm) or sum of squared errors (equivalently, the root of sum of squared errors ($\sum_{i=1}^{m} e_i^2)^{\frac{1}{2}}$, L_2-norm) or L_p-norm $\left((\sum_{i=1}^{m} e_i^p)^{\frac{1}{p}} \right)$. In general, the sum of squared errors or squared L_2-norm is minimized. This is called the least squares problem.

$$
\min \sum_{i=1}^{m} e_i^2 = \min \begin{bmatrix} e_1 & e_2 & \cdots & e_m \end{bmatrix} \begin{bmatrix} e_1 \\ e_2 \\ \vdots \\ e_m \end{bmatrix} = \min \left[(X\boldsymbol{\beta} - \mathbf{y})^T (X\boldsymbol{\beta} - \mathbf{y}) \right]
$$

$$
= \min \left[(\mathbf{y}^T - \boldsymbol{\beta}^T X^T)(X\boldsymbol{\beta} - \mathbf{y}) \right]
$$
$$
= \min \left[\mathbf{y}^T X\boldsymbol{\beta} - \boldsymbol{\beta}^T X^T X\boldsymbol{\beta} - \mathbf{y}^T \mathbf{y} + \boldsymbol{\beta}^T X^T \mathbf{y} \right]
$$
$$
= \min \left[-\boldsymbol{\beta}^T X^T X\boldsymbol{\beta} + 2\mathbf{y}^T X\boldsymbol{\beta} - \mathbf{y}^T \mathbf{y} \right] = \min f(\boldsymbol{\beta}) \qquad (3.9)
$$

We observe that the objective function for this optimization problem is a function of $\boldsymbol{\beta}$. Solving this optimization problem will result in a solution for $\boldsymbol{\beta}$. The solution to this optimization problem is obtained by differentiating $f(\boldsymbol{\beta})$ with respect to $\boldsymbol{\beta}$ and setting the differential to zero ($\nabla f(\boldsymbol{\beta}) = 0$) at $\boldsymbol{\beta} = \boldsymbol{\beta}^*$ (optimal solution). Solutions to optimization problems are described in detail in Chapter 4.

Differentiating $f(\boldsymbol{\beta})$ and setting the differential to zero results in

$$
\nabla f(\boldsymbol{\beta}^*) = 2(X^T X)\boldsymbol{\beta}^* - 2X^T y = 0 \implies (X^T X)\boldsymbol{\beta}^* = X^T y
$$

Assuming that all the columns are linearly independent (full column rank), $X^T X$ will be a full rank matrix and its inverse exists.

$$
\boldsymbol{\beta}^* = (X^T X)^{-1} X^T \mathbf{y} \qquad (3.10)
$$

While this solution $\boldsymbol{\beta}^*$ might not satisfy all the equations, this solution will ensure that the errors in all the equations are collectively minimized. This is an optimization view for case 2. The same equation can be derived using ideas of projections.

When the rank of the matrix $r < n$, then this becomes a case of no solution or infinite number of solutions. Since the columns are now dependent, multiple possible combinations of the columns to represent $\mathbf{y}$ are now possible, if a solution exists. It is also interesting to notice that equation 3.10 cannot be used any more because $X^T X$ is not invertible when the matrix is not full column rank.

Equation 3.10 can be generalized as

$$\boldsymbol{\beta}^* = X^{\dagger}\mathbf{y} \tag{3.11}$$

where $X^{\dagger}$ is referred to as "Moore-Penrose pseudo inverse" or simply "pseudo inverse". The computation of the pseudo-inverse will be described later when singular value decomposition is explained. Equation 3.11 identifies a solution for all cases.

3.5.3   Case 3: $m < n$

Let us again consider the full rank case, $r = m$. Since there are less equations than variables, there will be infinite number of solutions and one has to pick one solution. In this case, one can minimize the norm of $\boldsymbol{\beta}$ or $\min \boldsymbol{\beta}^T \boldsymbol{\beta}$. An optimization formulation for this case is $\min \boldsymbol{\beta}^T \boldsymbol{\beta}$ subject to $X\boldsymbol{\beta} = \mathbf{y}$. We will see this later as a constrained optimization problem (refer Chapter 4) and the solution to this problem is given by

$$\boldsymbol{\beta}^* = X^T (XX^T)^{-1}\mathbf{y} \tag{3.12}$$

Substitution of the expression $\boldsymbol{\beta}$ in $X\boldsymbol{\beta}$ will verify that $\boldsymbol{\beta}$ is a valid solution. When the rank $r < m$, this becomes a case of no solution or infinite number of solutions. The reason is the following. Since row rank is the same as the column rank, the number of independent columns is $r < m$. However $\mathbf{y} \in \mathbb{R}^m$, which means that unless $\mathbf{y}$ is in a subspace spanned by r basis vectors of the column space of X, there will not be a solution. If it is, then there are infinite number of solutions because multiple linear combinations can be identified. Similar to the previous case, if the matrix is rank deficient, then XX^T is not invertible and equation 3.12 can't be used. Equation 3.12 can be generalized as

$$\boldsymbol{\beta}^* = X^{\dagger}\mathbf{y} \tag{3.13}$$

Interestingly in the $r = n$ case, $X^{\dagger} = (X^T X)^{-1}X^T$ and in the $r = m$ case, $X^{\dagger} = X^T (XX^T)^{-1}$.

Example 3.19. *Solve the equations*

(a) $2u + v = 5$ and $u + v = 4$
(b) $2u + v = 5$ and $2u + v = 4$
(c) $2u + v = 5$ and $4u + 2v = 10$

Solution 3.19.

(a) The equations can be represented in matrix form as follows:

$$\begin{bmatrix} 2 & 1 \\ 1 & 1 \end{bmatrix} \begin{bmatrix} u \\ v \end{bmatrix} = \begin{bmatrix} 5 \\ 4 \end{bmatrix}$$

Since the matrix X is full rank, X^{-1} exists and thus the solution is given by

$$\begin{bmatrix} u \\ v \end{bmatrix} = \begin{bmatrix} 2 & 1 \\ 1 & 1 \end{bmatrix}^{-1} \begin{bmatrix} 5 \\ 4 \end{bmatrix} = \begin{bmatrix} 1 & -1 \\ -1 & 2 \end{bmatrix} \begin{bmatrix} 5 \\ 4 \end{bmatrix}$$

$$\begin{bmatrix} u^* \\ v^* \end{bmatrix} = \begin{bmatrix} 1 \\ 3 \end{bmatrix}$$

(b) The equations can be represented in matrix form as follows:

$$\begin{bmatrix} 2 & 1 \\ 2 & 1 \end{bmatrix} \begin{bmatrix} u \\ v \end{bmatrix} = \begin{bmatrix} 5 \\ 4 \end{bmatrix}$$

Since LHS of both equations are the same while RHS are different, the two equations are inconsistent. Thus, there is no solution to the given set of equations. In general, this can be identified by comparing the rank of X to the rank of the augmented matrix $[X \ y]$. If both the ranks are not the same, the given system of equations are inconsistent. In this problem, rank of X is 1 while rank of the augmented matrix is 2 indicating inconsistent equations.

(c) The equations can be represented in matrix form as follows:

$$\begin{bmatrix} 2 & 1 \\ 4 & 2 \end{bmatrix} \begin{bmatrix} u \\ v \end{bmatrix} = \begin{bmatrix} 5 \\ 10 \end{bmatrix}$$

Here, the second equation is twice the first equation and hence, both equations can be satisfied at the same time (consistent). This can also be concluded by comparing ranks of X and $[X \ y]$. For the given example, ranks of both X and $[X \ y]$ are the same (rank = 1). As rank is one, both equations represent the same relation between variables and there is only one equation ($2u + v = 5$). Consequently, there are infinite solutions to the given set of equations. For any u, v can be identified using the expression $v = 5 - 2u$. An optimal solution can be found using the expression 3.13.

$$\begin{bmatrix} u^* \\ v^* \end{bmatrix} = \begin{bmatrix} 2 \\ 1 \end{bmatrix}$$

Example 3.20. *Solve the equations: $2u + v = 5$, $u + v = 4$ and $2u + 4v = 11$.*

Solution 3.20. *The equations can be represented in matrix form as follows:*

$$\begin{bmatrix} 2 & 1 \\ 1 & 1 \\ 2 & 4 \end{bmatrix} \begin{bmatrix} u \\ v \end{bmatrix} = \begin{bmatrix} 5 \\ 4 \\ 11 \end{bmatrix}$$

There are more equations than variables. Solving using equation 3.10

$$\begin{bmatrix} u^* \\ v^* \end{bmatrix} = \begin{bmatrix} 1.5854 \\ 1.9756 \end{bmatrix}$$

3.6   ORTHOGONALITY, PROJECTIONS, AND HYPERPLANES

In this section, concepts related to orthogonality, projections, and hyperplanes are introduced.

3.6.1   Notion of Distance and Orthogonality

Until now $X\boldsymbol{\beta} = \mathbf{y}$ was largely interpreted as a set of equations where parameters $\boldsymbol{\beta}$ are calculated. $\boldsymbol{\beta}$ can also be thought of as a point in an n-dimensional
space. For example, if there are two parameters β_1 and β_2, $\boldsymbol{\beta}$ can be represented as a datapoint in a two-dimensional plane with β_1 and β_2 representing
the distances along β_1 and β_2 axes respectively. $\boldsymbol{\beta}$ can also be considered as a
vector between the origin $(\beta_1 = 0, \beta_2 = 0)$ and the datapoint. The length or
magnitude of this vector is $d = \sqrt{\beta_1^2 + \beta_2^2}$.

The same interpretations can be extended to any number of parameters or
a vector of variables. Although the fundamental mathematics remain the same
irrespective of the dimension of the vector, geometric concepts are difficult to
visualize in dimensions higher than 3-D.

Consider the case of 2 points in 2 dimensions: $\boldsymbol{\beta}^1 = \begin{bmatrix} \beta_1^1 \\ \beta_2^1 \end{bmatrix}$ and $\boldsymbol{\beta}^2 =$
$\begin{bmatrix} \beta_1^2 \\ \beta_2^2 \end{bmatrix}$. The vector from $\boldsymbol{\beta}^1$ to $\boldsymbol{\beta}^2$ is given by $\boldsymbol{\beta}^2 - \boldsymbol{\beta}^1$ and the magnitude of
this vector or the distance between the points can be calculated as

$$l = |\boldsymbol{\beta}^2 - \boldsymbol{\beta}^1|_2 \tag{3.14}$$

$$l = \sqrt{(\beta_1^2 - \beta_1^1)^2 + (\beta_2^2 - \beta_2^1)^2} \tag{3.15}$$

$$l = \sqrt{(\boldsymbol{\beta}^2 - \boldsymbol{\beta}^1)^T(\boldsymbol{\beta}^2 - \boldsymbol{\beta}^1)} \tag{3.16}$$

where $|\cdot|_2$ represents L_2-norm.

A unit vector is a vector with magnitude 1, *i.e.*, the distance from origin to
the end-point of the vector is 1. For example, vectors $\begin{bmatrix} 1 & 0 \end{bmatrix}^T$ and $\begin{bmatrix} 1/\sqrt{2} & 1/\sqrt{2} \end{bmatrix}^T$
are unit vectors. Unit vectors are used to define directions in a coordinate
system. Any vector in the direction of a unit vector can be written as a
product of that unit vector and the magnitude of the vector.

Two vectors are said to be orthogonal to each other when their dot product
is zero [42, 25]. Dot product or scalar product of two n-dimensional vectors
$\boldsymbol{\beta}^1$ and $\boldsymbol{\beta}^2$ is given by

$$\boldsymbol{\beta}^1 \cdot \boldsymbol{\beta}^2 = \sum_{j=1}^{n} \beta_j^1 \beta_j^2 = (\boldsymbol{\beta}^1)^T \boldsymbol{\beta}^2 \tag{3.17}$$

Thus, vectors $\boldsymbol{\beta}^1$ and $\boldsymbol{\beta}^2$ are orthogonal to each other if and only if

$$\boldsymbol{\beta}^1 \cdot \boldsymbol{\beta}^2 = (\boldsymbol{\beta}^1)^T \boldsymbol{\beta}^2 = 0 \tag{3.18}$$

Orthonormal vectors are orthogonal vectors with unit magnitude. Vectors
$\begin{bmatrix} 1 & -2 & 4 \end{bmatrix}^T$ and $\begin{bmatrix} 2 & 5 & 2 \end{bmatrix}^T$ are orthogonal vectors, but not orthonormal. However, these vectors can be made orthonormal by dividing them with their
magnitudes. Note that all orthonormal vectors are orthogonal.

There are several useful properties related to orthogonality of subspaces of matrices. Row space of any matrix X is orthogonal to null space of X and column space of X is orthogonal to left null space of X.

Example 3.21. *Find if vectors* $\mathbf{a} = \begin{bmatrix} 1 & -1 & 0 \end{bmatrix}^T$ *and* $\mathbf{b} = \begin{bmatrix} 1 & 1 & -\sqrt{2} \end{bmatrix}^T$ *are orthogonal? If orthogonal, find the corresponding orthonormal vectors.*

Solution 3.21. *Taking the dot product*

$$\mathbf{a} \cdot \mathbf{b} = 1 - 1 + 0 = 0$$

Since the dot product is zero, the vectors a and b are orthogonal.

$$|\mathbf{a}| = \sqrt{1+1} = \sqrt{2} \ and \ |\mathbf{b}| = \sqrt{1+1+2} = 2$$

Since the magnitudes of a and b are not 1, they are not orthonormal. The corresponding orthonormal set would be $\frac{\mathbf{a}}{|\mathbf{a}|}$ *and* $\frac{\mathbf{b}}{|\mathbf{b}|}$.

$$\frac{\mathbf{a}}{|\mathbf{a}|} = \frac{1}{\sqrt{2}} \begin{bmatrix} 1 & -1 & 0 \end{bmatrix} \ and \ \frac{\mathbf{b}}{|\mathbf{b}|} = \frac{1}{2} \begin{bmatrix} 1 & 1 & -\sqrt{2} \end{bmatrix}$$

The above two vectors are both orthogonal and orthonormal.

Example 3.22. *Find if vectors* $\mathbf{a} = \begin{bmatrix} 1 & -1 & 0 \end{bmatrix}^T$, $\mathbf{b} = \begin{bmatrix} 1 & 1 & -\sqrt{2} \end{bmatrix}^T$ *and* $\mathbf{c} = \begin{bmatrix} 1 & 1 & \sqrt{2} \end{bmatrix}^T$ *are orthogonal?*

Solution 3.22. *For a set of vectors to be orthogonal, each pair has to be orthogonal. From 3.21, we know that* **a** *and* **b** *are orthogonal. We need to see, whether the pairs* $\mathbf{a}, \mathbf{c}$ *and* $\mathbf{b}, \mathbf{c}$ *are orthogonal.*

$$\mathbf{a} \cdot \mathbf{c} = 1 - 1 + 0 = 0 \ and \ \mathbf{b} \cdot \mathbf{c} = 1 + 1 - 2 = 0$$

Hence the vectors are orthogonal.

3.6.2 Projection of Vectors onto Subspaces

The notion of projection of vectors onto subspaces is of great relevance in DS. A vector $\mathbf{y} \in \mathbb{R}^m$ can be projected on to a k-dimensional subspace in $\mathbb{R}^m$ $(k < m)$ [42]. Projections are generally used to reduce the number of attributes to be considered for visualization or further analysis.

Consider a 2D plane (subspace) in a three-dimensional space. Let us assume that the basis vectors of this plane are $\mathbf{x_1}$ and $\mathbf{x_2}$. Any vector in this plane can be written as a linear combination of $\mathbf{x_1}$ and $\mathbf{x_2}$. Consider a vector $\mathbf{y} \in \mathbb{R}^3$ that doesn't lie on this plane as shown in Figure 3.2. Projection refers to the identification of a vector $(\hat{\mathbf{y}})$ in the 2D plane (subspace) that best represents $\mathbf{y}$. In this case, $\hat{\mathbf{y}}$ can be written as a linear combination of $\mathbf{x_1}$ and $\mathbf{x_2}$. The best projection is when the distance between any point on the original vector

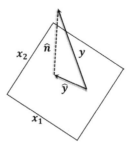

FIGURE 3.2 Projection of a vector onto a plane.

and the corresponding point on the projection is minimum. This will happen when $\mathbf{y} - \hat{\mathbf{y}}$ is normal to the plane (refer Figure 3.2).

$$\hat{\mathbf{y}} = \beta_1 \mathbf{x_1} + \beta_2 \mathbf{x_2} \tag{3.19}$$

β_1 and β_2 are yet to be determined. Using vector addition

$$\mathbf{y} = \hat{\mathbf{y}} + \hat{\mathbf{n}} = \beta_1 \mathbf{x_1} + \beta_2 \mathbf{x_2} + \hat{\mathbf{n}} \tag{3.20}$$

where $\hat{\mathbf{n}}$ is the normal vector to the plane as shown in Figure 3.2. Since $\hat{\mathbf{n}}$ is normal to the plane, $\hat{\mathbf{n}}^T \mathbf{x_1} = \mathbf{x_1}^T \hat{\mathbf{n}} = 0$ and $\hat{\mathbf{n}}^T \mathbf{x_2} = \mathbf{x_2}^T \hat{\mathbf{n}} = 0$. These relationships can be used to find β_1 and β_2. Using $\mathbf{x_1}^T \hat{\mathbf{n}} = 0$,

$$\mathbf{x_1}^T (\mathbf{y} - \beta_1 \mathbf{x_1} - \beta_2 \mathbf{x_2}) = 0 \implies \mathbf{x_1}^T \mathbf{y} - \beta_1 \mathbf{x_1}^T \mathbf{x_1} - \beta_2 \mathbf{x_1}^T \mathbf{x_2} = 0 \tag{3.21}$$

If the basis vectors are orthogonal, $\mathbf{x_1}^T \mathbf{x_2} = \mathbf{x_2}^T \mathbf{x_1} = 0$. Substituting in the above equations and rearranging,

$$\mathbf{x_1}^T \mathbf{y} = \beta_1 \mathbf{x_1}^T \mathbf{x_1} \implies \beta_1 = \frac{\mathbf{x_1}^T \mathbf{y}}{\mathbf{x_1}^T \mathbf{x_1}} \tag{3.22}$$

β_2 can be found following the same steps from $\mathbf{x_2}^T \hat{\mathbf{n}} = 0$. Substituting for β_1 and β_2 in the equation for $\hat{\mathbf{y}}$,

$$\hat{\mathbf{y}} = \frac{\mathbf{x_1}^T \mathbf{y}}{\mathbf{x_1}^T \mathbf{x_1}} \mathbf{x_1} + \frac{\mathbf{x_2}^T \mathbf{y}}{\mathbf{x_2}^T \mathbf{x_2}} \mathbf{x_2} \tag{3.23}$$

The foregoing analysis can be generalized to identify the projection of $\mathbf{y}$ onto a space spanned by k linearly independent vectors (not necessarily orthogonal).

$$\hat{\mathbf{y}} = \sum_{j=1}^{k} \beta_j \mathbf{x_j} \implies \hat{\mathbf{y}} = \begin{bmatrix} \mathbf{x_1} & \mathbf{x_2} & \cdots & \mathbf{x_k} \end{bmatrix} \begin{bmatrix} \beta_1 \\ \beta_2 \\ \vdots \\ \beta_k \end{bmatrix} = X\beta$$

Using orthogonality idea,

$$X^T(\mathbf{y} - \hat{\mathbf{y}}) = X^T(\mathbf{y} - X\beta) = 0 \implies X^T\mathbf{y} - X^TX\beta = 0$$
$$\beta = (X^TX)^{-1}X^T\mathbf{y} \tag{3.24}$$
$$\hat{\mathbf{y}} = X(X^TX)^{-1}X^T\mathbf{y} \tag{3.25}$$

This is the projection of $\mathbf{y}$ onto the subspace spanned by $\mathbf{x_1}$ to $\mathbf{x_k}$.

Example 3.23. *Find the projection of vectors* $\mathbf{a} = \begin{bmatrix} 1 & 1 & 1 \end{bmatrix}^T$ *and* $\mathbf{b} = \begin{bmatrix} 1 & -2 & 3 \end{bmatrix}^T$ *onto the space spanned by the vectors* $\mathbf{x_1} = \begin{bmatrix} 1 & 0 & 0 \end{bmatrix}^T$ *and* $\mathbf{x_2} = \begin{bmatrix} 0 & 1 & 0 \end{bmatrix}^T$.

Solution 3.23.

$$X = \begin{bmatrix} \mathbf{x_1} & \mathbf{x_2} \end{bmatrix} = \begin{bmatrix} 1 & 0 \\ 0 & 1 \\ 0 & 0 \end{bmatrix}$$

$$(X^TX)^{-1} = \left(\begin{bmatrix} 1 & 0 & 0 \\ 0 & 1 & 0 \end{bmatrix} \begin{bmatrix} 1 & 0 \\ 0 & 1 \\ 0 & 0 \end{bmatrix} \right)^{-1} = \begin{bmatrix} 1 & 0 \\ 0 & 1 \end{bmatrix}^{-1} = \begin{bmatrix} 1 & 0 \\ 0 & 1 \end{bmatrix}$$

$$X(X^TX)^{-1}X^T = \begin{bmatrix} 1 & 0 \\ 0 & 1 \\ 0 & 0 \end{bmatrix} \begin{bmatrix} 1 & 0 \\ 0 & 1 \end{bmatrix} \begin{bmatrix} 1 & 0 & 0 \\ 0 & 1 & 0 \end{bmatrix} = \begin{bmatrix} 1 & 0 \\ 0 & 1 \\ 0 & 0 \end{bmatrix} \begin{bmatrix} 1 & 0 & 0 \\ 0 & 1 & 0 \end{bmatrix} = \begin{bmatrix} 1 & 0 & 0 \\ 0 & 1 & 0 \\ 0 & 0 & 0 \end{bmatrix}$$

Projection of $\mathbf{a}$ *onto the space spanned by* $\mathbf{x_1}$ *and* $\mathbf{x_2}$ *is given by*

$$\hat{\mathbf{a}} = \begin{bmatrix} 1 & 0 & 0 \\ 0 & 1 & 0 \\ 0 & 0 & 0 \end{bmatrix} \begin{bmatrix} 1 \\ 1 \\ 1 \end{bmatrix} = \begin{bmatrix} 1 \\ 1 \\ 0 \end{bmatrix}$$

Similarly, the projection of $\mathbf{b}$ *onto the space spanned by* $\mathbf{x_1}$ *and* $\mathbf{x_2}$ *is given by*

$$\hat{\mathbf{b}} = \begin{bmatrix} 1 & 0 & 0 \\ 0 & 1 & 0 \\ 0 & 0 & 0 \end{bmatrix} \begin{bmatrix} 1 \\ -2 \\ 3 \end{bmatrix} = \begin{bmatrix} 1 \\ -2 \\ 0 \end{bmatrix}$$

3.6.3   Generating Orthogonal Vectors through Projections

In some of the DS applications, working with orthogonal vectors derived from linearly independent vectors possess some advantages. Consider n LI vectors $\mathbf{x_1} \ldots \mathbf{x_n}$ and the problem of deriving n orthogonal vectors $\mathbf{o_1} \ldots \mathbf{o_n}$ from this set. The procedure is outlined below (three steps are shown which can be similarly extended to n vectors).

$$\mathbf{o_1} = \mathbf{x_1} \text{ and } \mathbf{o_2} = \mathbf{x_2} - \frac{\mathbf{o_1}^T \mathbf{x_2}}{\mathbf{o_1}^T \mathbf{o_1}} \mathbf{o_1} \tag{3.26}$$

$$\mathbf{o_3} = \mathbf{x_3} - \frac{\mathbf{o_1}^T \mathbf{x_3}}{\mathbf{o_1}^T \mathbf{o_1}} \mathbf{o_1} - \frac{\mathbf{o_2}^T \mathbf{x_3}}{\mathbf{o_2}^T \mathbf{o_2}} \mathbf{o_2} \tag{3.27}$$

It can be verified that $\mathbf{o_1}^T \mathbf{o_2} = \mathbf{o_1}^T \mathbf{o_3} = \mathbf{o_2}^T \mathbf{o_3} = 0$. These orthogonal vectors are identified sequentially by removing the components of already identified orthogonal vectors (through projections) from the current LI vector under consideration. This process is called the Gram-Schmidt orthogonalization process. These orthogonal vectors can be made orthonormal through division by their magnitudes.

3.6.4 Understanding Noise Removal through Projections

The idea of projections onto subspaces can be applied in noise removal from data. Consider Figure 3.3, where a linear relationship between x_1 and x_2 is plotted. If there is no noise in the data, then all the points will fall on this line. However, due to noise, datapoints are away from the line. One noise removal approach is to project the datapoints onto the hyperplane (line in this 2D case) as shown in Figure 3.3. This idea will be described in more detail in the function approximation chapter.

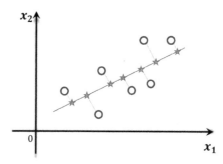

FIGURE 3.3 Projection of points onto a hyperplane.

3.6.5 Understanding Partitions through Hyperplanes and Half-spaces

The notion of hyperplanes and half-spaces is of considerable importance in classification tasks. Consider a generic linear equation $\hat{\mathbf{n}}^T \mathbf{x} + b = 0$, where $\hat{\mathbf{n}} = \begin{bmatrix} \beta_1, & \dots & , \beta_n \end{bmatrix}^T$. Figure 3.4 depicts one such line in two dimensions. Consider any two points $\mathbf{x}^1$ and $\mathbf{x}^2$ on this line; both the points will satisfy

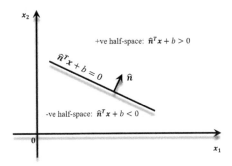

FIGURE 3.4 Concept of hyperplane and half-spaces.

the equation for the line. Hence,

$$\hat{\mathbf{n}}^T \mathbf{x}^1 + b = 0 \text{ and } \hat{\mathbf{n}}^T \mathbf{x}^2 + b = 0 \tag{3.28}$$

$$\implies \hat{\mathbf{n}}^T (\mathbf{x}^2 - \mathbf{x}^1) = 0 \tag{3.29}$$

From the definition of orthogonal vectors, it can be seen that vector $\hat{\mathbf{n}}$ is orthogonal to the vector $\mathbf{x}^2 - \mathbf{x}^1$. That is, $\hat{\mathbf{n}}$ is a normal vector to the line $\hat{\mathbf{n}}^T \mathbf{x} + b = 0$ as $\mathbf{x}^2 - \mathbf{x}^1$ is in the direction of the line.

Two independent equations in two dimensions represent a point (a 0D object), where the lines intersect. If the lines are parallel to each other, then there are no intersection points. If this were extended to three-dimenstional space, each equation $\hat{\mathbf{n}}^T \mathbf{x} + b = 0$ ($\beta_1 x_1 + \beta_2 x_2 + \beta_3 x_3 + b = 0$) represents a plane. $\hat{\mathbf{n}}$ is normal to the plane. Two equations in a three-dimensional space can be visualized as intersection of two planes. If the planes are not parallel, the intersection will be characterized by a line. This can be extended to an n-dimensional space and each equation will represent a geometric entity of dimension $n - 1$. This entity is called a hyperplane [42, 3, 25]. Hyperplanes for a 3D space are 2D planes and hyperplanes of a 2D space are 1D lines.

Similar to how a line would split the 2D space into two halves, every hyperplane would split the nD space into two halves. If the hyperplane is represented by the equation $\hat{\mathbf{n}}^T \mathbf{x} + b = 0$, then points in the half-space (open) in the direction of $\hat{\mathbf{n}}$ will satisfy $\hat{\mathbf{n}}^T \mathbf{x} + b > 0$ and points in the half-space (open) in the direction of $-\hat{\mathbf{n}}$ will satisfy $\hat{\mathbf{n}}^T \mathbf{x} + b < 0$ as shown in Figure 3.4.

Consider a 2D space with a hyperplane $x_1 + 3x_2 + 4 = 0$ (line). The normal to this hyperplane is $\hat{\mathbf{n}} = \begin{bmatrix} 1 & 3 \end{bmatrix}^T$. The half-spaces corresponding to this hyperplane are $x_1 + 3x_2 + 4 > 0$ (positive half-space) and $x_1 + 3x_2 + 4 < 0$ (negative half-space). Now, if we consider a point $\mathbf{x} = \begin{bmatrix} 1 & -1 \end{bmatrix}^T$, the value of $x_1 + 3x_2 + 4$ is 2 and hence this point belongs to the positive half-space. Similarly, if we take the datapoint $\mathbf{x} = \begin{bmatrix} 1 & -2 \end{bmatrix}^T$, the value of $x_1 + 3x_2 + 4$ is -1 and hence, this point belongs to the negative half-space.

This idea of hyperplane and half-spaces is of great importance in binary classification problems. Given the training dataset for binary classification (say, classes 0 and 1), one could find an optimal hyperplane ($\hat{\mathbf{n}}^T\mathbf{x} + b = 0$) that separates the two classes. Each half-space would represent one class. When a new datapoint is obtained, one can determine whether this point lies in Class-0 or Class-1 by evaluating the value of $\hat{\mathbf{n}}^T\mathbf{x} + b$.

Example 3.24. *Consider a 2D space with a hyperplane $-2x + 4y + 4 = 0$. Classify the following points based on the half-spaces they belong to.*

$$(-1, 1),\ (1, -1),\ (0, 1),\ (2, 5),\ (-3, -1),\ (5, 0)$$

Solution 3.24. *For $(-1, 1)$, $-2x+4y+4 = 10$. For $(1, -1)$, $-2x+4y+4 = -2$. For $(0, 1)$, $-2x + 4y + 4 = 8$. For $(2, 5)$, $-2x + 4y + 4 = 20$. For $(-3, -1)$, $-2x + 4y + 4 = 6$. For $(5, 0)$, $-2x + 4y + 4 = -6$. Hence, the following classification can be made.*

Along $\hat{\mathbf{n}} = \begin{bmatrix} -2 & 4 \end{bmatrix}^T$ *(Positive half-space)*	*Along $-\hat{\mathbf{n}} = \begin{bmatrix} 2 & -4^T \end{bmatrix}^T$* *(Negative half-space)*
(-1, 1)	*(1, -1)*
(0, 1)	*(5, 0)*
(2,5)	
(-3, -1)	

3.7   EIGENVALUES, EIGENVECTORS, AND SVD

A matrix X can also be viewed as a linear operator that transforms one vector to another.

$$\underset{m \times n}{X}\ \underset{n \times 1}{\boldsymbol{\beta}} = \underset{m \times 1}{\mathbf{y}} \tag{3.30}$$

In equation 3.30, the vector $\boldsymbol{\beta} \in \mathbb{R}^n$ is transformed to $\mathbf{y} \in \mathbb{R}^m$. An interesting transformation is given by equation 3.31. For a $n \times n$ square matrix,

$$\underset{n \times n}{X}\ \underset{n \times 1}{\boldsymbol{\beta}} = \lambda\ \underset{n \times 1}{\boldsymbol{\beta}} \tag{3.31}$$

In this case, the vector on being operated on by a square matrix X, is modified by just a scalar multiplication. This equation leads to the concept of eigenvalues and eigenvectors [42, 25].

3.7.1   Eigenvalues and Eigenvectors

In equation 3.31, λ is referred to as the eigenvalue and $\boldsymbol{\beta}$ is the eigenvector. Equation 3.31 can be rewritten as

$$(X - \lambda I)\boldsymbol{\beta} = 0 \tag{3.32}$$

The solution $\boldsymbol{\beta}$ has to be in the null space of $X - \lambda I$ (null space definition from before). For a non-trivial $\boldsymbol{\beta}$ to exist, determinant $|X - \lambda I| = 0$. A determinant is a scalar quantity calculated for a square matrix [42]. Two properties of determinants that are of great interest are: (i) If the columns (rows) are independent, then the determinant is non-zero, and (ii) If the columns (rows) are dependent, then it is zero. For non-trivial $\boldsymbol{\beta}$, columns of $X - \lambda I$ have to be dependent and hence, the determinant has to be zero. This is a nth order polynomial equation in λ ($P_n(\lambda)$) and will have n roots. The roots of the polynomial equation can be real or complex. In general, if the eigenvalues are distinct, there will be one eigenvector corresponding to each of the distinct eigenvalues. These eigenvectors are unique only up to a scaling constant. In the case of repeated eigenvalues, there may or may not be n LI eigenvectors. The number of times an eigenvalue repeats is called the algebraic multiplicity of the eigenvalue and the number of eigenvectors for a repeated eigenvalue is called the geometric multiplicity of the eigenvalue. It can be shown that geometric multiplicity is upper bounded by the algebraic multiplicity. When there are multiple eigenvectors corresponding to an eigenvalue, then any linear combination of these eigenvectors is also an eigenvector. For example, if $\boldsymbol{\beta}_1$ and $\boldsymbol{\beta}_2$ are eigenvectors for the same eigenvalue λ, then any $c_1\boldsymbol{\beta}_1 + c_2\boldsymbol{\beta}_2$ is also an eigenvector. Consequently, any vector in the subspace spanned by the eigenvectors of a specific eigenvalue is also an eigenvector for that eigenvalue. Eigenvalues and eigenvectors of square data matrices (numbers of variables and samples are the same) can be used to understand linear relationships among variables. From equation 3.32, the eigenvectors corresponding to zero eigenvalues are vectors in the null space of X since $(X - \lambda I)\boldsymbol{\beta} = 0$ will translate to $X\boldsymbol{\beta} = \mathbf{0}$. Since null space identifies the linear relationships among variables, eigenvectors corresponding to zero eigenvalue identify the relationships among variables. From this, the rank of the matrix (independent variables) can be calculated through rank-nullity theorem.

There are several properties of importance for symmetric matrices ($X = X^T$) that is of relevance in DS and ML. If X is a real symmetric matrix, then it can be shown that all eigenvalues will be real. Further, real symmetric matrices will always have "n" orthogonal eigenvectors (irrespective of distinct or repeated eigenvalues). Since eigenvectors are unique only upto a scaling constant, the eigenvectors of a symmetric matrix can be chosen to be orthonormal. These results are used extensively in optimization (Hessian matrix is symmetric) and Principal Component Analysis or PCA (covariance matrix is symmetric).

A matrix A is said to be positive definite if $\boldsymbol{\beta}^T A \boldsymbol{\beta} > 0 \quad \forall \quad \boldsymbol{\beta} \neq 0$. It can be shown that a matrix is positive definite if and only if (iff) all its eigenvalues are positive. A matrix A is said to be positive semi-definite if $\boldsymbol{\beta}^T A \boldsymbol{\beta} \geq 0 \ \forall \ \boldsymbol{\beta} \neq 0$. A matrix is positive semi-definite iff all its eigenvalues are greater than or equal to zero.

Any square symmetric matrix X can be written as

$$\underset{n\times n}{X} = \underset{n\times n}{Q} \underset{n\times n}{\Lambda} \underset{n\times n}{Q^T} \tag{3.33}$$

where the columns of Q are eigenvectors of X and Λ is a diagonal matrix with the eigenvalues of X as the diagonal elements. If λ_j is the jth diagonal element of Λ, then $\mathbf{q_j}$ (the jth column of Q) will be the eigenvector corresponding to the eigenvalue λ_j. This decomposition is referred to as spectral decomposition or eigenvalue decomposition.

Expanding the decomposition, we have

$$X = \begin{bmatrix} \mathbf{q_1} & \mathbf{q_2} & \cdots & \mathbf{q_n} \end{bmatrix} \begin{bmatrix} \lambda_1 & 0 & \cdots & 0 \\ 0 & \lambda_2 & \cdots & 0 \\ \vdots & \vdots & \ddots & \vdots \\ 0 & 0 & \cdots & \lambda_n \end{bmatrix} \begin{bmatrix} \mathbf{q_1^T} \\ \mathbf{q_2^T} \\ \vdots \\ \mathbf{q_n^T} \end{bmatrix} \tag{3.34}$$

$$X = \begin{bmatrix} \mathbf{q_1} & \mathbf{q_2} & \cdots & \mathbf{q_n} \end{bmatrix} \begin{bmatrix} \lambda_1 \mathbf{q_1^T} \\ \lambda_2 \mathbf{q_2^T} \\ \vdots \\ \lambda_n \mathbf{q_n^T} \end{bmatrix} \tag{3.35}$$

$$X = \lambda_1 \mathbf{q_1}\mathbf{q_1^T} + \lambda_2 \mathbf{q_2}\mathbf{q_2^T} + \ldots + \lambda_n \mathbf{q_n}\mathbf{q_n^T} \tag{3.36}$$

where each term is a rank 1 matrix. For a symmetric matrix, if the eigenvalues λ_{r+1} to λ_n are zero then the above relation can be simplified as,

$$X = \lambda_1 \mathbf{q_1}\mathbf{q_1^T} + \lambda_2 \mathbf{q_2}\mathbf{q_2^T} + \ldots + \lambda_r \mathbf{q_r}\mathbf{q_r^T} \tag{3.37}$$

The matrix X has a total of n^2 elements. Representation of X in the above form allows the the same information to be stored using $(n+1)r$ elements (eigenvectors corresponding to the r non-zero eigenvalues and the eigenvalues themselves). If $r << n$, good data compression can be achieved.

In DS applications, because of the noise in measurements, typically data matrices are unlikely to have zero eigenvalues, even if there are relationships between variables. In such cases one might choose to identify certain eigenvalues that are small (in comparison to others) and treat them as close to zero. As a result a symmetric matrix X may be decomposed as follows

$$\underset{n\times n}{X} = \underset{n\times n}{T} + \underset{n\times n}{U} \tag{3.38}$$

where

$$\underset{n\times n}{T} = \underset{n\times r}{Q_t} \underset{r\times r}{\Lambda_t} \underset{r\times n}{Q_t^T} \tag{3.39}$$

$$\underset{n\times n}{U} = \underset{n\times n-r}{Q_u} \underset{n-r\times n-r}{\Lambda_u} \underset{n-r\times n}{Q_u^T} \tag{3.40}$$

In the previous equation the first r eigenvalues (in descending order) are considered as important and the remaining $n-r$ small eigenvalues are ignored. The matrix now reconstructed with the r eigenvalues can be thought of as retaining the most relevant information from the X matrix.

$$\underset{n \times n}{X_{relevant}} = \underset{n \times r}{Q_t} \ \underset{r \times r}{\Lambda_t} \ \underset{r \times n}{Q_t^T} \tag{3.41}$$

Example 3.25. *Find the eigenvalues and eigenvectors of the matrix $X = \begin{bmatrix} 2 & 1 \\ 1 & 2 \end{bmatrix}$.*

Solution 3.25. *To find the eigenvalues, we solve $|X - \lambda I| = 0$ which implies*

$$|X - \lambda I| = \left| \begin{bmatrix} 2 - \lambda & 1 \\ 1 & 2 - \lambda \end{bmatrix} \right| = 0 \implies (2 - \lambda)^2 - 1 = 0 \implies 2 - \lambda = \pm 1$$

$$\lambda = 1, 3$$

Thus, the eigenvalues are 1 and 3. For $\lambda = 1$, $X\beta = \lambda\beta$ indicates,

$$\begin{bmatrix} 2 & 1 \\ 1 & 2 \end{bmatrix} \begin{bmatrix} \beta_1 \\ \beta_2 \end{bmatrix} = \begin{bmatrix} \beta_1 \\ \beta_2 \end{bmatrix} \implies 2\beta_1 + \beta_2 = \beta_1 \ \& \ \beta_1 + 2\beta_2 = \beta_2 \implies \beta_1 = -\beta_2$$

Hence, the eigenvector for $\lambda = 1$ is $\begin{bmatrix} -1 & 1 \end{bmatrix}^T$. For $\lambda = 3$, $X\beta = \lambda\beta$ indicates,

$$\begin{bmatrix} 2 & 1 \\ 1 & 2 \end{bmatrix} \begin{bmatrix} \beta_1 \\ \beta_2 \end{bmatrix} = 3 \begin{bmatrix} \beta_1 \\ \beta_2 \end{bmatrix} \implies 2\beta_1 + \beta_2 = 3\beta_1 \ \& \ \beta_1 + 2\beta_2 = 3\beta_2 \implies \beta_1 = \beta_2$$

Hence, the eigenvector for $\lambda = 3$ is $\begin{bmatrix} 1 & 1 \end{bmatrix}^T$.

Example 3.26. *Find the spectral decomposition of the matrix $\begin{bmatrix} 2 & 1 \\ 1 & 1 \end{bmatrix}$.*

Solution 3.26. *Eigenvalues of X are 0.3820 and 2.618. The corresponding eigenvectors are $\begin{bmatrix} 0.5257 & -0.8507 \end{bmatrix}^T$ and $\begin{bmatrix} -0.8507 & -0.5257 \end{bmatrix}^T$. Thus, the spectral decomposition of X is*

$$X = Q \Lambda Q^T \ \left(X \text{ is symmetric, i.e., } X = X^T \right)$$

$$X = \begin{bmatrix} 0.5257 & -0.8507 \\ -0.8507 & -0.5257 \end{bmatrix} \begin{bmatrix} 0.3820 & 0 \\ 0 & 2.618 \end{bmatrix} \begin{bmatrix} 0.5257 & -0.8507 \\ -0.8507 & -0.5257 \end{bmatrix}^T$$

Example 3.27. *Find X^{47} for the matrix $X = \begin{bmatrix} 4 & 5 & 1 \\ 5 & 4 & 1 \\ 1 & 1 & 1 \end{bmatrix}$.*

Solution 3.27. *Since X is a symmetric matrix, we can write*

$$X = Q \Lambda Q^T$$

$$where \ Q = \begin{bmatrix} -0.7071 & -0.1196 & -0.6969 \\ 0.7071 & -0.1196 & -0.6969 \\ -0.0000 & 0.9856 & -0.1691 \end{bmatrix} \ and \ \Lambda = \begin{bmatrix} -1.0000 & 0 & 0 \\ 0 & 0.7574 & 0 \\ 0 & 0 & 9.2426 \end{bmatrix}.$$

$$X^{47} = Q\Lambda^{47}Q^T$$

$$= \begin{bmatrix} -0.7071 & -0.1196 & -0.6969 \\ 0.7071 & -0.1196 & -0.6969 \\ -0.0000 & 0.9856 & -0.1691 \end{bmatrix} \times \begin{bmatrix} -1 & 0 & 0 \\ 0 & 2.1296 \times 10^{-06} & 0 \\ 0 & 0 & 2.4684 \times 10^{45} \end{bmatrix}$$

$$\times \begin{bmatrix} -0.7071 & 0.7071 & -0.0000 \\ -0.1196 & -0.1196 & 0.9856 \\ -0.6969 & -0.6969 & -0.1691 \end{bmatrix}$$

$$= 10^{45} \begin{bmatrix} 1.1989 & 1.1989 & 0.2909 \\ 1.1989 & 1.1989 & 0.2909 \\ 0.2909 & 0.2909 & 0.0706 \end{bmatrix}$$

Note: If X was asymmetric and has a full set of linearly independent eigenvectors, then $X^n = Q\Lambda^n Q^{-1}$.

Example 3.28. *Solve the equations $2u + v = 5$ and $u + v = 4$ using spectral decomposition.*

Solution 3.28. *The given equations can be represented as $X\beta = y$ where $X = \begin{bmatrix} 2 & 1 \\ 1 & 1 \end{bmatrix}$ and $y = \begin{bmatrix} 5 \\ 4 \end{bmatrix}$. Note that X is symmetric. Hence,*

$$X = Q\,\Lambda\,Q^T \quad \Longrightarrow \quad X^{-1} = Q\,\Lambda^{-1}\,Q^T$$

$$\beta^* = X^{-1}y = Q\,\Lambda^{-1}\,Q^T y$$

$$= \begin{bmatrix} 0.5257 & -0.8507 \\ -0.8507 & -0.5257 \end{bmatrix} \begin{bmatrix} \frac{1}{0.382} & 0 \\ 0 & \frac{1}{2.618} \end{bmatrix} \begin{bmatrix} 0.5257 & -0.8507 \\ -0.8507 & -0.5257 \end{bmatrix} \begin{bmatrix} 5 \\ 4 \end{bmatrix} = \begin{bmatrix} 1 \\ 3 \end{bmatrix}$$

Example 3.29. *Find the eigenvalues of the following matrices and comment on the relation between rank and eigenvalues.*

$$a) \ X_1 = \begin{bmatrix} 1 & 1 & 0 \\ 1 & 2 & 0 \\ 0 & 0 & 0 \end{bmatrix}; \quad b) \ X_2 = \begin{bmatrix} 1 & 0 & 2 \\ 0 & 0 & 2 \\ 0 & 0 & 0 \end{bmatrix}; \quad c) \ X_3 = \begin{bmatrix} 1 & 1 & 0 \\ 1 & 1 & 0 \\ 0 & 0 & 2 \end{bmatrix};$$

Solution 3.29. *a) For the given symmetric matrix, eigenvalues are 0, 0.382, and 2.618. The rank of the given matrix is 2. For the given matrix, rank of the matrix is equal to the number of non-zero eigenvalues.*
b) For the given asymmetric matrix, eigenvalues are 1, 0, and 0. Rank of this matrix is 2. Here, rank is not equal to the number of non-zero eigenvalues.
c) For the given symmetric matrix of rank 2, eigenvalues are 0, 2, and 2. Number of non-zero eigenvalues is equal to the rank of the matrix.

Remark. *For a real and symmetric matrix A, rank is equal to the number of non-zero eigenvalues (counted with repeats).*

3.7.2   Singular Value Decomposition (SVD)

In general, data matrices are rectangular and non-symmetric (even if they are square). Many of the interesting and useful results in the last section are for symmetric square matrices. It is useful to understand how these results can be utilized in working with rectangular matrices that are of relevance to DS. Singular Value Decomposition (SVD) is an important technique that is used in DS and the key ideas in SVD are built from the results for square symmetric matrices.

For any rectangular matrix, SVD of X is,

$$\underset{m \times n}{X} = \underset{m \times m}{Q_1} \ \underset{m \times n}{\Sigma} \ \underset{n \times n}{Q_2^T} \tag{3.42}$$

where Q_1 and Q_2 has orthonormal columns i.e., $Q_1 Q_1^T = Q_1^T Q_1 = I$ and $Q_2 Q_2^T = Q_2^T Q_2 = I$. Σ is a diagonal matrix whose diagonal values are called as singular values. Singular values are the square root of eigenvalues of XX^T (or $X^T X$). Columns of Q_1 and Q_2 are eigenvectors of XX^T and $X^T X$ respectively [42, 3]. When compared to the eigenvalue decomposition (equation 3.33), it can be seen that the form is the same except that there are two matrices Q_1 and Q_2 (only Q in spectral decomposition).

Now,

$$XX^T = Q_1 \Sigma Q_2^T Q_2 \Sigma^T Q_1^T \implies XX^T = Q_1 \Lambda_1 Q_1^T$$

where $\Lambda_1 = \Sigma\Sigma^T$. This is the spectral decomposition of XX^T. Similarly,

$$X^T X = Q_2 \Lambda_2 Q_2^T \tag{3.43}$$

where $\Lambda_2 = \Sigma\Sigma^T$. Eigenvalues of XX^T and $X^T X$ are by construction non-negative, since these are guaranteed to be positive semi-definite. This is because $y^T XX^T y = (X^T y)^T X^T y$, which is guaranteed to be non-negative for all y. Similar arguments hold for $X^T X$. Thus, SVD of a rectangular matrix X is based on the spectral decompositions of XX^T and $X^T X$ utilizing all the results pertaining to symmetric matrices.

Consider the example $X = \begin{bmatrix} 1 & 1 & 0 \\ 2 & 3 & 1 \\ 4 & 5 & 1 \\ 3 & 7 & 5 \end{bmatrix}$. The rank of this matrix is 3. SVD of this matrix is given below (computed using a programming platform).

$$X = \begin{bmatrix} -0.11 & 0.25 & -0.51 & -0.82 \\ -0.32 & 0.22 & 0.83 & -0.41 \\ -0.54 & 0.71 & -0.2 & 0.41 \\ -0.78 & -0.62 & -0.13 & 0 \end{bmatrix} \begin{bmatrix} 11.55 & 0 & 0 \\ 0 & 2.74 & 0 \\ 0 & 0 & 0.077 \\ 0 & 0 & 0 \end{bmatrix} \begin{bmatrix} -0.45 & 0.62 & -0.64 \\ -0.79 & 0.055 & 0.61 \\ -0.41 & -0.79 & -0.46 \end{bmatrix}$$
$$\underset{Q_1}{} \qquad\qquad \underset{\Sigma}{} \qquad\qquad \underset{Q_2}{}$$

$$\tag{3.44}$$

It can be seen that the Σ contains the square root of eigenvalues of XX^T and $X^T X$ arranged in the decreasing order. Q_1 has the eigenvectors of XX^T and Q_2 has the eigenvectors of $X^T X$.

As alluded to in the section on solving linear equations, SVD can be used in finding solutions to all cases of $X\beta = \mathbf{y}$. Since $X = Q_1 \Sigma Q_2^T$, pseudo-inverse of X can be written as

$$X^\dagger = (Q_1 \Sigma Q_2^T)^\dagger = (Q_2^T)^\dagger \Sigma^\dagger Q_1^\dagger$$

Since Q_1 and Q_2 are orthonormal, $Q_1^\dagger = Q_1^T$ and $(Q_2^T)^\dagger = Q_2$. Thus $X^\dagger = Q_2 \Sigma^\dagger Q_1^T$. Since Σ is a rectangular diagonal matrix of size $m \times n$, the pseudo-inverse $\Sigma^\dagger$ will be a rectangular diagonal matrix of size $n \times m$ with inverse of non-zero diagonal elements of Σ as its diagonal entries. The solution to $X\beta = \mathbf{y}$ can thus be found using $\beta = X^\dagger \mathbf{y}$. This is particularly useful when the size of the matrix X is very large. This inverse of any matrix is called the Moore-Penrose inverse.

Example 3.30. *Perform SVD of the matrix* $X = \begin{bmatrix} 1 & 2 & 3 \\ 6 & 5 & 4 \end{bmatrix}$.

Solution 3.30.

$$XX^T = \begin{bmatrix} 1 & 2 & 3 \\ 6 & 5 & 4 \end{bmatrix} \begin{bmatrix} 1 & 6 \\ 2 & 5 \\ 3 & 4 \end{bmatrix} = \begin{bmatrix} 14 & 28 \\ 28 & 77 \end{bmatrix}$$

Eigenvalues of XX^T are 87.6456 and 3.3544. The corresponding eigenvectors are $\begin{bmatrix} 0.3554 & 0.9347 \end{bmatrix}^T$ and $\begin{bmatrix} -0.9347 & 0.3554 \end{bmatrix}^T$, respectively.

$$X^T X = \begin{bmatrix} 1 & 6 \\ 2 & 5 \\ 3 & 4 \end{bmatrix} \begin{bmatrix} 1 & 2 & 3 \\ 6 & 5 & 4 \end{bmatrix} = \begin{bmatrix} 37 & 32 & 27 \\ 32 & 29 & 26 \\ 27 & 26 & 25 \end{bmatrix}$$

Eigenvalues of $X^T X$ are 87.6456, 3.3544, and 0. The corresponding eigenvectors are $\begin{bmatrix} 0.637 & 0.5751 & 0.5133 \end{bmatrix}^T$, $\begin{bmatrix} 0.6539 & -0.0505 & -0.7549 \end{bmatrix}^T$, and $\begin{bmatrix} 0.4082 & -0.8165 & 0.4082 \end{bmatrix}^T$, respectively. Singular values are the square root of eigenvalues of XX^T and $X^T X$. The singular values for the given matrix are 9.3619 and 1.8315. Thus, we get the following SVD for the matrix X.

$$X = Q_1 \Sigma Q_2^T = \begin{bmatrix} 0.3554 & -0.9347 \\ 0.9347 & 0.3554 \end{bmatrix} \begin{bmatrix} 9.3619 & 0 & 0 \\ 0 & 1.8315 & 0 \end{bmatrix} \begin{bmatrix} 0.637 & 0.6539 & 0.4082 \\ 0.5751 & -0.0505 & -0.8165 \\ 0.5133 & -0.7549 & 0.4082 \end{bmatrix}^T$$

Evaluating SVD using inbuilt packages in a programming platform, we get,

$$X = \begin{bmatrix} 0.3554 & 0.9347 \\ 0.9347 & -0.3554 \end{bmatrix} \begin{bmatrix} 9.3619 & 0 & 0 \\ 0 & 1.8315 & 0 \end{bmatrix} \begin{bmatrix} 0.637 & -0.6539 & 0.4082 \\ 0.5751 & 0.0505 & -0.8165 \\ 0.5133 & 0.7549 & 0.4082 \end{bmatrix}^T$$

Note that some of the columns of Q_1 and Q_2 identified using a programming platform are the negatives of the corresponding columns that were identified by our calculations. But, both the representations would result in correct decomposition of X. The difference is because of the fact that if β is an eigenvector, $c\beta$ is also an eigenvector for any value of c. Hence, SVD is not unique for any matrix. Σ is unique (with singular values in the descending order) while Q_1 and Q_2 are unique upto sign changes in columns or the orientation of the eigenvectors.

Example 3.31. *Perform SVD of the matrix* $\begin{bmatrix} 2 & 1 \\ 1 & 1 \end{bmatrix}$.

Solution 3.31. *Using inbuilt packages in various programming platforms, we can get the SVD of any matrix.*

$$Q_1 = \begin{bmatrix} -0.8507 & -0.5257 \\ -0.5257 & 0.8507 \end{bmatrix} \text{ and } \Sigma = \begin{bmatrix} 2.618 & 0 \\ 0 & 0.3820 \end{bmatrix} \text{ and } Q_2 = \begin{bmatrix} -0.8507 & -0.5257 \\ -0.5257 & 0.8507 \end{bmatrix}$$

See that, $Q_1 \Sigma Q_2^T = \begin{bmatrix} 2 & 1 \\ 1 & 1 \end{bmatrix} = X$

Example 3.32. *Solve* $2u + v = 5$ *and* $u + v = 4$ *using SVD*

Solution 3.32. *Given* $X\beta = \mathbf{y}$, $\beta^* = X^{-1}\mathbf{y}$. *Using SVD,* $X = Q_1 \Sigma Q_2^T$,

$$\beta^* = (Q_1 \Sigma Q_2^T)^{-1}\mathbf{y} = (Q_2^T)^{-1} \Sigma^{-1} (Q_1)^{-1})\mathbf{y} = Q_2 \Sigma^{-1} Q_1^T \mathbf{y}$$

Substituting Q_1, Q_2, *and* Σ *values obtained in Example 3.31, we get* $\beta^* = \begin{bmatrix} 1 & 3 \end{bmatrix}^T$. *Thus the solution to the given equations is* $u = 1$ *and* $v = 3$.

3.7.3 Understanding Data Spread, Significant Directions, Linear Relationships and Noise Removal through SVD

For the rectangular data matrix X, the number of samples is generally greater than the number of variables. Hence, the rank is decided by the linearly independent variables. It can be shown that the rank of X and $X^T X$ are equal. Consider the null spaces of X and $X^T X$.

$$X\beta = 0 \implies X^T X\beta = 0 \tag{3.45}$$

$$X^T X\beta = 0 \implies \beta^T X^T X\beta = 0 \implies (X\beta)^T X\beta = 0 \implies X\beta = 0 \tag{3.46}$$

Hence $\beta \in \mathcal{N}(X)$, iff $\beta \in \mathcal{N}(X^T X)$. Consequently, the dimensions of null spaces of X and $X^T X$ are the same.

$$\mathcal{N}(X^T X) = \mathcal{N}(X) \tag{3.47}$$

Now, rank-nullity theorem for X states that

$$r_1 + Dim(\mathcal{N}(X)) = n \tag{3.48}$$

where r_1 is the rank of X. Similarly, rank-nullity theorem for $X^T X$ can be written as

$$r_2 + Dim(\mathcal{N}(X^T X)) = n \tag{3.49}$$

Since $Dim(\mathcal{N}(X)) = Dim(\mathcal{N}(X^T X))$, we have $r = r_1 = r_2$.

Since the rank of $X^T X$ and X are equal, the number of non-zero eigenvalues of $X^T X$ (*i.e.*, the number of non-zero singular values counting repeats) is equal to the rank of the matrix X. Since the $n - r$ eigenvectors in Q_2 corresponding to the zero eigenvalues form the null space of $X^T X$ and $\mathcal{N}(X^T X) = \mathcal{N}(X)$, they also form the nullspace of X. As a result, vectors in $\mathcal{N}(X^T X)$ provide the relationships between variables. The remaining r eigenvectors in Q_2 provide the subspace in which the data (stored in X) is located. The eigenvector corresponding to the largest eigenvalue would represent the direction of maximum spread in the data or the dominant direction in the data.

$X = Q_1 \Sigma Q_2^T$ will translate to

$$X = \sigma_1 \mathbf{q_1^1} \mathbf{q_2^{1T}} + \ldots + \sigma_n \mathbf{q_1^n} \mathbf{q_2^{nT}} \tag{3.50}$$

where each of the matrix $\mathbf{q_1^n} \mathbf{q_2^{nT}}$ is a rank 1 matrix. σ_i is the ith singular value in Σ. If r is the rank of X, then the r terms in this expansion correspond to the non-zero singular values. $\sigma_{r+1}, \ldots, \sigma_n$ are all zeros. Thus the above equation can be simplified as

$$X = \sigma_1 \mathbf{q_1^1} \mathbf{q_2^{1T}} + \ldots + \sigma_r \mathbf{q_1^r} \mathbf{q_2^{rT}} \tag{3.51}$$

This can be written as

$$\underset{m \times n}{X} = \underset{m \times r}{Q_1} \underset{r \times r}{\Sigma} \underset{r \times n}{Q_2^T} \tag{3.52}$$

This is referred to as compact SVD. Compact SVD representation helps in dimensionality reduction.

In the presence of uncertainty, true data matrices are unlikely to possess zero singular values even when there are relationships among variables. One has to make a judgement about relevant information and this usually translates to retaining or dropping of terms in the SVD expansion based on the magnitude of singular values. If terms corresponding to r largest singualr values of the matrix are retained in the compact SVD form, the resultant matrix can be thought of as the denoised data matrix. The notional noise component can be obtained using the terms for the remaining $n - r$ singular values.

$$\underset{m \times n}{X} = \underset{m \times r}{Q_{1t}} \underset{r \times r}{\Sigma_t} \underset{r \times n}{Q_{2t}^T} + \underset{m \times n-r}{Q_{1u}} \underset{n-r \times n-r}{\Sigma_u} \underset{n-r \times n}{Q_{2u}^T} \tag{3.53}$$

The first term in the RHS represents T (true data) and the second term corresponds to U (uncertain data).

$$\underset{m \times n}{T} = \underset{m \times r}{Q_{1t}} \underset{r \times r}{\Sigma_t} \underset{r \times n}{Q_{2t}^T} \tag{3.54}$$

All of these ideas can be illustrated through a simple 2D dataset shown in Figure 3.5. Corresponding to this dataset, there will be two singular values. Spread of data in the direction of the solid line drawn in the figure is provided by the highest singular value while the spread along the perpendicular direction (shown by dashed line) is provided by the other singular value. If the second singular value is deemed to be small, then the first vector in Q_2 will provide the direction (solid line) and the second vector in Q_2 will define the relationship (through the normal to the line). Projection of

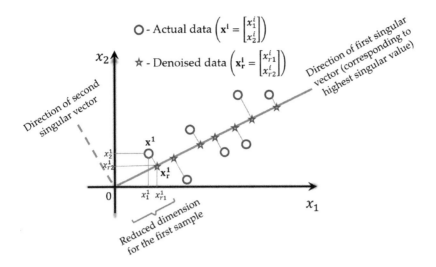

FIGURE 3.5 Data spread and significance of eigenvalues.

any point onto the line will provide noise removed datapoint. Retaining only
the first term in the expanded SVD form (equation 3.51) will provide the
denoised matrix. Dimensionality reduction is achieved through the use of the
one-dimensional variable, the coordinate value corresponding to any point in
the direction of maximum spread.

Example 3.33. *Find the direction of maximum spread in the data matrix*

$$X = \begin{bmatrix} 2.6 & 8.03 \\ 3.4 & 11.11 \\ 3.8 & 11.55 \\ 1.4 & 5.03 \\ 2.7 & 8.64 \end{bmatrix}$$

Solution 3.33.

$$X^T X = \begin{bmatrix} 42.01 & 132.912 \\ 132.912 & 421.266 \end{bmatrix}$$

*Eigenvalues and eigenvectors can be computed using computational tools like
MATLAB or Python. Eigenvalues are 463.207 and 0.069. Since 463.207 is
the highest eigenvalue, the eigenvector corresponding to this eigenvalue will
be the direction of maximum spread. Eigenvector of $\lambda_1 = 463.207$ is $\nu_1 = \begin{bmatrix} 0.3009 & 0.9536 \end{bmatrix}^T$.*

*The line joining $(0,0)$ and $(0.3009, 0.9536)$ will be the direction of maximum
spread in the data. This is also evident from Figure 3.6. Note that the datapoints lie along the arrow which shows the identified direction of maximum
spread.*

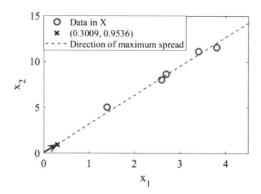

FIGURE 3.6 Data given in Example 3.33. Direction of maximum spread based on eigenvector is also shown using the arrow.

Example 3.34. *Find the linear relations between variables in the data matrix* $X = \begin{bmatrix} 1.1 & 2.05 \\ 2 & 4.01 \end{bmatrix}$ *using SVD.*

Solution 3.34. *From 3.35, we know that the insignificant eigenvalue is* 0.0616 *and the corresponding eigenvector is* $\begin{bmatrix} -0.8920 & 0.4520 \end{bmatrix}^T$. *Hence the corresponding linear relation is:*

$$\begin{bmatrix} x_1 & x_2 \end{bmatrix} \begin{bmatrix} -0.8920 \\ 0.4520 \end{bmatrix} = 0 \implies x_2 = 1.9735 x_1$$

Example 3.35. *Find the compact SVD for the data matrix,* $X = \begin{bmatrix} 1.1 & 2.05 \\ 2 & 4.01 \end{bmatrix}$.

Solution 3.35. *Taking SVD, we have* $X = Q_1 \Sigma Q_2^T$ *where*

$$Q_1 = \begin{bmatrix} -0.4607 & -0.8876 \\ -0.8876 & 0.4607 \end{bmatrix}; \qquad Q_2^T = \begin{bmatrix} -0.4520 & -0.8920 \\ -0.8920 & 0.4520 \end{bmatrix}$$

$$\Sigma = \begin{bmatrix} 5.0486 & 0 \\ 0 & 0.0616 \end{bmatrix}$$

From Σ, *we know that the singular values (square roots of eigenvalues) are* 5.05 *and* 0.06. *Since the second eigenvalue is close to zero in comparison the other, we can retain only the first eigenvalue. Then, the corresponding compact SVD will be as follows:*

$$X \approx Q_{1t} \Sigma_t Q_{2t}^T$$

where

$$Q_{1t} = \begin{bmatrix} -0.4607 \\ -0.8876 \end{bmatrix} \text{ and } Q_{2t}^T = \begin{bmatrix} -0.4520 & -0.8920 \end{bmatrix} \text{ and } \Sigma_t = \begin{bmatrix} 5.0486 \end{bmatrix}$$

EXERCISE PROBLEMS

Exercise 3.1. *Find the rank and nullity of the following matrices*

(a) $A = \begin{bmatrix} 3.7 & 1 \\ 7.4 & 2 \end{bmatrix}$

(b) $A = \begin{bmatrix} 3.7 & 1 \\ 7.4 & 4 \end{bmatrix}$

(c) $A = \begin{bmatrix} 3.7 & 1 & 2.4 \\ 4.6 & 5 & 1.5 \\ 8.3 & 6 & 3.9 \end{bmatrix}$

(d) $A = \begin{bmatrix} 3.7 & 1 & 2.4 \\ 4.6 & 5 & 1.5 \\ 2.3 & 7 & 1.9 \end{bmatrix}$

(e) $A = \begin{bmatrix} 1 & 4 & 9 \\ 4 & 5 & 2 \\ 3 & 3 & 1 \end{bmatrix}$

(f) $A = \begin{bmatrix} 10 & 11 & 4 \\ 4 & 5 & 2 \\ 3 & 3 & 1 \end{bmatrix}$

Exercise 3.2. *Find the dimensions of the four fundamental subspaces of the following matrices*

(a) $A = \begin{bmatrix} 3.7 & 1 \\ 7.4 & 2 \end{bmatrix}$

(b) $A = \begin{bmatrix} 3.7 & 1 \\ 7.4 & 4 \end{bmatrix}$

(c) $A = \begin{bmatrix} 3.7 & 1 & 2.4 \\ 4.6 & 5 & 1.5 \\ 8.3 & 6 & 3.9 \end{bmatrix}$

(d) $A = \begin{bmatrix} 3.7 & 1 & 2.4 \\ 4.6 & 5 & 1.5 \\ 2.3 & 7 & 1.9 \end{bmatrix}$

(e) $A = \begin{bmatrix} 1 & 4 & 9 \\ 4 & 5 & 2 \\ 3 & 3 & 1 \end{bmatrix}$

(f) $A = \begin{bmatrix} 10 & 11 & 4 \\ 4 & 5 & 2 \\ 3 & 3 & 1 \end{bmatrix}$

Exercise 3.3. *Find the basis set for the row-space and column space of the following matrices:*

(a) $A = \begin{bmatrix} 3.7 & 1 \\ 7.4 & 2 \end{bmatrix}$

(b) $A = \begin{bmatrix} 3.7 & 1 \\ 7.4 & 4 \end{bmatrix}$

(c) $A = \begin{bmatrix} 3.7 & 1 & 2.4 \\ 4.6 & 5 & 1.5 \\ 8.3 & 6 & 3.9 \end{bmatrix}$

(d) $A = \begin{bmatrix} 3.7 & 1 & 2.4 \\ 4.6 & 5 & 1.5 \\ 2.3 & 7 & 1.9 \end{bmatrix}$

(e) $A = \begin{bmatrix} 1 & 4 & 9 \\ 4 & 5 & 2 \\ 3 & 3 & 1 \end{bmatrix}$

(f) $A = \begin{bmatrix} 10 & 11 & 4 \\ 4 & 5 & 2 \\ 3 & 3 & 1 \end{bmatrix}$

Exercise 3.4. *Find the LU decomposition of the following matrices:*

(a) $A = \begin{bmatrix} 3.7 & 1 \\ 7.4 & 2 \end{bmatrix}$

(b) $A = \begin{bmatrix} 3.7 & 1 \\ 7.4 & 4 \end{bmatrix}$

(c) $A = \begin{bmatrix} 3.7 & 1 & 2.4 \\ 4.6 & 5 & 1.5 \\ 8.3 & 6 & 3.9 \end{bmatrix}$

(d) $A = \begin{bmatrix} 3.7 & 1 & 2.4 \\ 4.6 & 5 & 1.5 \\ 2.3 & 7 & 1.9 \end{bmatrix}$

(e) $A = \begin{bmatrix} 1 & 4 & 9 \\ 4 & 5 & 2 \\ 3 & 3 & 1 \end{bmatrix}$

(f) $A = \begin{bmatrix} 10 & 11 & 4 \\ 4 & 5 & 2 \\ 3 & 3 & 1 \end{bmatrix}$

Exercise 3.5. *Is there any linear relation between variables in the following data matrices? If yes, find those linear relations.*

(a) $A = \begin{bmatrix} 3.7 & 1 \\ 7.4 & 2 \end{bmatrix}$

(b) $A = \begin{bmatrix} 3.7 & 1 \\ 7.4 & 4 \end{bmatrix}$

(c) $A = \begin{bmatrix} 3.7 & 1 & 2.4 \\ 4.6 & 5 & 1.5 \\ 8.3 & 6 & 3.9 \end{bmatrix}$

(d) $A = \begin{bmatrix} 3.7 & 1 & 2.4 \\ 4.6 & 5 & 1.5 \\ 2.3 & 7 & 1.9 \end{bmatrix}$

(e) $A = \begin{bmatrix} 1 & 4 & 9 \\ 4 & 5 & 2 \\ 3 & 3 & 1 \end{bmatrix}$

(f) $A = \begin{bmatrix} 10 & 11 & 4 \\ 4 & 5 & 2 \\ 3 & 3 & 1 \end{bmatrix}$

Exercise 3.6. *Find whether the following vectors are orthogonal? If orthogonal, find the corresponding orthonormal vectors.*

(i) $\mathbf{a} = \begin{bmatrix} 2 & -3 & 0 \end{bmatrix}^T$ and $\mathbf{b} = \begin{bmatrix} 3 & 2 & -4 \end{bmatrix}^T$

(ii) $\mathbf{a} = \begin{bmatrix} 2 & -3 & 0 \end{bmatrix}^T$, $\mathbf{b} = \begin{bmatrix} 3 & 2 & -4 \end{bmatrix}^T$, and $\mathbf{c} = \begin{bmatrix} 1.5 & 1 & 1 \end{bmatrix}^T$

(iii) $\mathbf{a} = \begin{bmatrix} 2 & -3 & 1 \end{bmatrix}^T$, $\mathbf{b} = \begin{bmatrix} 4 & 2 & -2 \end{bmatrix}^T$, and $\mathbf{c} = \begin{bmatrix} -5 & -1 & 7 \end{bmatrix}^T$

Exercise 3.7. *Consider a 2D space with a hyperplane $5x + 2y + 7z + 4 = 0$. Classify the following points based on the half-spaces they belong to.*

$$(-1, 2, 4), \ (2, 7, -1), \ (3, 0, 1), \ (2, 5, -7), \ (-3, -1, 1), \ (5, 0, 0)$$

Exercise 3.8. *Solve the following equations using matrix inverse ideas:*

(a) $3x + 4y = 11$ *and* $x + y = 3$
(b) $3x + 6y = 12$ *and* $x + 2y = 4$
(c) $2x + y + 3z = 8$, $x + 2y + 3z = 5$ *and* $4x + 5y + 6z = 40$
(d) $2x + y + 3z = 8$, $x + 2y + 3z = 5$ *and* $4x + 5y + 9z = 40$
(e) $3x + 4y = 11$, $x + y = 3$ *and* $2x + 5y = 7$

Exercise 3.9. *Solve the following equations using spectral decomposition:*

(a) $3x + 4y = 11$ *and* $x + y = 3$
(b) $3x + 6y = 12$ *and* $x + 2y = 4$
(c) $2x + y + 3z = 8$, $x + 2y + 3z = 5$ *and* $4x + 5y + 6z = 40$
(d) $2x + y + 3z = 8$, $x + 2y + 3z = 5$ *and* $4x + 5y + 9z = 40$
(e) $3x + 4y = 11$, $x + y = 3$ *and* $2x + 5y = 7$

Exercise 3.10. *Solve the following equations using SVD:*

(a) $3x + 4y = 11$ *and* $x + y = 3$
(b) $3x + 6y = 12$ *and* $x + 2y = 4$
(c) $2x + y + 3z = 8$, $x + 2y + 3z = 5$ *and* $4x + 5y + 6z = 40$
(d) $2x + y + 3z = 8$, $x + 2y + 3z = 5$ *and* $4x + 5y + 9z = 40$
(e) $3x + 4y = 11$, $x + y = 3$ *and* $2x + 5y = 7$

Exercise 3.11. *Find the direction of maximum spread in the data matrix*

$$A = \begin{bmatrix} 2.6 & 8.03 & 7.4 \\ 3.4 & 11.11 & 5.18 \\ 3.8 & 11.55 & 6.39 \\ 1.4 & 5.03 & 4.3 \\ 2.7 & 8.64 & 2.7 \\ 1.78 & 3.65 & 3.39 \\ 4.84 & 7.31 & 9.25 \end{bmatrix}$$

Exercise 3.12. *Find A^{72} for the matrix $A = \begin{bmatrix} 4 & 5 & 1 \\ 5 & 4 & 1 \\ 3 & 1 & 1 \end{bmatrix}$ using*

(a) Spectral decomposition
(b) Cayley-hamilton theorem

Exercise 3.13. *Find the compact SVD for the data matrix $A = \begin{bmatrix} 3.02 & 9.21 & 1.07 \\ 4.11 & 11.89 & 1.23 \\ 5.94 & 18.13 & 2.02 \end{bmatrix}$ and find the linear relations between the variables, if any. Find the denoised data such that 95% variance is captured.*

Exercise 3.14. *Verify Cayley-Hamilton theorem for the matrix $A = \begin{bmatrix} 5 & 7 & 4 \\ 1 & 3 & 2 \\ 6 & 3 & 1 \end{bmatrix}$ and find A^{-1}.*

Exercise 3.15. *Calculate the Jordan block for the following matrices, and comment on the algebraic and geometric multiplicities of the eigenvalues of both matrices.*

a) $A = \begin{bmatrix} -3 & 0 & 4 & 0 \\ -2 & 0 & 0 & 1.1 \\ -1.2 & 1 & 0 & 0.6 \\ 0.6 & 0 & 1 & 0.8 \\ 2.4 & 0 & 0 & 2 \end{bmatrix}$; *b)* $A = \begin{bmatrix} 0 & 0 & 1 & 0 \\ 0 & 0 & 1 & 1 \\ 0 & 0 & 0 & 0 \\ 0 & 0 & 0 & 0 \end{bmatrix}$

Exercise 3.16. *Identify whether the matrix $\begin{bmatrix} 3.1 & 4.23 & 5.01 \\ 7.35 & 8.12 & 4.93 \\ 5.71 & 0.59 & 1.66 \end{bmatrix}$ is diagonal-izable.*

Exercise 3.17. *Perform QR factorization of the matrix $\begin{bmatrix} 3.1 & 4.23 & 5.01 \\ 7.35 & 8.12 & 4.93 \\ 5.71 & 0.59 & 1.66 \end{bmatrix}$ using Gram-Schmidt orthogonalization procedure.*

Exercise 3.18. *Given a list of products*

Products	Electricity Cost	Production Cost	Raw Material Cost	Labour Cost
A	275	110	53	10
B	290	120	76	15
C	352	85	96	20
D	452	130	120	12
E	140	70	45	11
F	185	53	72	8
G	220	110	40	6
H	276	135	87	13
I	312	145	97	12
J	276	129	65	5

1. *Mean center the above data matrix X and perform SVD (USV^T).*
2. *Construct the best rank 2 approximation of the matrix X ($X_{reduced}$)*
3. *How does the constructed $X_{reduced}$ compare with X?*

4 Optimization for DS and ML

Optimization is at the core of data science. While ML algorithms can be viewed from different perspectives, the derivation of the algorithm itself can be ultimately formalized as an optimization problem. An example of this is the well-known Hebbian learning [17], where the original arguments were inspired from biological conceptualization; however, the same learning can be derived as a solution to an optimization problem. Similarly, while the original clustering algorithms and the corresponding learning rules were intuitively derived, these rules can be shown to result from an optimization formulation. A fundamental understanding of optimization algorithms is critical for a deeper understanding of ML algorithms [46].

4.1 ELEMENTS OF AN OPTIMIZATION FORMULATION

Every optimization problem has three main components. The first is the objective function to be maximized or minimized. The identification of this function is critical as it identifies and formalizes the quantity of interest. The second is the decision variables that can be assigned values to optimize the objective function. The third is a set of constraints that describe the rules that the decision variables should follow [36, 45, 8]. Mathematically, an optimization problem is formulated as follows

$$
\begin{aligned}
\min_{\mathbf{x}} \quad & f(\mathbf{x}) \\
\text{subject to} \quad & \mathbf{g}(\mathbf{x}) \leq 0 \\
& \mathbf{h}(\mathbf{x}) = 0
\end{aligned}
\tag{4.1}
$$

In the formulation, $\mathbf{x}$ is a vector of variables (decision variables), $f(\mathbf{x})$ is a scalar function (objective function) and $\mathbf{g}(\mathbf{x})$ is a vector of functions that the decision variables have to satisfy (inequality constraints), and $\mathbf{h}(\mathbf{x})$ is also a vector of functions that the decision variables have to satisfy (equality constraints). Optimization problems are classified into various categories based on the nature of the decision variables and the objective and constraint functions. In what follows, different types of optimization functions will be briefly described and the most important optimization formulation from an ML viewpoint will be identified.

The objective function can be a linear or nonlinear function of the decision variables. The decision variables can be of different types. These variables could be continuous or discrete. When the variables are discrete, they could

DOI: 10.1201/b23276-4

TABLE 4.1

Types of Optimization Problems

Decision Variables	Objective Function	Constraints	Type
Continuous	Nonlinear	None	Unconstrained NLP
Continuous	Nonlinear	Nonlinear	NLP
Continuous	Quadratic	Linear	QP
Continuous	Linear	Linear	LP
Binary	Linear	Linear	BLP
Integer	Linear	Linear	ILP
Continuous and Integer	Linear	Linear	MILP
Continuous and Integer	Nonlinear	Nonlinear	MINLP

NLP - Nonlinear Programming
QP - Quadratic Programming
LP - Linear Programming
BLP - Binary Linear Programming
ILP - Integer Linear Programming
MILP - Mixed Integer Linear Programming
MINLP - Mixed Integer Nonlinear Programming

either take binary values (0 or 1) or take values from a set of integers. The constraint functions ($\mathbf{h}(\mathbf{x})$, $\mathbf{g}(\mathbf{x})$) could be linear or nonlinear. Common optimization formulations are described in Table 4.1.

While the table is self-explanatory, one point of clarification is that for a problem to be considered NLP, it is enough if one of either the objective function or the constraints is nonlinear [11, 28]. While all these problem types are of importance for more advanced data science solutions, the most important class of problems as far as ML is concerned is the unconstrained NLP [46]. Many learning rules in ML algorithms are derived as some modification of a gradient descent algorithm, used to solve NLPs. Constrained NLPs are solved through the Karush-Kuhn-Tucker (KKT) [20, 24] conditions. Familiarity with these conditions is critical to the understanding of ML algorithms.

Example 4.1. *For the optimization problems given below, identify whether the problem is univariate/multivariate, constrained/unconstrained and NLP/LP/QP/MINLP/BLP/ILP. Explain with reasons.*

a) $\min x^4$
b) $\min x^2 + y^2$
c) $\min x^4$ s.t. $-5 \leq x \leq 2$
d) $\min x^2 + y^2$ s.t. $-5 \leq x \leq 2$; $-4 \leq y \leq -1$
e) $\min x^2 + y^2$ s.t. $-5 \leq x \leq 2$; $-4 \leq x^3 + y^3 \leq 81$

f) $\min x^2 + y^2$ s.t. $-5 \leq x \leq 2$; $-4 \leq x^3 + y^3 \leq 81$; $y \in \mathbb{Z}$ *(integer space)*

Solution 4.1. *a) Univariate, unconstrained, NLP.*
(Only one variable x, no constraints, nonlinear objective function)
b) Multivariate, unconstrained, QP
(Since 2 variables (x&y), no constraints, quadratic objective function)
c) Univariate, constrained, NLP
d) Multivariate, constrained, QP
e) Multivariate, constrained, NLP
f) Multivariate, constrained, MINLP

4.2 DISCUSSION ON OBJECTIVE FUNCTIONS FOR CLASSIFICATION AND FUNCTION APPROXIMATION PROBLEMS

Unconstrained optimization problems comprise of only two (objective function and decision variables) of the three components of a general optimization problem. Before optimization solutions are discussed in detail, commonly used objective functions for function approximation and classification are outlined (one for each).

4.2.1 Function Approximation Objective Function

A typical function approximation problem will process data of the form shown in Table 4.2, for a single output and multiple inputs. The output prediction ($\hat{y}^i$) from an ML algorithm with a particular functional form is shown as an additional column. $\mathbf{p}$ denotes the vector of the required parameters. x_j^i is the ith sample of the jth variable and m represents the number of samples available for function approximation.

Different ML algorithms choose different functional forms. The goal of learning is to identify model parameter values that best fit the data. This would require the definition of a function that measures how well the model (with the chosen parameters) fits the data. Once such a function is defined, then the parameters can be used as decision variables for optimization.

TABLE 4.2

Data for Function Approximation Problem

Output	Input 1	...	Input n	Predicted $\hat{y}^i$
y^1	x_1^1	...	x_n^1	$f_{ML}(x_1^1, \ldots, x_n^1, \mathbf{p})$
$\vdots$	$\vdots$	...	$\vdots$	$\vdots$
y^m	x_1^m	...	x_n^m	$f_{ML}(x_1^m, \ldots, x_n^m, \mathbf{p})$

Sum of squared errors (SSE) shown in equation 4.2 [31] is a common objective function that is used.

$$\text{Error or Loss Function} = \sum_{i=1}^{m}(y^i - f_{ML}(x_1^i, \ldots, x_n^i, \mathbf{p}))^2 \qquad (4.2)$$

It can be seen that there are m terms, one for each of the samples. The aim of an ML algorithm is to minimize this error. The objective is a scalar error function that collectively minimizes the error for all the samples. As the objective function is a sum of squares, the lowest value that it can take is zero. When the parameters are chosen such that the objective function is zero, the identified model is an exact match to the data. This seldom occurs in real problems. The learning rule of an ML algorithm is an outcome of solving the minimization problem as shown in equation 4.3.

$$\min_{\mathbf{p}} \sum_{i=1}^{m}(y^i - f_{ML}(x_1^i, \ldots, x_n^i, \mathbf{p}))^2 \qquad (4.3)$$

The extension of the objective function to multiple output problems is rather straightforward and is shown in equation 4.4 for a problem with l outputs. One can notice that there is a superscript denoting a sample number and a subscript denoting the output identifier. $f_{ML,k}$ denotes the function corresponding to the kth output. An optimization-based ML rule can be setup for function approximation problems with multiple outputs.

$$\text{Error or Loss Function} = \sum_{i=1}^{m}\sum_{k=1}^{l}(y_k^i - f_{ML,k}(x_1^i, \ldots, x_n^i, \mathbf{p}))^2 \qquad (4.4)$$

Example 4.2. *Consider the following data.*

x	1	2	3
y	3	6	9

If we are interested in fitting the given data using a strict linear model $(\hat{y} = f_{ML}(x) = mx)$, *write the optimization problem to find the unknown parameter m.*

Solution 4.2. *Optimization problem:*

$$\min_{m} \sum_{i=1}^{3}\left(y^i - \hat{y}^i\right)^2 \Rightarrow \min_{m} \sum_{i=1}^{3}\left(y^i - mx^i\right)^2$$

$$\Rightarrow \min_{m} \quad (3 - m)^2 + (6 - 2m)^2 + (9 - 3m)^2$$

This is an unconstrained quadratic programming problem.

4.2.2   Classification Objective Function

Let us consider classification problems. Table 4.3 depicts typical data for a binary classification problem. There are some differences between data for classification and function approximation problems. For each input vector, the output takes a real value in the case of function approximation, whereas a class label (C_1 or C_2) is assigned in classification problems. For illustration purpose, data for a typical binary classification problem has been presented. The class information can be converted to numeric information by assigning $C_1 = -1$ and $C_2 = 1$.

An ML approach identifies a decision function $h(\mathbf{x})$, where $\mathbf{x}$ is a vector of input values, and the output of the decision function is converted to class information [31]. This is usually achieved through

$$h(\mathbf{x}) \leq 0 \quad \mathbf{x} \in C_1 \tag{4.5}$$
$$h(\mathbf{x}) > 0 \quad \mathbf{x} \in C_2 \tag{4.6}$$

As before, $h(\mathbf{x})$ is either chosen *a priori* or specified through some hyper-parameter tuning. The ML rule provides an approach to identify the best parameters for correct classification of all the samples. Since there is an output value for each input, it might be tempting to use the same objective function as in equation 4.2. However, a more appropriate objective function for this problem is given by equation 4.7.

$$\text{Error or Loss Function} = \sum_{i=1}^{m} e^{-h(x_1^i, \ \ldots, \ x_n^i, \ \mathbf{p})C_i} \tag{4.7}$$

Let us understand equation 4.7. It can be noticed that the loss function has a term corresponding to each of these samples. Whenever a datapoint is classified correctly, the term corresponding to that datapoint should be small. This would ensure that the benefits accrued by tweaking the parameters for these datapoints is minimal. On the other hand, for incorrectly classified datapoints, the loss term should be high, incentivizing the optimizer to tweak the parameters such that the datapoint is classified correctly. Let us consider a term in equation 4.7 to verify if the loss function exhibits this behavior. Assume datapoint k is correctly classified. If it belongs to class C_1, then $C_k = -1$ and $h(x_1^k, \ldots, x_n^k, \mathbf{p}) \leq 0$, which would make $e^{-h(x_1^k, \ldots, x_n^k, \ \mathbf{p})C_k}$ a small number (the farther away h is from 0 in the negative direction, the smaller this number will be). If this point were incorrectly classified ($h(x_1^k, \ldots, x_n^k, \mathbf{p}) > 0$), then it is obvious that the same term will be large. Similar arguments can be made if the datapoint k belongs to class C_2 (readers to verify).

Example 4.3. *Consider the classification problem given below.*

x_1	1	2	3	1	2	3
x_2	1	1	2	3	5	7
Class	-1	-1	-1	1	1	1

If a linear separator of the form $x_2 - mx_1 = 0$ is used to classify the data, write the corresponding optimization problem that can be used to find the unknown parameter 'm'.

TABLE 4.3

Data for Classification Problems

Class	Input 1	...	Input n	Predicted Class
C_1	x_1^1	...	x_n^1	$h_{ML}(x_1^1,\ldots,x_n^1,\mathbf{p})$
$\vdots$	$\vdots$	...	$\vdots$	$\vdots$
C_1	x_1^m	...	x_n^m	$h_{ML}(x_1^m,\ldots,x_n^m,\mathbf{p})$
C_2	$x_1^{(m+1)}$	...	$x_n^{(m+1)}$	$h_{ML}(x_1^{(m+1)},\ldots,x_n^{(m+1)},\mathbf{p})$
$\vdots$	$\vdots$	...	$\vdots$	$\vdots$
C_2	$x_1^{(m+1)}$	...	$x_n^{(m+l)}$	$h_{ML}(x_1^{(m+l)},\ldots,x_n^{(m+l)},\mathbf{p})$

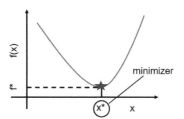

FIGURE 4.1 Single variable problem with a single minimum.

Solution 4.3. *The objective would be minimization of loss function* $=$ $\sum_{i=1}^{l} e^{-h(x_1^i, x_2^i, m) \times C_i}$.

$$\min_{m} \sum_{i=1}^{6} e^{-\left(x_2^i - m x_1^i\right) C_i}$$

$$\implies \min_{m} e^{(1-m)} + e^{(1-2m)} + e^{(2-3m)} + e^{-(3-m)} + e^{-(5-2m)} + e^{-(7-3m)}$$

4.3   FIRST- AND SECOND-ORDER ANALYTICAL CONDITIONS FOR OPTIMALITY OF UNCONSTRAINED NLPS

We now describe the key ideas in solving unconstrained NLP problems. Figure 4.1 shows a single variable function with a minimum point x^*, the point at which the function takes its minimum value f^*. The function depicted in Figure 4.1 is simple and is called a unimodal function (only one optimum point) [36].

In many other cases, functions are likely to have multiple optimum values as depicted in Figure 4.2; here, two minima and a maximum point can be observed. The two minima points are shown as x^{1*} and x^{2*}. Clearly, at x^{2*}, the function attains a lower value than at x^{1*} and hence x^{2*} is a global minimum

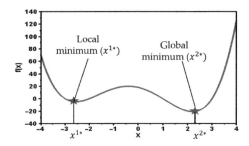

FIGURE 4.2 Single variable problem with multiple minima.

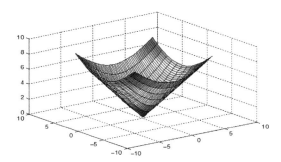

FIGURE 4.3 Two variables problem with a single minimum.

point and x^{1*} is the local minimum point. Interestingly, the problem of multiple local minima and the inability of numerical algorithms for optimization in identifying the global minima has been a concern in ML algorithms for a long time and continues to be so. The lull in activity in the field of neural networks in the late 1990s and early 2000s can be partially attributed to this problem. One could similarly visualize optimization problems in two decision variables; these will be three-dimensional plots with the third dimension used to denote the value of the function as shown in Figure 4.3. Similar to a single variable case, multivariable optimization problems could also have multiple optima, both local and global. [18, 14]. Figure 4.4 depicts one such function, popularly known as the Rastrigin function.

It becomes clear then that one should have a procedure to identify the values of the decision variables at which the objective function takes an optimum value (optimum point). It is also apparent that once such points are identified, they should be labeled as a minimum, maximum or neither. Further, one would also like to label a point as a global minimum point or a local minimum point. In what follows, we will see that it is possible to identify the optima points analytically, classify them as minima, maxima, or saddle points and finally also learn that it is not possible, in general, to identify if

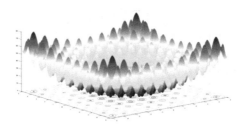

FIGURE 4.4 Two variables problem with a number of optimum points.

a point is a global or a local optimum, except for functions that have certain characteristics that can be established *a priori.*

We will now describe the analytical conditions for optimum solutions to unconstrained NLP problems. This can then be extended to constrained NLP problems. Formal proofs for these conditions are not provided in this book and an appeal to intuition is made to explain the basic ideas. There are a large number of excellent textbooks that cover the field of optimization in considerable detail [36, 28, 8, 4, 6]. Numerical techniques [45, 12] used to find solutions to these analytical conditions will also be described. The most popular among these is the gradient descent solution approach that is at the heart of many ML algorithms. Several modifications to this technique that result in different learning algorithms will also be outlined.

Consider equation 4.1, without any constraints $\mathbf{g}(\mathbf{x})$ and $\mathbf{h}(\mathbf{x})$. As discussed before, $f(\mathbf{x})$ is a scalar function in n variables $\mathbf{x} = [x_1 \ldots x_n]$. Equation 4.1 is posed as a minimization problem without any loss of generality. All maximization problems can be converted to minimization problems as $\max_{\mathbf{x}} f(\mathbf{x})$ is equivalent to $\min_{\mathbf{x}} -f(\mathbf{x})$. A solution to the minimization problem should provide the optimum value of the decision variables $\mathbf{x}^* = [x_1^* \ldots x_n^*]$ and $f(\mathbf{x}^*)$ will provide the minimum value that the objective function can take. This implies that one would need n equations to identify the values for the n variables. These n equations are provided by equation 4.8, also termed as the first-order necessary conditions [36, 45] for an optimum solution of $f(\mathbf{x})$,

$$\boldsymbol{\nabla} f = \begin{bmatrix} \partial f/\partial x_1 \\ \vdots \\ \partial f/\partial x_n \end{bmatrix} = 0 \qquad (4.8)$$

Several important comments need to be made regarding equation 4.8. When there are n nonlinear equations in n variables, all possibilities exist (no solutions, any number of solutions (1, 2, 3 and so on), infinite number of solutions). Any solution that satisfies the first order necessary conditions is an optimum point. We need to verify another condition to ascertain if the solution is a minimum, maximum or a saddle point. It might be worthwhile to contrast this with n linear equations in n variables (which we studied in Chapter 3), where the possibilities are no solution, unique solution or an infinite number of solutions.

Without the theoretical details, the conditions for a point being a local minimum, or maximum or a saddle point depends on a matrix that is computed at the optimum point, called Hessian matrix and denoted by $H(\mathbf{x}^*)$ [36, 6, 8].

$$H(x^*) = \begin{bmatrix} \frac{\partial^2 f}{\partial x_1^2} & \cdots & \frac{\partial^2 f}{\partial x_1 \partial x_n} \\ \vdots & \ddots & \vdots \\ \frac{\partial^2 f}{\partial x_n \partial x_1} & \cdots & \frac{\partial^2 f}{\partial x_n^2} \end{bmatrix}_{x^*} = 0 \qquad (4.9)$$

Given a multivariable function f, the matrix H computed using equation 4.9 will be a function of all the variables. Once we find a $\mathbf{x}^*$ that satisfies equation 4.8, then H at $\mathbf{x}^*$ can be computed. This is a numerical matrix, which is the Hessian matrix at the point $\mathbf{x}^*$. From here on it is easy to decide if the point $\mathbf{x}^*$ is a minimum point, maximum point or a saddle point. The first thing to notice is that the Hessian matrix is symmetric. This is an important observation, which guarantees that the eigenvalues are all real numbers (positive, negative or zero). If the Hessian matrix is positive definite (all eigenvalues strictly greater than zero), then $\mathbf{x}^*$ is a minimum point. If the matrix is negative definite (all eigenvalues strictly less than zero), then $\mathbf{x}^*$ is a maximum point. If there are both positive and negative eigenvalues, then $\mathbf{x}^*$ is a saddle point. Zero eigenvalues in the multivariate case and the second derivative becoming zero in the univariate case will require a little more analysis. Interested readers can consult several resources for a detailed analysis of these cases [36, 28, 8, 4, 6].

Consider Figure 4.2, which depicts a function with two minimum points. At these points, the first derivative is zero and the second derivative is positive. Using these conditions, it is not possible to determine if a point is a local or a global minimum. A determination about the point being a global minimum can only be made by enumerating all the minima points and finding the point(s) at which the function takes the least value. However, most algorithms will terminate when the first order necessary conditions are satisfied. Consequently, most ML algorithms are local algorithms.

Example 4.4. *Find the stationary optimum points and identify whether each of the stationary points corresponds to a maximum, minimum, or saddle point.*

a) $f(x) = (x-3)^2 - 7$
b) $f(x, y) = (x-3)^2 + (y+5)^2 - 7$
c) $f(x, y) = -(x-3)^2 - (y+5)^2 - 7$
d) $f(x, y) = (x-3)^2 - (y+5)^2 - 7$
e) $f(x, y) = \left(x - \frac{5}{3}\right)^2 + x^3 + (y+5)^2 - 7$

Solution 4.4. *a)* $\frac{df}{dx} = 0$ *for a stationary point. So,*

$$2(x-3) = 0 \Rightarrow x^* = 3$$

To find whether x^* *is a minimum/maximum/saddle point, we find the second derivative.*

$$f''(x)|_{x^*} = 2 > 0 \Rightarrow x^* = 3 \text{ is a minimum.}$$

b)

$$\nabla f = \begin{bmatrix} \frac{\partial f}{\partial x} \\ \frac{\partial f}{\partial y} \end{bmatrix} = \begin{bmatrix} 2(x-3) \\ 2(y+5) \end{bmatrix} = 0 \Rightarrow x^* = 3, \ y^* = -5$$

$$H = \begin{bmatrix} 2 & 0 \\ 0 & 2 \end{bmatrix}$$

The Hessian matrix has repeated eigenvalues at 2 (> 0). Thus, $\begin{bmatrix} 3 & -5 \end{bmatrix}^T$ *is a minimum.*

c)

$$f(x,y) = -(x-3)^2 - (y+5)^2 - 7$$

$$\nabla f = \begin{bmatrix} -2(x-3) \\ -2(y+5) \end{bmatrix} = 0 \Rightarrow x^* = 3; \ y^* = -5$$

$$H(x,y) = \begin{bmatrix} -2 & 0 \\ 0 & -2 \end{bmatrix}$$

Eigenvalues, $\lambda = -2, -2 < 0 \Longrightarrow \Rightarrow \mathbf{x}^* = \begin{bmatrix} 3 & -5 \end{bmatrix}^T$ *is a maximum.*

d)

$$f(x,y) = (x-3)^2 - (y+5)^2 - 7$$

$$\nabla f = \begin{bmatrix} 2(x-3) \\ -2(y+5) \end{bmatrix} = 0 \Rightarrow x^* = 3; \ y^* = -5$$

$$H(x,y) = \begin{bmatrix} 2 & 0 \\ 0 & -2 \end{bmatrix} \Longrightarrow \lambda = 2, -2$$

One of the eigenvalues is positive and one is negative. Hence, $\mathbf{x}^* = \begin{bmatrix} 3 & -5 \end{bmatrix}^T$ *is neither a maximum, nor a minimum. It is a saddle point.*

e)

$$f(x,y) = \left(x - \frac{5}{3}\right)^2 + x^3 + (y+5)^2 - 7$$

$$\nabla f = \begin{bmatrix} 2\left(x - \frac{5}{3}\right) + 3x^2 \\ 2(y+5) \end{bmatrix} = 0 \Rightarrow y^* = -5$$

$$3x^2 + 2x - \frac{10}{3} = 0 \Longrightarrow \Rightarrow x^* = \frac{-2 \pm \sqrt{4+40}}{2 \times 3} \Longrightarrow = -1.438 \text{ or } 0.772$$

So we have two stationary solutions: $\begin{bmatrix} -1.438 & -5 \end{bmatrix}^T$ *and* $\begin{bmatrix} 0.772 & -5 \end{bmatrix}^T$.

$$H\left(x, y\right) = \begin{bmatrix} 2 + 6x & 0 \\ 0 & 2 \end{bmatrix}$$

For $\begin{bmatrix} -1.438 & -5 \end{bmatrix}^T$, $H = \begin{bmatrix} -6.628 & 0 \\ 0 & 2 \end{bmatrix}$, $\lambda = -6.628, 2 \Rightarrow$ *Saddle point*

For $\begin{bmatrix} 0.772 & -5 \end{bmatrix}^T$, $H = \begin{bmatrix} 6.632 & 0 \\ 0 & 2 \end{bmatrix}$, $\lambda = 6.632, 2 \Rightarrow$ *Minimum*

4.4 NUMERICAL APPROACHES TO SOLVING OPTIMIZATION PROBLEMS

Equation 4.8 provides n equations in n variables that can be solved to identify local minima points. While theoretically this is sufficient, practically, solving n nonlinear equations in n variables is not an easy task, with analytical solutions not possible in many cases. Numerical approaches are required to identify solutions to the first-order necessary conditions, and such approaches have been investigated thoroughly in optimization literature. These are commonly referred to as "search techniques". When an optimization approach is used to solve an ML problem, the search techniques are referred to as the "learning algorithms" or "learning laws". While several algorithms have been investigated, gradient descent for optimization is the most important algorithm [45, 12]. The importance of this algorithm cannot be overstated, and many ML algorithms are variants of the gradient descent approach. We start with numerical algorithms for unconstrained univariate problems.

4.4.1 Univariate Problems

Consider a scalar function $f(x)$ that needs to be minimized. The necessary condition for a minimum point is $f'(x) = 0$. There are several ways of solving this problem. Of course, if equation $f'(x) = 0$ is analytically solvable, numerical algorithms are not required. In cases where the equation is not analytically solvable, there are two major categories of solutions: gradient based and bracketing methods. There are several variants in these categories. Two techniques, one for each of these categories, are described to explain the basic concepts. From an ML perspective, this univariate optimization is useful in multivariate problems to identify the "learning rate" (called "step length" in optimization literature) automatically.

4.4.1.1 Gradient-Based Approach for Univariate Optimization Problems

Taylor's series approximation is the fundamental idea that forms the basis for most of the optimization approaches. Equation 4.10 is a well-known approximation of any function around a point x^0 under certain mild assumptions (hot is an abbreviation for higher order terms).

$$f(x) = f(x^0) + f'(x^0)(x - x^0) + f''(x^0)\frac{(x - x^0)^2}{2!} + hot \qquad (4.10)$$

A first order approximation retains two terms of the expression on the LHS of equation 4.10. A numerical procedure for solving equation $f(x) = 0$ is as follows. Choose any arbitrary point x^0 and if $f(x^0) = 0$, then the problem is solved. If not, a new point x^1 is selected using the Taylor's series approximation for the function (denoted as f^a in equation 4.11). Assuming that x^1 solves this approximate equation, we have

$$f^a(x^1) = f(x^0) + f'(x^0)(x^1 - x^0) = 0$$
$$x^1 = x^0 - \frac{f(x^0)}{f'(x^0)} \qquad (4.11)$$

If this solution also satisfies the original equation $f(x) = 0$, then it is the desired solution [45, 12]. The key observation is that we now have a procedure to generate a sequence of improving solutions that could lead to the final solution. This can be represented as a recursion shown in equation 4.12, which can be terminated based on some criteria that characterizes the goodness of the current solution x^{k+1}.

$$x^{k+1} = x^k - \frac{f(x^k)}{f'(x^k)} \qquad (4.12)$$

This approach can be adopted for minimizing $f(x)$. In this case, we are interested in a solution to the equation that characterizes the minimum point, which is $f'(x) = 0$. Substituting $f'(x)$ for $f(x)$ in equation 4.12, we get the solution strategy for univariate optimization problems as in equation 4.13.

$$x^{k+1} = x^k - \frac{f'(x^k)}{f''(x^k)} \qquad (4.13)$$

Example 4.5. *Find a minimum of the function $f(x) = (x - 3)^2 - 7$ using gradient search method starting from $x = -5$.*

Solution 4.5. *To see if a value of x is a solution or not, we check the value of f' at that x value.*

$$f'(x) = 2(x - 3) \implies f'(x_0) = -16 \neq 0 \text{ and } f''(x) = 2 \qquad (4.14)$$

So, $x^1 = x^0 - \frac{f'(x^0)}{f''(x^0)} = -5 + 8 = 3$. To find whether x^1 is a minimum,

$$f'(x^1) = 0 \text{ and } f''(x^1) = 2 \qquad (4.15)$$

Since the first derivative is zero and the second derivative is positive, $x = 3$ is a minimum point of the given function. The corresponding function value is $f(x^1) = -7$.

4.4.1.2 Bracketing Methods

In bracketing methods, gradient computations are replaced by function evaluations. The basic concept used is the elimination of regions based on a determination that the optimum point will not be present in that region. Elimination of regions at every iteration will result in the identification of progressively smaller regions where the optimum will be found. When this region becomes small enough, the algorithm is terminated with the minimum point point being bracketed in that region. Eliminating large regions in each iteration will lead to tight bracketing of the minimum point in a fewer number of iterations [36].

A simple bracketing algorithm is described now to make the key ideas concrete. It is important to mention here that all of these bracketing approaches work under the assumption that the underlying function is a unimodal function in the region of interest. Here, unimodal means that there is only one minimum in the region of interest. An example of a unimodal function can be seen in Figure 4.1 and an example of a function that is not unimodal is shown in Figure 4.2. With this in the background, let us now consider Figure 4.5. The bracketing algorithm starts with three points x^1, x^2, x^3 such that two conditions, viz, $f(x^3) < f(x^1)$ and $f(x^3) < f(x^2)$ are satisfied. The minimum is bracketed between x^1 and x^2. The position of x^3 can be on either side of the minimum; it is not possible to decide which side it is on just based on the two conditions. Generate two points x^4 and x^5 on either side of x^3. There are four possibilities described in Table 4.4. The first case is not possible for unimodal functions in the interval. In other cases, the three new points and the reduction in the length of the interval are mentioned. If this process is continued with the three new points, after multiple iterations, the region will be smaller than a pre-specified threshold. The minimum will be bracketed in this small region.

TABLE 4.4

Decisions for Bracketing

Scenario	Possible	New Points	Reduction in Length
$f(x^4) < f(x^3), f(x^5) < f(x^3)$	No	NA	NA
$f(x^4) > f(x^3), f(x^5) > f(x^3)$	Yes	x^4, x^3, x^5	$(x^4 - x^1) + (x^2 - x^5)$
$f(x^4) < f(x^3), f(x^5) > f(x^3)$	Yes	x^1, x^4, x^3	$x^2 - x^3$
$f(x^5) < f(x^3), f(x^4) > f(x^3)$	Yes	x^3, x^5, x^2	$x^3 - x^1$

Example 4.6. *Find a minimum of the function $f(x) = (x - 3)^2 - 7$ using bracketing method given that the minimum is in the range $\begin{bmatrix} -5 & 7 \end{bmatrix}$.*

Solution 4.6. *Picking a point between -5 and 7, $x^3 = 0$. We have three points: $x^1 = -5$, $x^3 = 0$, and $x^2 = 7$. The corresponding function values are $f(x^1) = 57, f(x^3) = 2$, and $f(x^2) = 9$.*

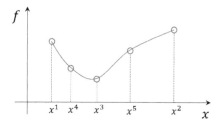

FIGURE 4.5 Bracketing concept.

Pick two points x^4 and x^5: $x^4 = -3$ and $x^5 = 4$. Correspondingly, $f(x^4) = 29$ and $f(x^5) = -6$. $f(x^5) < f(x^3) < f(x^2) < f(x^4) < f(x^1)$ while $x^1 < x^4 < x^3 < x^5 < x^2$. Hence, the new region is $\begin{bmatrix} 0 & 7 \end{bmatrix}$.

Picking two more points between x^3 and x^2: $x^6 = 1.5$ and $x^7 = 5$. Corresponding function values are $f(x^6) = -4.75$ and $f(x^7) = -3$.

Here, since $f(x^6) > f(x^5)$ and $f(x^7) > f(x^5)$, the new region is $\begin{bmatrix} x^6 & x^7 \end{bmatrix}$, i.e., $\begin{bmatrix} 1.5 & 5 \end{bmatrix}$. We pick two points in between 1.5 and 5: $x^8 = 3$ and $x^9 = 4.5$. Corresponding function values are $f(x^8) = -7$ and $f(x^9) = -4.75$. The new range is $\begin{bmatrix} x^6 & x^5 \end{bmatrix}$, i.e., $\begin{bmatrix} 1.5 & 4 \end{bmatrix}$. We could continue further by considering 2 additional points between points x^6 and x^5. Otherwise, we could terminate and pick the point corresponding to the lowest function value as the minimum point. Accordingly, $x^* = x^8 = 3$ is the optimum solution.

4.4.2 Multivariate Problems

For multivariate unconstrained nonlinear optimization problems, equation 4.8 represents a system of nonlinear equations. Many of the numerical approaches to solving a system (vector) of equations $\nabla f(x_1, \dots, x_n)$ follow the recursion process. An arbitrary initial point x^0 is chosen; this step is called the initialization step. A recursive equation of the form

$$\mathbf{x}^{k+1} = \mathbf{g}(\mathbf{x}^k) \tag{4.16}$$

is derived, where k is an iteration counter. From this, a sequence of points starting from $\mathbf{x}^0 \to \mathbf{x}^1 \to \dots \to \mathbf{x}^N$ are generated. The algorithm is terminated when $\mathbf{x}^N$ satisfies the system of equations to a pre-specified tolerance level.

We will now turn our attention to how the search equation 4.16 is defined. $\mathbf{g}(\mathbf{x}^k)$ is usually chosen as $\mathbf{g}(\mathbf{x}^k) = \mathbf{x}^k + \alpha^k \mathbf{s}^k$. $\mathbf{s}^k$ is referred to as the direction of search and α^k is the step length. At every iteration, the direction of search $\mathbf{s}^k$ and the step length α^k (how far should the point be moved in that direction) are identified.

$$\mathbf{x}^{k+1} = \mathbf{x}^k + \alpha^k \mathbf{s}^k \tag{4.17}$$

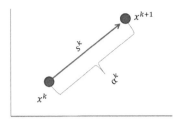

FIGURE 4.6 Directional search concept.

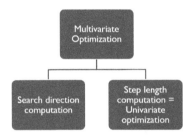

FIGURE 4.7 Solving multivariate optimization problems as univariate problems.

This process generates a sequence of points that will converge to a local optimum. This process is depicted in Figure 4.6. The search direction is chosen such that the decrease in $f(\mathbf{x})$ is rapid in that direction and α^k is chosen such that the decrease in $f(\mathbf{x})$ is maximum or equivalently $f(\mathbf{x}^k + \alpha^k \mathbf{s}^k)$ is minimized. Once $\mathbf{s}^k$ is identified any univariate method (numerically or analytically using equation 4.8) can be used to solve this problem. As a result, multivariate optimization problems can potentially be solved using univariate optimization as shown in Figure 4.7. We have already discussed how a univariate optimization can be performed and hence α^k can be identified using these approaches. It is worth pointing out that in many ML algorithms, the "step length" or "learning rate", is used as a tunable parameter.

4.4.2.1 Steepest Descent Algorithm

From the discussions until now, optimization of multivariate systems has been reduced to the identification of a search direction. There are several approaches to the identification of a search direction. However, the most common of these (and also the most relevant) is the steepest descent algorithm. Consider Figure 4.8a. The figure shows a two variable optimization problem. Notice that even for a two variable optimization problem, a three-dimensional visualization is needed to represent the additional objective function value. Every point in the

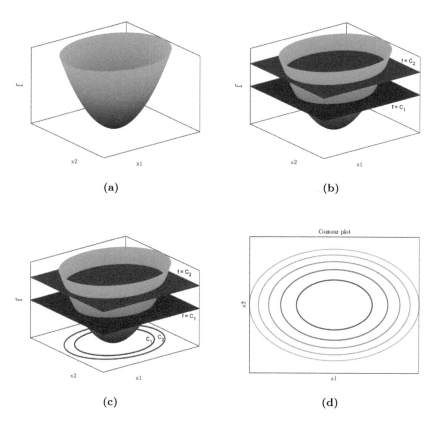

(a)   (b)

(c)   (d)

FIGURE 4.8 Idea of level-set and contours in multivariate problems.

figure is a three-dimensional vector $\left[x_1, x_2, z = f(x_1, x_2)\right]$. One linear equation in three variables is a plane. A nonlinear equation represents a curved surface as shown in Figure 4.8a.

We now connect equation 4.17 to solving multivariable optimization problems for functions such as the one shown in Figure 4.8a. The first thing to notice is that the recursion equation is relevant only in the space of the decision variables, which is the x_1-x_2 plane or any plane parallel to the x_1-x_2 plane. These planes can be represented by the equation $z = f(x_1, x_2) = C$ (constant). Depending on the constant value $(C_1, C_2,$ and so on), the planes will move up or down as shown in Figure 4.8b. At the current iteration k, $z = f(x_1, x_2) = C = f(x_1^k, x_2^k)$, which represents a curve as shown in Figure 4.8c. It is clear that there are many points with the same value of the objective function as (x_1^k, x_2^k). A collection of all these points is called the level-set. Points in the level-set obey both the equations $z = f(x_1, x_2)$ (points should be on the curved surface) and $z = f(x_1^k, x_2^k) = C$ (points on a plane parallel to the x_1-x_2 plane with the same objective function value). These two equations represent a curved line or contour (Figure 4.8c).

Since equation 4.17 is defined in the x_1-x_2 plane, the contours are projected down from different heights; the height of the contour from the x_1-x_2 plane are labeled as C_1, C_2 and so on. In summary, each of these contours represent level-sets, where any point in the contour will result in the same objective function value ($C_1, C_2, \ldots$, depending on which contour the point is located on).

Consider Figure 4.9a. Let $\mathbf{x}^k$ be the current point in the process of optimizing a function $f(\mathbf{x})$. Since contours are level-sets, if $\mathbf{x}^{k+1}$ is any point on the same contour, there would be no progress made in minimizing the function. The new point $\mathbf{x}^{k+1}$ must be located away from the contour, the question is where. Figure 4.9a shows three directions that can be chosen from $\mathbf{x}^k$. If one traverses along these directions for the same amount of distance, the new point that is reached is also depicted. Each of these points have their own level-sets, which are the contours on which the points fall. From the figure, it is clear that the level-sets are different and hence the objective function values are also different. Moving in the direction in the middle is the best option for maximum improvement (decrease) in the objective function for the same amount of movement (step length). It can be shown that locally, the best direction among all directions is given by the negative of the normal to the function at that point. Hence, the best $\mathbf{s}^k$ will be equal to $-\nabla f(\mathbf{x}^k)$; leading to the steepest descent equation

$$\mathbf{x}^{k+1} = \mathbf{x}^k - \alpha^k \nabla f(\mathbf{x}^k) \tag{4.18}$$

Figure 4.9b describes the impact of the step length on the steepest descent algorithm. It is very important that the correct length be chosen; three cases are shown in the figure, which are improvement in the objective function, no improvement in the objective function and worsening of the objective function, while moving in the best direction. Steepest descent is a "greedy search" or a "local search", where one moves in the locally optimal or best direction. This is also apparent because if one moves away from $\mathbf{x}^k$ marginally, the new point will be in a new contour and the normal to the contour at that point will be different from the normal used in the steepest descent algorithm. Figure 4.9c shows that at $\mathbf{x}^k$, there exists a direction that can lead to the optimum solution with the correct step-length. Unfortunately, even though one is certain that there is a "globally" optimal direction at every point, this cannot be uncovered with the local information. For the sake of illustration, Figure 4.10 shows the sequence of iterations to optimality when the steepest descent algorithm is applied on a prototypical function.

Example 4.7. *Find a minimum of the function* $f(x_1, x_2) = x_1^2 + x_2^2$ *using steepest descent search starting from* $x^0 = \begin{bmatrix} 1 & 1 \end{bmatrix}^T$ *and using a constant* $\alpha, \alpha^k = 0.3$. *Let the tolerance on* ∇f *be 0.075.*

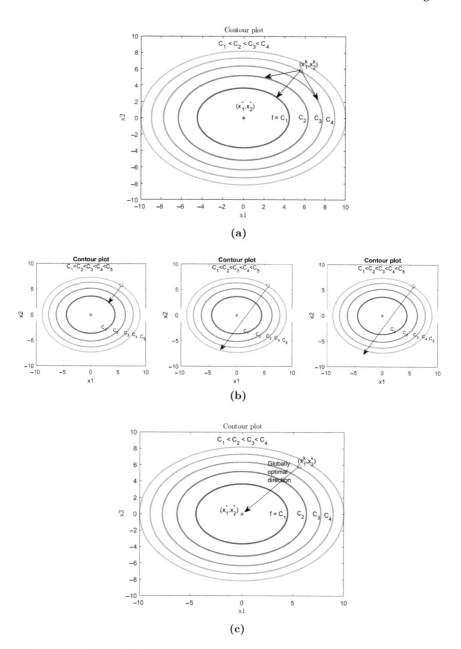

FIGURE 4.9 Understanding the working of steepest descent algorithm.

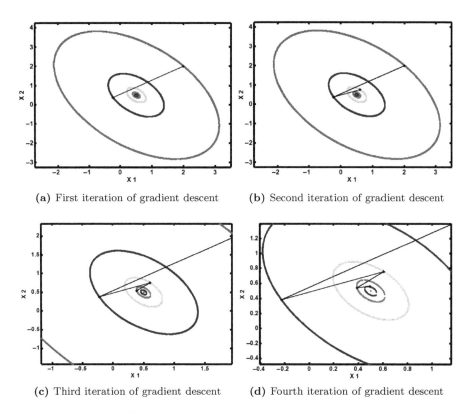

(a) First iteration of gradient descent   **(b)** Second iteration of gradient descent

(c) Third iteration of gradient descent   **(d)** Fourth iteration of gradient descent

FIGURE 4.10 Gradient descent algorithm.

Solution 4.7.

$$f(x_1, x_2) = x_1^2 + x_2^2; \quad \nabla f = \begin{bmatrix} 2x_1 \\ 2x_2 \end{bmatrix}; \quad H = \begin{bmatrix} 2 & 0 \\ 0 & 2 \end{bmatrix}$$

Since H is positive definite, the solution to $\nabla f = 0$ will provide a minimum of $f(x_1, x_2)$.

$$\mathbf{x}^1 = x^0 - 0.3 \nabla f_{\mathbf{x}^0} = \begin{bmatrix} 1 \\ 1 \end{bmatrix} - 0.3 \begin{bmatrix} 2 \\ 2 \end{bmatrix} = \begin{bmatrix} 0.4 \\ 0.4 \end{bmatrix} \Rightarrow \nabla f(\mathbf{x}^1) = \begin{bmatrix} 0.8 \\ 0.8 \end{bmatrix}$$

$$\mathbf{x}^2 = \begin{bmatrix} 0.4 \\ 0.4 \end{bmatrix} - 0.3 \begin{bmatrix} 0.8 \\ 0.8 \end{bmatrix} = \begin{bmatrix} 0.16 \\ 0.16 \end{bmatrix} \Rightarrow \nabla f(\mathbf{x}^2) = \begin{bmatrix} 0.32 \\ 0.32 \end{bmatrix}$$

$$\mathbf{x}^3 = \begin{bmatrix} 0.16 \\ 0.16 \end{bmatrix} - 0.3 \begin{bmatrix} 0.32 \\ 0.32 \end{bmatrix} = \begin{bmatrix} 0.064 \\ 0.064 \end{bmatrix} \Rightarrow \nabla f(\mathbf{x}^3) = \begin{bmatrix} 0.128 \\ 0.128 \end{bmatrix}$$

$$\mathbf{x}^4 = \begin{bmatrix} 0.064 \\ 0.064 \end{bmatrix} - 0.3 \begin{bmatrix} 0.128 \\ 0.128 \end{bmatrix} = \begin{bmatrix} 0.0256 \\ 0.0256 \end{bmatrix} \Rightarrow \nabla f(\mathbf{x}^4) = \begin{bmatrix} 0.0512 \\ 0.0512 \end{bmatrix}$$

Since $|\nabla f(\mathbf{x}^4)| < \epsilon$ *(0.075) we can terminate the iterations. Thus,* $\mathbf{x}^4 = \begin{bmatrix} 0.0256 & 0.0256 \end{bmatrix}^T$. *Note that the true solution is* $\mathbf{x}^* = \begin{bmatrix} 0 & 0 \end{bmatrix}^T$. *By decreasing the tolerance and proceeding with large number of iterations, a point close to the optimum can be reached.*

Example 4.8. *Find the minima of the function* $f(x_1, x_2) = x_1^2 + x_2^2$ *using steepest descent search starting from* $\mathbf{x}^0 = \begin{bmatrix} 1 & 1 \end{bmatrix}^T$ *and using optimum step length identified through univariate optimization.*

Solution 4.8. $f(x_1, x_2) = x_1^2 + x_2^2;$ $\nabla f = \begin{bmatrix} 2x_1 \\ 2x_2 \end{bmatrix};$

$$H(x_1, x_2) = \begin{bmatrix} 2 & 0 \\ 0 & 2 \end{bmatrix} \Rightarrow Positive\ definite$$

$$\mathbf{x}^{k+1} = \mathbf{x}^k + \alpha^k \mathbf{s}^k; \quad \mathbf{s}^k = -\nabla f(\mathbf{x}^k); \quad \mathbf{x}^0 = \begin{bmatrix} 1 \\ 1 \end{bmatrix}$$

$$\mathbf{x}^1 = \mathbf{x}^0 - \alpha^0 \begin{bmatrix} 2x_1^0 \\ 2x_2^0 \end{bmatrix} = \begin{bmatrix} 1 \\ 1 \end{bmatrix} - \alpha^0 \begin{bmatrix} 2 \\ 2 \end{bmatrix} = \begin{bmatrix} 1 - 2\alpha^0 \\ 1 - 2\alpha^0 \end{bmatrix}$$

$$f(\mathbf{x}^1) = 2(1 - 2\alpha^0)^2$$

To find α^0 *that minimized* $f(\mathbf{x}^1)$ *the most,*

$$\frac{df(\mathbf{x}^1)}{d\alpha^0} = 4(1 - 2\alpha^0) \times -2 = 0 \Rightarrow \alpha^0 = \frac{1}{2}$$

$$\mathbf{x}^1 = \begin{bmatrix} 1 - 2\alpha^0 \\ 1 - 2\alpha^0 \end{bmatrix} = \begin{bmatrix} 0 \\ 0 \end{bmatrix}$$

For $\mathbf{x}^1$, $f(\mathbf{x}^1) = 0$, $\nabla f = \begin{bmatrix} 0 \\ 0 \end{bmatrix}$. *Hence,* $\mathbf{x}^* = \mathbf{x}^1$ *is the optimum. Note that we arrived at the solution in one step.*

4.4.2.2   Newton's Method

It would be worthwhile to also describe an alternate approach to identifying the direction for the recursion given by equation 4.17 for multivariable optimization. Taylor's series expansion up to a second order term for a multivariable function (around a point $\mathbf{x}^k$) is given by equation 4.19.

$$f^a(\mathbf{x}) = f(\mathbf{x}^k) + \nabla f(\mathbf{x}^k)(\mathbf{x} - \mathbf{x}^k) + \frac{1}{2}(\mathbf{x} - \mathbf{x}^k)^T H(\mathbf{x}^k)(\mathbf{x} - \mathbf{x}^k) \quad (4.19)$$

It is now possible to optimize this approximate function f^a using equation 4.8 to find $\mathbf{x}^*$ and denote this solution as $\mathbf{x}^{k+1}$. This is called the Newton's method.

$$\nabla f^a(\mathbf{x}) = \nabla f(\mathbf{x}^k) + H(\mathbf{x}^k)(\mathbf{x} - \mathbf{x}^k) = 0 \qquad (4.20)$$
$$\mathbf{x}^{k+1} = \mathbf{x}^* = \mathbf{x}^k - H(\mathbf{x}^k)^{-1}\nabla f(\mathbf{x}^k) \qquad (4.21)$$

A generalization (or combination) of both the approaches given by equations 4.21 and 4.17 is the one shown in equation 4.22. To retrieve equation 4.18, one needs to set $M = I$ and $\mathbf{s}^k = \nabla f(\mathbf{x}^k)$ and to retrieve equation 4.21 one needs to set $M = H(\mathbf{x}^k)^{-1}$, $\alpha^k = 1$ and $\mathbf{s}^k = \nabla f(\mathbf{x}^k)$. Another viewpoint that is also useful is that in steepest descent, variables are changed from iteration to iteration only based on the gradient of the function with respect to the corresponding variable, whereas in Newton's method and methods where $M \neq I$, changes in variables through iterations require the gradient of the function with respect to all variables.

$$\mathbf{x}^{k+1} = \mathbf{x}^k - \alpha^k M^k \mathbf{s}^k \qquad (4.22)$$

Example 4.9. *Find a minimum of the function* $f(x_1, x_2) = x_1^2 + x_2^2$ *using Newton's method starting from* $\mathbf{x}^0 = \begin{bmatrix} 1 & 1 \end{bmatrix}^T$.

Solution 4.9.

$$\nabla f(x_1, x_2) = \begin{bmatrix} 2x_1 \\ 2x_2 \end{bmatrix}; \quad H = \begin{bmatrix} 2 & 0 \\ 0 & 2 \end{bmatrix}; \quad H^{-1} = \begin{bmatrix} 0.5 & 0 \\ 0 & 0.5 \end{bmatrix}$$

$$\mathbf{x}^{k+1} = \mathbf{x}^k - H\left(\mathbf{x}^k\right)^{-1} \nabla f\left(\mathbf{x}^k\right) = \begin{bmatrix} x_1^k \\ x_2^k \end{bmatrix} - \begin{bmatrix} 0.5 & 0 \\ 0 & 0.5 \end{bmatrix} \begin{bmatrix} 2x_1^k \\ 2x_2^k \end{bmatrix} = \begin{bmatrix} 0 \\ 0 \end{bmatrix}$$

$$\Rightarrow \nabla f = \begin{bmatrix} 0 \\ 0 \end{bmatrix} \Rightarrow \mathbf{x}^* = \mathbf{x}^1$$

Irrespective of what your $\mathbf{x}^0$ *is, we reach* $\mathbf{x}^* = \begin{bmatrix} 0 & 0 \end{bmatrix}^T$ *in one iteration for this function.*

Example 4.10. *Find a minimum of the function* $f(x_1, x_2) = x_1^2 + 10x_2^2$ *using steepest descent (using optimum step length identified through univariate optimization) and Newton's method. Start from* $\begin{bmatrix} 2 & 5 \end{bmatrix}^T$.

Solution 4.10. $f(x_1, x_2) = x_1^2 + 10x_2^2 \Rightarrow \nabla f = \begin{bmatrix} 2x_1 \\ 20x_2 \end{bmatrix} \Rightarrow H = \begin{bmatrix} 2 & 0 \\ 0 & 20 \end{bmatrix}$

Steepest descent: $\mathbf{x}^{k+1} = \mathbf{x}^k - \alpha^k \nabla f^k$. *Hence,*

$$\mathbf{x}^1 = \mathbf{x}^0 - \alpha^0 \begin{bmatrix} 4 \\ 100 \end{bmatrix} = \begin{bmatrix} 2 - 4\alpha^0 \\ 5 - 100\alpha^0 \end{bmatrix}$$

$$f(\mathbf{x}^1) = (2 - 4\alpha^0)^2 + 10(5 - 100\alpha^0)^2$$

$$\frac{df(\mathbf{x}^1)}{d\alpha^0} = (2(2 - 4\alpha^0) \times -4) + 20(5 - 100\alpha^0)(-100)$$

$$\implies -16 + 32\alpha^0 - 10^4 + 2 \times 10^5 \alpha^0 = 0 \implies 200032\alpha^0 = 10016$$

$$\implies \alpha^0 = \frac{10016}{200032} \approx 0.05$$

$$\mathbf{x}^1 = \begin{bmatrix} 2 - 4 \times 0.05 \\ 5 - 100 \times 0.05 \end{bmatrix} = \begin{bmatrix} 1.8 \\ 0 \end{bmatrix} \Rightarrow f(\mathbf{x}^1) = 1.8^2 \text{ and } \nabla f^1 = \begin{bmatrix} 3.6 \\ 0 \end{bmatrix}$$

$$\mathbf{x}^2 = \begin{bmatrix} 1.8 \\ 0 \end{bmatrix} - \alpha^1 \begin{bmatrix} 3.6 \\ 0 \end{bmatrix} = \begin{bmatrix} 1.8 - 3.6\alpha^1 \\ 0 \end{bmatrix}$$

$$f(\mathbf{x}^2) = (1.8 - 3.6\alpha^1)^2 + 0$$

$$\frac{df(\mathbf{x}^2)}{d\alpha^1} = 0 \Rightarrow 2 \times (-3.6)(1.8 - 3.6\alpha^1) = 0 \Rightarrow \alpha = \frac{1.8}{3.6} = \frac{1}{2}$$

$$\mathbf{x}^2 = \begin{bmatrix} 0 \\ 0 \end{bmatrix} \Rightarrow \nabla f(\mathbf{x}^2) = \begin{bmatrix} 0 \\ 0 \end{bmatrix} \Rightarrow \mathbf{x}^* = \mathbf{x}^2$$

Newton's Method

$$\mathbf{x}^1 = \mathbf{x}^0 - H(\mathbf{x}^0)^{-1} \nabla f(\mathbf{x}^0) = \begin{bmatrix} 2 \\ 5 \end{bmatrix} - \begin{bmatrix} 2 & 0 \\ 0 & 20 \end{bmatrix}^{-1} \begin{bmatrix} 4 \\ 100 \end{bmatrix}$$

$$= \begin{bmatrix} 2 \\ 5 \end{bmatrix} - \begin{bmatrix} 0.5 & 0 \\ 0 & 0.05 \end{bmatrix} \begin{bmatrix} 4 \\ 100 \end{bmatrix} = \begin{bmatrix} 0 \\ 0 \end{bmatrix}$$

$$\nabla f(\mathbf{x}^1) = \begin{bmatrix} 0 \\ 0 \end{bmatrix} \Rightarrow \mathbf{x}^* = \mathbf{x}^1$$

4.4.2.3 Algebraic Derivation of Steepest Descent Method

Consider the first-order approximation of Taylor's series expansion, $f^a(\mathbf{x}) = f(\mathbf{x}^k) + \nabla f(\mathbf{x}^k)(\mathbf{x} - \mathbf{x}^k)$. This approximation will work well when the first-order term dominates all the other terms. This will happen when $\mathbf{x}$ is close to $\mathbf{x}^k$ because the higher order terms comprise of higher powers of the distance between $\mathbf{x}$ and $\mathbf{x}^k$. It is apparent that $(\mathbf{x} - \mathbf{x}^k)$ must be in the direction of $-\nabla f(\mathbf{x}^k)$ to reduce the objective function the most. This will result in a dot product between two vectors in the opposite direction ($\nabla f(\mathbf{x}^k)$ and $(\mathbf{x} - \mathbf{x}^k)$).

Example 4.11. *Solve the unconstrained optimization problem described in Example 4.2 using gradient descent algorithm starting from $m = 0$*

Solution 4.11. $f(m) = (3-m)^2 + (6-2m)^2 + (9-3m)^2$
To find the optimum m,

$$\frac{df}{dm} = -2(3-m) - 4(6-2m) - 6(9-3m)$$

$$m^{k+1} = m^k + \alpha^k s^k \text{ and } s^k = -\frac{df}{dm}\Big|_{m=m^k}$$

Initial guess: $m^0 = 0$.
Iteration 1:

$$m^1 = 0 - \alpha^0 \left[-2(3-m^0) - 4(6-2m^0) - 6(9-3m^0)\right] = 84\alpha^0$$

$$f(m^1) = (3 - 84\alpha^0)^2 + (6 - 2 \times 84\alpha^0)^2 + (9 - 3 \times 84\alpha^0)^2$$

To find optimum value of α^0, $\frac{df(m^1)}{d\alpha^0} = 0$.

$$\implies 2 \times (-84)(3 - 84\alpha^0) + 2 \times (-2 \times 84)(6 - 2 \times 84\alpha^0)$$
$$+ 2 \times (-3 \times 84)(9 - 3 \times 84\alpha^0) = 0$$

$$\alpha^0 = \frac{42}{14 \times 84} = \frac{1}{28} \implies m^1 = 84 \times \frac{1}{28} = 3$$

To check whether m^1 *is optimum,* $\frac{df}{dm}\Big|_{m^1} = -2 \times 0 - 4 \times 0 - 6 \times 0 = 0$. *Hence,* $m^* = 3$ *is the optimum solution. Note that for a dataset with three samples, the objective function is a sum of three terms with each term a function of the decision variables. Since one unknown parameter was used in the proposed model, there was only one decision variable. Hence, the derivative of the three terms in the objective function with respect to one decision variable had to be computed. If the data had 10,000 samples, the objective function would be a sum of 10,000 terms with each term being a function of unknown parameters. If there were three unknown parameters or decision variables (eg: $ax^2 + bx + c$ as the fitting function), then the derivative needs to be computed with respect to all three unknown variables. All of these simultaneous computations can potentially render the gradient descent algorithm computationally complex when there is a large dataset.*

4.5 DESCRIPTION OF STOCHASTIC GRADIENT DESCENT

In the previous section, the description of local methods also pointed out that while there is a globally optimal direction at all times, the steepest descent will not uncover this direction (or something close to this direction) until the current point under consideration is close to the optimum solution. Stochastic gradient descent approaches provide random fluctuations to the steepest descent at every iteration with the hope that better directions are found (statistically) for faster convergence. It is important to note that theoretical guaranties—that a solution will be found within a reasonable amount of time—are still not possible. While we described that there exist directions traversing which, one can get to a local optimum in one iteration, the same idea is true for global optimization as well.

Consider Figure 4.11a. In this figure, a two-variable function with local and global optimum points is plotted. The contour plots for this function are shown in Figure 4.11b. In this figure, two prototypical points, one in each valley, have been highlighted. For each of these points, one can see two directions, which will reach the local and global optimum points and one direction, which might reach neither, but the best direction based on the local information at that point. The key observation is that at any point, directions for local and global optimum points are available but cannot be uncovered. Stochastic gradient descent algorithms try to uncover these directions by leveraging randomization.

To understand stochastic gradient descent, let us go back to equation 4.17. We assume that the $(k + 1)^{th}$ iteration is completed when all the decision variables are updated from β^k to β^{k+1} using all the samples. To keep the description general, let us define a loss function $lf(\mathbf{z}^i, \beta)$ which can be computed for every sample (M samples), where β is a vector of decision variables and $\mathbf{z}^i$ is a sample point.

$$f(\beta) = \sum_{i=1}^{M} lf(\mathbf{z}^i, \beta) \qquad (4.23)$$

Appropriate loss functions could be chosen based on the task at hand (classification or function approximation). For example, if we are building a linear model $y = ax + b$ between y and x and denote $\mathbf{z} = [y \; x]^T$ and $\beta = [a \; b]^T$, then the loss function $lf(\mathbf{z}^i, \beta) = (y^i - (ax^i + b))^2$. Similar loss functions can be conceptualized for different classification and function approximation problems.

If one were to apply the steepest descent algorithm to identify the decision variables, equation 4.17 will lead to

$$\beta^{k+1} = \beta^k - \alpha^k \sum_{i=1}^{M} \nabla lf(\mathbf{z}^i, \beta)|_{\beta^k} \qquad (4.24)$$

Equation 4.24 specifies the point at which the gradient is evaluated, which happens to be β^k. Given β^k, this step will produce a gradient that is deterministic, in other words, when the algorithm is initialized at β^0, the sequence of solutions at all iterations are fixed and deterministic. As discussed before, variations in these directions at each iteration might lead to the exploration of directions that are not purely greedy, leading to faster convergence and possible attainment of global optimum.

This stochastic behavior with the steepest descent framework is achieved through the following idea. In equation 4.24, the objective function is constructed with all the samples in the same batch. This strategy could be modified by considering batches of samples sequentially. Let us consider the simplest case, where we split the data into two batches $i \in [1 \; (M/2)]$ and

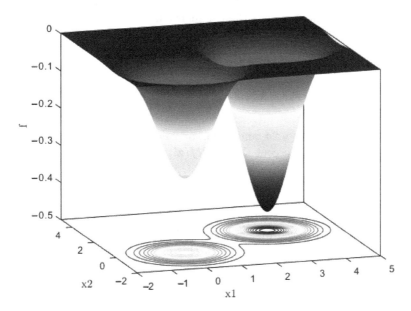

(a) Function with two optimum points, one local and one global

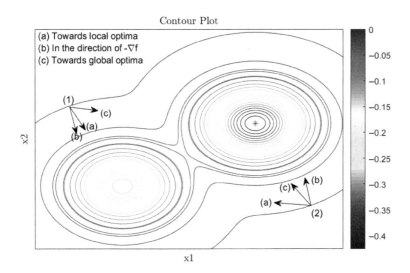

(b) Level sets or contour plots

FIGURE 4.11 Directions for local and global optimum.

$i \in [(M/2)\ M]$. If one generates, at the kth iteration, new values for β using only the first batch of data and then sequentially using the second batch of data, then the steepest descent equations will become

$$\beta^{k,1} = \beta^k - \alpha^{k,0} \sum_{i \in [1\ M/2]} \nabla l f(\mathbf{z}^i, \beta)|_{\beta^k}$$

$$\beta^{k+1} = \beta^{k,1} - \alpha^{k,1} \sum_{i \in [(M/2)+1\ M]} \nabla l f(\mathbf{z}^i, \beta)|_{\beta^{k,1}} \qquad (4.25)$$

Comparing equations 4.24 and 4.25, one can see that the gradient in the standard steepest descent $\Sigma_{i=1}^{i=M} \nabla l f(\mathbf{z}^i, \beta)|_{\beta^k}$ has been replaced by $\Sigma_{i \in [1\ M/2]} \nabla l f(\mathbf{z}^i, \beta)|_{\beta^k} + \Sigma_{i \in [M/2\ M]} \nabla l f(\mathbf{z}^i, \beta)|_{\beta^{k,1}}$ in the stochastic gradient descent (if we choose $\alpha^{k,0} = \alpha^{k,1}$). The gradient can be made stochastic by randomly changing the order in which the batches are processed. In the case of two batches, equation 4.25 will return different gradients based on whether batch 1 of data is processed first or second. The limiting case of this is when each batch contains only one sample. In this case, randomizing the order in which the samples are processed in a stochastic gradient descent algorithm will lead to different gradients. When all the data is processed at the same time, we recover the original deterministic steepest descent algorithm.

Another approach to bringing in stochasticity is to update the decision variables in batches but using all the sample points in the objective function. A completely equivalent derivation for such a version of stochastic steepest descent can then be derived. This approach is also called as "alternating optimization" and is quite popular in ML literature. Clustering algorithms use this notion of alternating optimization. Finally, one could sub-select both variables and samples in batches and carry out steepest descent in batches to generate stochastic behavior. The underlying expectation, as described before, is that these stochastic modifications to a purely greedy search has the possibility of enhancing convergence speeds and the ability to identify globally optimum solutions. The ideas of sub-selecting samples and variables for randomization is used in a very popular ML algorithm, the random forest algorithm.

Example 4.12. *Solve the unconstrained optimization problem described in Example 4.2 using stochastic gradient descent algorithm, starting from $m = 0$. Take 2 batches: batch 1 with samples $1, 2$ and batch 2 with sample 3.*
Solution 4.12.

$$l f(\mathbf{z}^i, m) = (y^i - mx^i)^2$$

$$m^{k,1} = m^k - \alpha^{k,0} \sum_{i=1,2} \nabla l f|_{m^k}; \quad m^{k+1} = m^{k,1} - \alpha^{k,1} \sum_{i=3} \nabla l f|_{m^{k,1}}$$

Iteration 1:

$$m^{1,1} = m^0 - \alpha^{1,0} \left[\tfrac{\partial}{\partial m}(3-m)^2 + \tfrac{\partial}{\partial m}(6-2m)^2 \right]\Big|_{m^0}$$

$$= 0 - \alpha^{1,0} \left[-2(3-m) - 4(6-2m) \right]\big|_{m^0} = 30\alpha^{1,0}$$

$$f\left(m^{1,1}\right) = \sum_{i=1}^{2} lf\left(\mathbf{z}^i, m\right) = (3-m)^2 + (6-2m)^2 = (3-m)^2 + 4(3-m)^2$$
$$= 5(3-m)^2 = 5\left(3 - 30\alpha^{1,0}\right)^2$$

$$\frac{df\left(m^{1,1}\right)}{d\alpha^0} = 10\left(3 - 30\alpha^{1,0}\right) \times -30 = 0 \Rightarrow 30\alpha^{1,0} = 3$$
$$\alpha^{1,0} = 0.1 \Rightarrow m^{1,1} = 3$$
$$m^1 = m^{1,1} - \alpha^{1,0}\left[\frac{d}{dm}\left(9 - 3m\right)^2\right]\Big|_{m^{1,1}}$$
$$= 3 - \alpha^{1,0}\left[2\left(9 - 3m\right) \times -3\right]\big|_{m^{1,1}} = 3 - 0 = 3$$
$$\frac{df}{dm}\left(m^1\right) = -6\left(9 - 3m\right) = 0$$

$\Rightarrow m^1 = 3$ *is the optimum solution.*

4.6 ALTERNATE LEARNING ALGORITHMS

The algorithms that were described until now are all based on the computation of gradients and the local minima problems associated with these greedy searches were described. One approach to address these problems is through stochastic gradients. Another approach to address the global optimization problem is the use of multiple solutions that are updated simultaneously without the computation of gradients. In both the regular and stochastic versions of the steepest descent algorithm, starting with a solution, a final solution is identified through a sequence of steps. In contrast, in alternate algorithms such as Genetic Algorithms (GA) and Particle Swarm Optimization (PSO), one starts with a population of solutions. Similar to how a single solution is modified to a new solution in traditional algorithms, a population of solutions is converted into a new population in an iteration (also called a generation) in GA, for example. Several operations are performed to convert one population to another. The key idea is that these transformations use only the evaluation of the objective function for different members of the population to identify the new population and the gradient information is typically not used. Hence, these approaches are readily applicable to problems where gradients might not be defined or difficult to compute. Some of these transformations use randomization and hence these algorithms are inherently stochastic in nature. Further, these algorithms always terminate with a pool of promising solutions, which could address the problem of global optimization to some extent. The use of GA and PSO in ML has been extensively explored [15, 22]. Detailed texts on GA and PSO can be consulted for a better understanding of these algorithms [23].

4.7   IMPACT OF NON-CONVEXITY ON ML ALGORITHMS

It can be noticed that all the algorithms described are in the category of local algorithms—that is they guarantee that a local optimum point will be identified. The non-gradient based alternate algorithms have been demonstrated in some cases to be better at identifying globally optimum solutions or at least many locally optimum solutions. However, these algorithms cannot provide theoretical guaranties on convergence to global optimum. The idea of global optimum is of critical importance in ML algorithms—this can in fact be the difference between success and failure of applications that utilize ML. To understand this, consider a loss function that captures the errors between predicted and actual values. The best model should have zero errors and this is the global minimum. Now an ML model will try to achieve this by tweaking the parameters in the model through an optimization routine. If this error function has multiple optima, then the ML algorithm could get stuck at a local optimum, where the error is greater than zero. From an optimization viewpoint, this solution for parameters will satisfy all the local conditions and standard algorithms will not find ways to improve the solution. In many cases of parsimonious representation (very few parameters in the ML model), it will not be clear if zero error is even achievable and one cannot accurately determine if the solution is a local or a global solution.

Two approaches evaluated by the ML community to address this problem have been discussed already. These are the stochastic gradient descent algorithms and the evolution of multiple solutions toward global optimality. However, these algorithms can't provide any theoretical guarantees regarding the quality of the identified solution. Approaches to provide theoretical guaranties are if the objective function itself is shown to have some special properties or if a specific global optimization approach is employed.

The best case scenario is if one could theoretically establish that the objective function is a convex function. Convex functions are also unimodal functions that have only one optimum (minimum) value. Examples of convex functions can be seen in Figures 4.1 and 4.3. Non-convex functions are shown in Figures 4.2 and 4.4 (this is an extreme case). Since local optimum is the same as the global optimum for convex functions, once a local greedy algorithm converges, one need not look for better solutions. A function is convex if it satisfies the following condition

$$f(\alpha \mathbf{x} + (1 - \alpha)\mathbf{y}) \le \alpha f(\mathbf{x}) + (1 - \alpha)f(\mathbf{y}) \ \forall \ \mathbf{x}, \mathbf{y} \qquad (4.26)$$

An equivalent condition is that a function is convex if it is twice differentiable and the Hessian matrix $H(\mathbf{x})$ is positive semi-definite $\forall \ \mathbf{x}$. Convex functions can be shown to follow several properties, using which convexity properties of composite functions can be established. An important point to note here is that for an optimization problem to be convex, the objective function must be convex, and the constraints should also form a convex set. We do not describe this here as constraints have not been introduced as yet.

More details on this can be found in many excellent textbooks on optimization [26, 36]. When the optimization problem is not a convex problem, then algorithms that specifically address non-convex problems with theoretical guaranties can be explored. Convex underbounding algorithms such as the α-bb algorithm provide guaranties on identifying globally optimum solution within an ϵ neighborhood [43].

4.8 HANDLING CONSTRAINTS

Similar to the discussion on unconstrained nonlinear problems, we will start with theoretical necessary and sufficiency conditions for constrained optimization. Understanding these conditions will allow one to understand some of the advanced ML algorithms. There are several numerical approaches (sequential quadratic programming, active-set methods, reduced gradients, and so on) that can be employed to identify solutions that satisfy the necessary and sufficiency conditions. However, these are very detailed and not critically required for understanding ML. There are several excellent books on optimization [45, 46, 12] that describe these numerical methods in considerable detail.

4.8.1 Equality Constraints

We know that the necessary and sufficiency conditions for an optimum point in the unconstrained case is given by equation 4.8 and the second-order sufficiency condition is $H(\mathbf{x}^k)$ is positive definite. Equivalent necessary and sufficiency conditions need to be identified in the case of constrained optimization problems. It is worthwhile to geometrically understand how constraints change the necessary conditions before the conditions are introduced. Consider the function $2x_1^2 + 4x_2^2$. A simple computation will reveal that $(0,0)$ is the point at which the function is minimized. The level-set contours and this unconstrained minimum point are shown in Figure 4.12a. Now consider the case where the solution should also satisfy the linear constraint $h(\mathbf{x}) : 3x_1 + 2x_2 = 12$, this line is shown in Figure 4.12b along with the level-set contours. It is evident that the original optimum point is not a valid solution anymore as the solution has to lie on the line to satisfy the equality constraint. However, the question is where in the line should the solution lie.

Let us perform a thought-experiment and start moving the unconstrained optimum point towards the constraint. As this solution is being moved, every point will have a corresponding level-set where the objective function value will be the same. As we move away from origin, the objective function increases (level-set contours with increasing objective function values). From a minimization viewpoint, we want to move the solution as little as possible, while satisfying the constraint. This will be the exact point $(\mathbf{x}^*)$ where the constraint becomes a tangent to a level-set, shown as Contour C in Figure 4.12b. This point represents the best trade-off on the objective function value while satisfying the constraint. Any less trade-off (shown by contour (C_-)) will violate feasibility and with any more (contour C_+), the trade-off is more than what is required. The mathematical equivalent of this geometric intuition is that, at $\mathbf{x}^*$, $\boldsymbol{\nabla} h(\mathbf{x}^*)$, and $\boldsymbol{\nabla} f(\mathbf{x}^*)$ have to be in the same or opposite direction for $\mathbf{x}^*$ to be an optimum point.

This result must be generalized to nonlinear equality constraints and for cases where there are more constraints than one. The conditions for the general case, where there are m linear equality constraints $\mathbf{h}(\mathbf{x})$ is given by (where $\boldsymbol{\lambda}$ is a $m \times 1$ vector)

$$l(\mathbf{x}, \boldsymbol{\lambda}) = f(\mathbf{x}) + \boldsymbol{\lambda}^T \mathbf{h}(\mathbf{x})$$
$$\boldsymbol{\nabla} l(\mathbf{x}) = 0 \tag{4.27}$$
$$\mathbf{h}(\mathbf{x}) = 0 \tag{4.28}$$

Equations 4.27 and 4.28 when solved together will lead to an optimal solution $\mathbf{x}^*$. It can be noticed that m new variables $\boldsymbol{\lambda}$ have been added, one each for the constraints. As a result, the number of variables become $n+m$. Necessary conditions, taken together, yield $n + m$ equations (n gradient equations and m equality constraints) to solve for the variables. We do not describe the second-order conditions here as they are a little more involved and from an ML perspective, the first order necessary conditions are enough to generating an initial understanding of ML algorithms. Numerical approaches to solving these equations are also involved. Purely gradient based searches will not work because a new point identified by the steepest descent algorithm might not satisfy the equality constraints. One needs to additionally ensure that the solutions considered also satisfy the equality constraints. This is achieved by either searching in a space of solutions where all the equality constraints are satisfied or by ensuring that an unconstrained gradient solution is modified to satisfy equality constraints as the algorithm proceeds through multiple descent steps. An interested reader is referred to [26] for understanding these ideas in detail.

Example 4.13. *Solve the following problem by direct substitution:*

$$\min_{x_1, x_2} \quad f(x_1, x_2) = 4x_1^2 + 5x_2^2$$
$$\text{s.t.} \quad 2x_1 + 3x_2 = 6$$

Solution 4.13.

$$2x_1 + 3x_2 = 6 \Rightarrow x_1 = \frac{6 - 3x_2}{2}$$

$$f(x_1, x_2) = 4x_1^2 + 5x_2^2 = 4\left(\frac{6 - 3x_2}{2}\right)^2 + 5x_2^2$$
$$= 36 + 9x_2^2 - 36x_2 + 5x_2^2 = 14x_2^2 - 36x_2 + 36$$

*Thus, the given optimization problem can be rewritten as the following uncon-
strained optimization problem.*

$$\min_{x_2} \ 14x_2^2 - 36x_2 + 36 = f_2(x_2)$$

$$f_2'(x_2) = 28x_2 - 36; \qquad f_2''(x_2) = 28$$

The above problem can be solved by

$$f_2'(x_2)\Big|_{x_2^*} = 0 \Rightarrow 28x_2 = 36 \Rightarrow x_2 = \frac{36}{28} = \frac{9}{7}$$

Since $f_2''(x_2^*) > 0$, x_2^* is a minimum to the function f_2. Now,

$$x_1^* = \frac{6 - 3x_2^*}{2} = \frac{42 - 27}{14} = \frac{15}{14}$$

$$f(x_1^*, x_2^*) = 4\left(\frac{15}{14}\right)^2 + 5\left(\frac{9}{7}\right)^2 = \frac{90}{7}$$

Example 4.14. *Solve the following optimization problem using the concept
of Lagrangian.*

$$\min_{x_1, x_2} \quad f(x_1, x_2) = 4x_1^2 + 5x_2^2$$

$$\text{s.t.} \quad\quad 2x_1 + 3x_2 = 6$$

Solution 4.14.

$$L(x_1, x_2, \lambda) = 4x_1^2 + 5x_2^2 + \lambda(2x_1 + 3x_2 - 6)$$

At the optimum $(x_1^*, x_2^*, \lambda^*)$,

$$\frac{\partial L}{\partial x_1} = 0 \Rightarrow 8x_1^* + 2\lambda^* = 0; \qquad \frac{\partial L}{\partial x_2} = 0 \Rightarrow 10x_2^* + 3\lambda^* = 0$$

$$\frac{\partial L}{\partial \lambda} = 0 \Rightarrow 2x_1^* + 3x_2^* - 6 = 0$$

We have three linear equations in three unknowns. From $\frac{\partial L}{\partial x_1}$, $-2\lambda^* = 8x_1^* \Rightarrow \lambda^* = -4x_1^*$. Substituting in $\frac{\partial L}{\partial x_1}$, $10x_2^* - 12x_1^* = 0 \Rightarrow x_2^* = 1.2x_1^*$. Substituting in $\frac{\partial L}{\partial x_1}$, $2x_1^* + 3.6x_1^* - 6 = 0 \Rightarrow 5.6x_1^* = 6$.

$$x_1^* = \frac{6}{5.6} = \frac{60}{56} = \frac{15}{14} \ \text{and} \ x_2^* = 1.2 \times \frac{15}{14} = \frac{9}{7} \implies f(x_1^*, x_2^*) = \frac{90}{7}$$

4.8.2 Inequality Constraints

The next question that needs to be answered is, what happens when the constraint becomes an inequality constraint. There are two possibilities $l(\mathbf{x}) \leq 0$

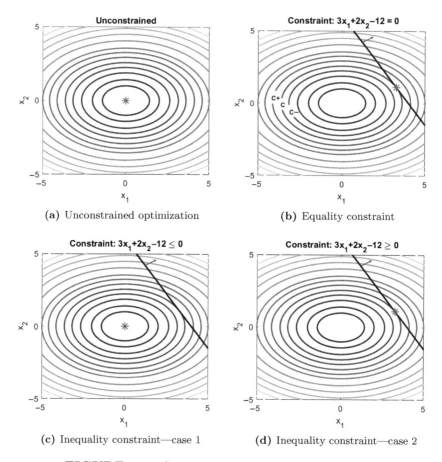

(a) Unconstrained optimization   **(b)** Equality constraint

(c) Inequality constraint—case 1   **(d)** Inequality constraint—case 2

FIGURE 4.12 Geometric understanding of constraints.

and $l(\mathbf{x}) \geq 0$. In Figure 4.12c, the direction of normal to the line is indicated. When, $l(\mathbf{x}) \leq 0$, any point on the half-plane going away from the direction of the normal will satisfy the constraint. This essentially means that the original unconstrained solution is the optimum solution in this case. When $l(\mathbf{x}) \geq 0$, then the half-plane in the direction of the normal is the feasible region, where points will satisfy the equality constraint. In this case, as shown in Figure 4.12d, the solution with equality constraint is the solution. The same arguments that were made in terms of trade-offs in the equality case are valid here also.

The key insight from this discussion is that in the inequality $(\leq, \geq)$ constraint case, for an optimum point under consideration, one has to determine if the constraint is satisfied as an equality $(=)$ or a strict inequality $(<, >)$. If it is satisfied as an equality (the constraint is then called as an active constraint), then the same approach as the solution to equality constrained optimization problem is valid, except that we must ensure that there is no possibility of improving the objective function locally by moving away from the optimum point under consideration. If it is satisfied as a strict inequality, then one must check if the solution is in the correct side of the inequality, that is, the strict inequality constraint is satisfied by the point (feasibile point or not). Mathematically, given an objective function $f(\mathbf{x})$ and vector of m equality constraints $\mathbf{h}(\mathbf{x})$, and a vector of p inequality constraints $\mathbf{g}(\mathbf{x})$ (without loss of generality we always assume $\leq$, $\geq$ inequalities can be converted to $\leq$ inequalities through multiplication by -1), the necessary conditions are

$$l(\mathbf{x}, \boldsymbol{\lambda}, \boldsymbol{\mu}) = f(\mathbf{x}) + \boldsymbol{\lambda}^T \mathbf{h}(\mathbf{x}) + \boldsymbol{\mu}^T \mathbf{g}(\mathbf{x})$$

$$\nabla l(\mathbf{x}, \boldsymbol{\lambda}, \boldsymbol{\mu}) = 0 \tag{4.29}$$

$$\mathbf{h}(\mathbf{x}) = 0 \tag{4.30}$$

$$\mu_i g_i(\mathbf{x}) = 0 \ \forall \ i \in [1:p] \tag{4.31}$$

$$\mu_i \geq 0 \ \forall \ i \in [1:p] \tag{4.32}$$

$$\mathbf{g}(\mathbf{x}) \leq 0 \tag{4.33}$$

where $\boldsymbol{\mu}$ is a $p \times 1$ vector.

It can be observed that there are more conditions than the main gradient condition (equation 4.29) and the equality condition (equation 4.30). We discussed that inequality constraints can either be active or inactive and equation 4.31 reflects this. For every $i \in [1:p]$, either μ_i is zero or $g_i(\mathbf{x})$ is zero. If μ_i is zero, the constraint is inactive and will not participate in the gradient equation (equation 4.29). When $g_i(\mathbf{x})$ is zero, this means that the constraint is active and the optimum solution is on the constraint surface. Whenever this happens, it can be shown that if $\mu_i > 0$ (equation 4.32), then there is no scope for improvement in the objective function in the local region. Of course, equation 4.33 ensures that all the inequalities are satisfied for the optimum point under consideration. Numerical approaches for solving inequality constrained problems build on the solution to equality constrained problems. One idea is to guess which constraints might be active and which ones might be inactive and once this choice is made, the problem then becomes one of equality constrained optimization problem and solution approaches for solving these types of problems can then be leveraged. There are several details that need to be addressed to make such algorithms work [45, 12].

To understand multiple inequality constraints consider the following optimization problem

$$\min_{x_1, x_2} \quad 2x_1^2 + 4x_2^2$$
$$\text{s.t.} \quad 3x_1 + 2x_2 \leq 12$$
$$2x_1 + 5x_2 \geq 10$$
$$x_1 \leq 1$$

The problem has two decision variables with three inequality constraints. For the sake of illustration, one of these inequalities has been chosen to be a $\geq$ inequality. The necessary conditions (equations 4.29, 4.30, 4.31, 4.32, 4.33) for this problem are as follows:

$$4x_1 + 3\mu_1 - 2\mu_2 + \mu_3 = 0$$
$$8x_2 + 2\mu_1 - 5\mu_2 = 0$$
$$\mu_1(3x_1 + 2x_2 - 12) = 0$$
$$\mu_2(10 - 2x_1 - 5x_2) = 0$$
$$\mu_3(x_1 - 1) = 0$$
$$\mu_1 \geq 0$$
$$\mu_2 \geq 0$$
$$\mu_3 \geq 0$$
$$3x_1 + 2x_2 \leq 12$$
$$2x_1 + 5x_2 \geq 10$$
$$x_1 \leq 1$$

It can be noticed that the $\geq$ constraint has been converted to a $\leq$ constraint as seen from the fourth condition in the list of conditions above. As explained above, to solve for these conditions, one could proceed by choosing constraints as either being active (A) or inactive (I). Since there are three constraints, there are eight possibilities. Table 4.5 provides the analytical computations for each of these combinations and a geometric interpretation of these results is provided in Figure 4.13.

Example 4.15. *Solve the following optimization problem.*

$$\begin{aligned} \min \quad & x_1 + x_2 \\ \text{s.t.} \quad & x_1^2 + x_2^2 \leq 2 \end{aligned}$$

Solution 4.15.

$$L(x_1, x_2, \mu) = x_1 + x_2 + \mu\left(x_1^2 + x_2^2 - 2\right)$$

KKT conditions:

$$\frac{\partial L}{\partial x_1} = 0 \Rightarrow 1 + 2\mu x_1 = 0$$
$$\frac{\partial L}{\partial x_2} = 0 \Rightarrow 1 + 2\mu x_2 = 0$$
$$\mu\left(x_1^2 + x_2^2 - 2\right) = 0$$
$$\mu \geq 0$$
$$x_1^2 + x_2^2 - 2 \leq 0$$

TABLE 4.5

Combinatorial Analysis of All Cases

Sl.no	Active (A) /Inactive (I) constraints			Solution (x, μ)	Possible optima (Y/N)	Remark
	(a)	(b)	(c)			
1	A	A	A	Infeasible	N	3 equations and 2 variables. No feasible solution.
2	A	A	I	$x = [3.6364 \quad 0.5455]$ $\mu = [-5.2 \quad -1.45 \quad 0]$	N	$x_1 \le 1$ is not satisfied, $\mu_1 < 0$, $\mu_2 < 0$
3	A	I	A	$x = [1 \quad 4.5]$ $\mu = [-18 \quad 0 \quad 50]$	N	$\mu_1 < 0$
4	I	A	A	$x = [1 \quad 1.6]$ $\mu = [0 \quad 2.56 \quad 1.12]$	Y	All constraints and KKT conditions satisfied
5	A	I	I	$x = [3.27 \quad 1.09]$ $\mu = [-4.36 \quad 0 \quad 0]$	N	$x_1 \le 1$ is not satisfied, $\mu_1 < 0$
6	I	A	I	$x = [1.21 \quad 1.51]$ $\mu = [0 \quad 2.45 \quad 0]$	N	$x_1 \le 1$ is not satisfied
7	I	I	A	$x = [1 \quad 0]$ $\mu = [0 \quad 0 \quad -4]$	N	$\mu_3 < 0$
8	I	I	I	$x = [0 \quad 0]$ $\mu = [0 \quad 0 \quad 0]$	N	$2x_1 + 5x_2 \ge 10$ is not satisfied

Case 1: Active constraint

$$x_1^2 + x_2^2 - 2 = 0 \ and \ \mu \ge 0$$

$$x_1 = \frac{-1}{2\mu}; \quad x_2 = \frac{-1}{2\mu}$$

$$x_1^2 + x_2^2 = 2 \Rightarrow \frac{1}{4\mu^2} + \frac{1}{4\mu^2} = 2 \quad \frac{1}{2\mu^2} = 2 \Rightarrow \mu^2 = \frac{1}{4} \Rightarrow \mu = \pm\frac{1}{2}$$

As μ has to be greater than or equal to 0, $\mu = \frac{1}{2}$.

$$x_1^* = -1; \quad x_2^* = -1; \quad f(x_1^*, x_2^*) = -2$$

Case 2: Inactive Constraint

$$x_1^2 + x_2^2 - 2 < 0 \ and \ \mu = 0$$

$\frac{\partial L}{\partial x_1} = 0$ *and* $\frac{\partial L}{\partial x_1} = 0$ *cannot be satisfied as* $\mu = 0$ *and thus, this case cannot be the optimum solution.*

So, the optimum solution is $\begin{bmatrix} -1 & -1 \end{bmatrix}^T$.

4.9   DYNAMIC PROGRAMMING

Dynamic programming (DP) is an optimization technique that was developed by Richard Bellman in the 1940s [5]. Before discussing dynamic programming, one needs to understand recursion.

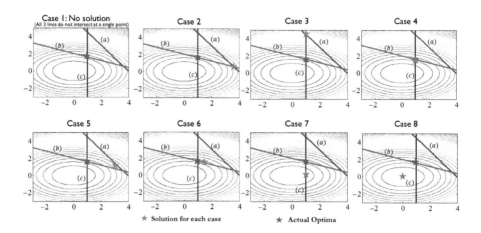

FIGURE 4.13 Pictorial depiction of combinatorial cases.

4.9.1   Recursion

Recursion is a method to approach a large class of problems. The original problem is decomposed into levels of subproblems until subproblems that can be solved easily are reached. These trivially solved subproblems are known as the base cases of the recursion. To understand recursion, let us consider a very simple example - to find the sum of the first n positive integers (this is only for demonstration, the best approach is to use the result $(n(n + 1)/2)$).

Let $S(n)$ represent the sum of the first n positive integers. To solve this, we will make use of the observation that $S(n) = n + S(n - 1)$ as the sum of the first n positive integers is the sum of the first $(n - 1)$ positive integers plus the nth positive integer. Here we can see that problem can be solved by using the result of a smaller subproblem. Now consider $S(1)$. The value of $S(1)$ is clearly 1. $S(1) = 1$ is the base case of the recursion. Hence, our strategy to solve the problem would be to break the problem S(n) into smaller subproblems until we reach $S(1) = 1$. Finally, the solutions to the subproblems are used to solve the original larger problem.

Listing 4.1: Sum of the first n positive integers

```
1   def  S(n):
2        if  (n == 1):
3            return  1
4        else:
5            return  (n + S(n − 1))
```

Let us look at one more example to understand recursion. The Fibonacci sequence is a very well-known sequence where a given number in the sequence is the sum of the previous two numbers in the sequence and the first two

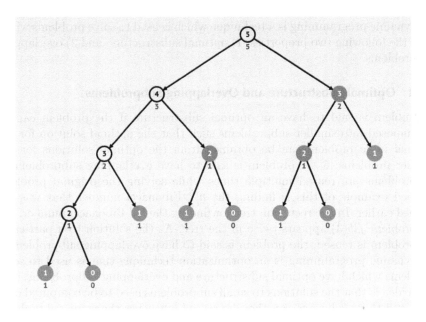

FIGURE 4.14 Recursion Tree for Fibonacci.

numbers are 0 and 1. The first 10 numbers of the Fibonacci sequence are 0, 1, 1, 2, 3, 5, 8, 13, 21, 34. Our aim is to calculate the nth number in the Fibonacci sequence. Let $fib(n)$ represent the nth number in the sequence. From the definition of the sequence, $fib(1) = 0; fib(2) = 1; fib(n) = fib(n-1) + fib(n-2)$ for $n > 2$. Clearly the problem of finding the nth Fibonacci can be split into two smaller problems, finding the $(n-1)^{th}$ number in the sequence and finding the $(n-2)^{th}$ number in the sequence with $fib(1) = 0$ and $fib(2) = 1$ being the base cases. A recursion tree (refer Figure 4.14) illustrates how the problem is split into subproblems.

Listing 4.2: Recursion for Fibonacci sequence

```
1   def fib(n):
2       if (n == 1):
3           return 0
4       elif (n == 2):
5           return 1
6       else:
7           return (fib(n − 1) + fib(n − 2))
```

Dynamic programming is a technique which is used to solve problems which have the following two properties: 1) optimal substructure, and 2) overlapping subproblems.

4.9.2 Optimal Substructure and Overlapping Subproblems

A problem is said to have an optimal substructure if the problem can be decomposed into smaller subproblems such that the optimal solution for the original large problem can be obtained from the optimal solutions for the smaller problems [5]. A problem is said to have overlapping subproblems if subproblems are reused multiple times while solving the original problem. A good example of this is finding the nth Fibonacci number that was described earlier. In the recursion tree for finding the 5th Fibonacci number, the subproblem $fib(3)$ appears twice in the tree. As the solution to a particular subproblem is reused, the problem is said to have overlapping subproblems.

Dynamic programming is an optimization technique that is used to solve problems which have optimal substructure and overlapping subproblems. The main idea is that the solutions to small subproblems need to be computed only once and these solutions can then be reused whenever the same subproblem appears in the recursion. Optimal substructure is a necessary property as this technique only works if the optimal solution to the overall problem depends on the optimal solution to the subproblems. Otherwise, there is no use in storing the optimal solutions to subproblems. Also notice that if the problem does not have overlapping subproblems, then storing optimal solutions to the subproblems does not achieve anything since the subproblems are never reused. In the Fibonacci tree, it is clear that the subproblem $fib(3)$ needs to be solved only once. Hence, we will apply DP to the problem by storing the solution of each subproblem. Whenever a new subproblem is encountered, we first check if that problem has already been solved. If yes, then the stored answer to that subproblem is directly used.

Listing 4.3: DP for Fibonacci sequence

```
1   import numpy as np
2   n = 100
3   storefib = np.full((n,1),-1)   #Creates an nx1 matrix and ...
        fills every cell with 1 . Here 1 acts as an ...
        indicator which tells us that a particular subproblem ...
        is unsolved
4   storefib[1][0] = 0
5   storefib[2][0] = 1
6   def fib(n):
7       if(storefib[n][0]!=-1):  #If subproblem is already ...
            solved , then use the stored solution
8           return storefib[n][0]
9       else:
10          storefib[n][0] = fib(n-1) + fib(n-2) #Otherwise compute ...
                and store
11          return storefib[n][0]
```

4.9.3 2D Dynamic Programming Example

Consider a $m \times n$ rectangular grid with the point $(1,1)$ at the bottom left corner and (m,n) at the top right corner as shown in Figure 4.15. Now, let us say that a robot is placed at cell $(1,1)$ and has to reach the cell (m,n) in a sequence of moves. We define a move as going either one cell to the right or one cell up. The robot cannot move to the left or downwards. We want to find out the number of different ways in which a robot can reach cell (m,n).

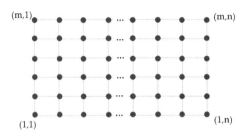

FIGURE 4.15 2D Dynamic Programming Example

Let $ans(i,j)$ be the number of different ways to reach the cell (i,j) given that the robot starts at cell $(1,1)$. Therefore, the question boils down to finding $ans(m,n)$. To solve this, we are going to make an interesting observation. Let us say that we are trying to find $ans(i,j)$, that is the number of ways to reach the cell (i,j), where i and j are greater than 1. One move before the robot reaches the cell (i,j), it is at either $(i,j-1)$ or $(i-1,j)$ because the robot can only move up or to the right. This implies that the number of ways to reach (i,j) is the sum of the number of ways to reach $(i,j-1)$ and $(i-1,j)$. Formally,

$$ans(i,j) = ans(i,j-1) + ans(i-1,j), \text{ where } i,j > 1 \qquad (4.34)$$

Let us now look at the base case of the solution, when $i = 1$ or $j = 1$. If both i and j are 1, then the number of ways to reach the cell is 1 (which is just arriving at the cell). Also, the number of ways to reach any cell $(n,1)$ or $(1,n)$ is only 1, either moving only right or up. Therefore,

$$ans(i,j) = 1, \text{ if } i = 1 \text{ or } j = 1 \qquad (4.35)$$

Thus, we arrive at a solution to the problem. To find the number of ways to reach a cell (i,j), we use the results to subproblems $ans(i,j-1)$ and $(i-1,j)$ with the base case being equation 4.35. Recursion tree for this grid problem is shown in Figure 4.16.

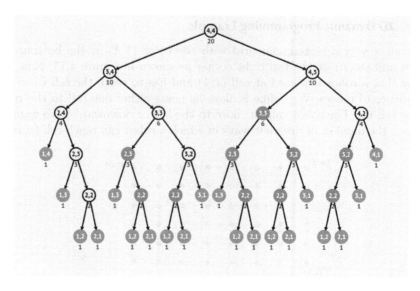

FIGURE 4.16 Recursion Tree for Grid Problem.

Listing 4.4: 2D Dynamic Programming Example

```
1  def ans(m, n):
2      storeans =[[None](n + 1) for i in range(m + 1)]
3      storeans[1][1] = 1
4      for i in range(m + 1):
5          storeans[i][1] = 1
6      for j in range(n+1):
7          storeans[1][j] = 1
8      for i in range(2, m + 1):
9          for j in range(2, n + 1):
10             storeans[i][j] = storeans[i 1][j] + ...
                   storeans[i][j 1]
11
12     return storeans[m][n]
```

EXERCISE PROBLEMS

Exercise 4.1. *For the optimization problems given below, identify whether the problem is univariate/multivariate, constrained/unconstrained and NPL/LP/QP/MINLP/BLP/ILP. Explain with reasons.*

a) $\min x^3$
b) $\min 4x^2 + 3y^2$
c) $\min 4x^3 + 3y^2$
d) $\min 4x + 3y$ s.t. $-4 \le x \le 8$
e) $\min 4x + 3y$ s.t. $-5 \le x \le 2;\quad -4 \le y \le -1$
f) $\min 4x + 3y$ s.t. $-5 \le x \le 2;\quad -4 \le y^2 \le 81$

g) $\min 4x + 3y$ s.t. $-5 \le x \le 2;$ $-4 \le y^2 \le 81;$ $y \in \mathbb{Z}$ *(integer space)*
h) $\min 4x^3 + 3y$ s.t. $-5 \le x \le 2;$ $-4 \le y \le 8;$
i) $\min 3x^2 + 2y^2 + z^2$ s.t. $-5 \le x \le 6;$ $-3 \le y \le 9;$ $-4 \le z \le 7$
j) $\min 3x + 2y + z$ s.t. $-5 \le x \le 6;$ $-3 \le y \le 9;$ $-4 \le z \le 7$
k) $\min 3x^2 + 2y^2 + z^2$ s.t. $-5 \le x \le 6;$ $-3 \le x^3 + y^3 \le 81;$ $-4 \le$
   $x^2 + z^2 \le 49$

Exercise 4.2. *Consider the data provided in the table.*

x	1	2	3
y	5	8	11

If we are interested in fitting the data using a linear model $\hat{y} = f_{ML}(x) = mx + c$, write the corresponding optimization problem to find the unknown parameters m and c. Solve this optimization using

(a) Newton's method
(b) Stochastic gradient descent method

Exercise 4.3. *Consider the classification problem given below.*

x	5	-1	-4	2	3	3
y	4	2	6	3	-5	7
class	-1	1	-1	1	-1	1

If a linear separator of the form $y - ax - b = 0$ is used to classify the data, write the corresponding optimization problem that can be used to find the unknown parameters a and b.

Exercise 4.4. *Find the stationary optimum points and identify whether each of the stationary points corresponds to a maximum, minimum or saddle point.*

a) $f(x) = (2x - 5)^2 - 4$
b) $f(x, y) = (2x - 5)^2 + (6 - 3y)^2 - 4$
c) $f(x, y) = (2x - 5)^2 + 3y - 4$
d) $f(x, y) = (2x - 5)^2 + 3y^2 - 9y - 4$
e) $f(x, y) = (2x - 5)^3 + (-9 + y)^3$
f) $f(x_1, x_2) = x_1^4 + 2x_1 x_2^2 + 5x_1 x_2 - 3x_1^2 x_2 + 7$

Exercise 4.5. *Find a minimum of the function $f(x) = (2x - 5)^2 - 4$ using gradient search method starting from -5.*

Exercise 4.6. *Find a minimum of the function $f(x) = (2x - 5)^2 - 4$ using bracketing method given that the minimum is in the range $\begin{bmatrix} -5 & 10 \end{bmatrix}$.*

Exercise 4.7. *Find a minimum of the function $f(x_1, x_2) = (2x_1 - 5)^2 + 9x_2^2 - 4$ using steepest descent algorithm starting from $x^0 = \begin{bmatrix} -5 & 10 \end{bmatrix}^T$. Use tolerance on ∇f as 0.02.*

a) Use $\alpha = 0.4$.

b) Use α identified using univariate optimization.

Exercise 4.8. *Find a minimum of the function $f(x_1, x_2) = (2x_1-5)^2+9x_2^2-4$ using Newton's method starting from $x^0 = \begin{bmatrix} -5 & 10 \end{bmatrix}^T$.*

Exercise 4.9. *Solve the optimization problem*

$$\min_{x_1, x_2} \quad (2x_1 - 5)^2 + 9x_2^2 - 4$$

$$\text{s.t.} \quad 3x_1 - 2x_2 = 5$$

using direct substitution and Lagrangian methods.

Exercise 4.10. *Solve the following optimization problems*

a)

$$\min_{x_1, x_2} \quad (2x_1 - 5)^2 + 9x_2^2 - 4$$

$$\text{s.t.} \quad 3x_1 - 2x_2 \leq 5$$

c)

$$\min_{x_1, x_2} \quad (2x_1 - 5)^2 + 9x_2^2 - 4$$

$$\text{s.t.} \quad 3x_1 - 2x_2 \geq 5$$

b)

$$\min_{x_1, x_2} \quad (2x_1 - 5)^2 + 9x_2^2 - 4$$

$$\text{s.t.} \quad 3x_1 - 2x_2 \leq 5$$
$$x_1 \leq 2.5$$
$$x_2 \geq 3$$

d)

$$\min_{x_1, x_2} \quad (2x_1 - 5)^2 + 9x_2^2 - 4$$

$$\text{s.t.} \quad x_1 \leq 2.5$$
$$x_2 \geq -2$$

Exercise 4.11. *Check the convexity of the following optimization problem and find the global minimum, if it exists.*

$$\min \ f(x_1, x_2, x_3) = 3x_1^2 + 9x_2^2 + 2x_3^2 - 5x_1x_2 + 2.5x_1x_3 - 4x_1 - 2x_2 - 10$$

Exercise 4.12. *Consider the following optimization problems. Interpret the problems geometrically and determine the optimal solution based on geometry. Sketch the feasible set.*

(a) $\min_x \ x_1 + x_2$ *s.t.* $x_1^2 + x_2^2 - 2 = 0$
(b) $\min_x \ x_1 + x_2$ *s.t.* $x_1^2 + x_2^2 \leq 2$
(c) $\min_x \ (x_1 - 1.5)^2 + (x_2 - 1/8)^2$ *s.t.* $x_1 + x_2 \leq 1$, $x_1 - x_2 \leq 1$, $-x_1 + x_2 \leq 1$, and $-x_1 - x_2 \leq 1$

Exercise 4.13. *Identify whether the following sets are convex or not. Explain why.*

(a) Region constructed by the constraints $x_2 \geq 1 - x_1$, $x_2 \leq 1 + 0.5x_1$, $x_1 \leq 2$ and $x_2 \geq 0$.

(b) Region constructed by the constraints $2x_2 - 4 - x_1 \leq 0$, $x_1 + x_2 - 5 \leq 0$, $x_1 \leq 5$, $-x_1 \leq 0$, $-x_2 \geq 0$, and $(x_1 - 1.5)^2 + 0.5(x_2 - 1)^2 \leq 1$.

5 Statistical Foundations for DS and ML

The aim of this chapter is to introduce those core ideas of statistics that are relevant for data science. As the field of statistics is vast, concepts that lay the foundation for data science have been carefully chosen and elaborated here. Every effort has been made to maintain the presentation as intuitive as possible; however, rigorous mathematical analysis has been included as and when appropriate.

5.1 DECOMPOSITION OF A DATA MATRIX INTO MODEL AND UNCERTAINTY MATRICES

The characterization of uncertainty starts with the notion of probability. Let us begin with an example—suppose one is doing an experiment to demonstrate Newton's second law of motion ($F_{net} = ma$) using a ball that weighs 0.5 kg. The experimental set-up consists of the ball, a device to generate force, a device to measure the acceleration of the ball (how these devices work is not considered to be relevant for the purpose of this example), and a smooth surface that is generated to negate the effects of friction. Since the surface is smooth, the net force on the ball will be equal to the externally applied force, which can be controlled. If the experiment is repeated several times with force values 1 N, 2 N, 3 N, 4N, etc., and the corresponding values of acceleration are measured, two sets of values, one being net force (F) on the ball, and the other being the corresponding acceleration can be generated. Considering that the applied force values are 1 N, 2 N, 3 N, etc. and the mass of the ball is 0.5 kg, it is expected that the values of acceleration in m/s^2 would be 2, 4, 6, etc. The corresponding data is shown in Table 5.1.

It is critical to note that the values above are the "expected" measurements of acceleration given the force values and mass of the ball, within the framework of Newton's second law. However, if the same experiment were to be repeated multiple times, will the measurements of acceleration always

TABLE 5.1

Data for Force and Acceleration

$F(N)$	1	2	3	4	5	6	7
$a(m/s^2)$	2	4	6	8	10	12	14

DOI: 10.1201/b23276-5

TABLE 5.2

Data for Force and Acceleration

$F(N)$	1	2	3	4	5	6	7
$a(m/s^2)$	1.98	4.03	5.95	7.90	10.07	11.88	14.06

yield 2, 4, 6, etc.? In practice, this is impossible. The fundamental reason is simple—any measurement is always accompanied with an error. The mass of the ball was considered to be 0.5 kg, but a more precise scale might measure it to weigh 0.498 kg, and an even better one may measure 0.4985 kg. The same can be said about the devices used to generate force and measure acceleration. Even if one sets the external force to 1 N, the device may generate 0.999 N one time, and 1.001 N the next. Similarly, acceleration measurements may not be consistent across experimental runs, and absolutely friction free smooth surfaces are practically impossible to generate. Therefore, obtaining the expected values, or even data with the same values, is very unlikely. In other words, no two measurements will be the same; every measurement will be associated with an unavoidable error or uncertainty. This of course does not mean that Newton's second law of motion is not valid, all properly taken measurements will have values that are "close" to the expected values. But how "close" is close enough? How does one extract Newton's second law from a group of similar, but not identical data? From this example, it is clear that handling uncertainty/error is fundamental to deriving insights from collected data, and thus, fundamental to data science.

Suppose one obtains measurements with the set-up mentioned above, and data shown in Table 5.2 is obtained. The aim is to try and extract Newton's second law from this data. This data can be organized in a matrix form. For this, let us assume that the error is additive in nature, *i.e.*, $X = T + U$, where X is the data matrix obtained from the experiment (where the first column is the force and the second column is the acceleration), T is a matrix containing true, error-removed data, and U is the "uncertainty" matrix, which contains the error values, or the uncertain component of the experimental data. The additive model assumes that the experimentally obtained data matrix can be decomposed into the sum of an invariant "true data" matrix, and an error matrix that varies from one experimental run to the next. For the data shown in Table 5.2, the decomposition can be written as follows.

$$
\begin{bmatrix} 1 & 1.98 \\ 2 & 4.03 \\ 3 & 5.95 \\ 4 & 7.90 \\ 5 & 10.07 \\ 6 & 11.88 \\ 7 & 14.06 \end{bmatrix} = \begin{bmatrix} 1 & 2 \\ 2 & 4 \\ 3 & 6 \\ 4 & 8 \\ 5 & 10 \\ 6 & 12 \\ 7 & 14 \end{bmatrix} + \begin{bmatrix} 0 & -0.02 \\ 0 & 0.03 \\ 0 & -0.05 \\ 0 & 0.10 \\ 0 & 0.07 \\ 0 & -0.12 \\ 0 & 0.06 \end{bmatrix} \tag{5.1}
$$

It can be observed that the first matrix obtained upon decomposition, or T, matches the expected values from the experiment, and the other matrix provides the error or uncertainty associated with the measurement. Different experimental runs will thus yield different data matrices that can be similarly broken down into T and U matrices, and U will vary while T will remain invariant. A closer inspection of this example reveals that the decomposition of the data matrix was made possible as the form of Newton's second law was known *a priori*, the mass is precisely known, and force measurements are deviod of errors. However, without these assumptions, there are infinite ways of decomposing any data matrix. If such a relationship between the variables ($F_{net} = ma$) is unknown, how can one achieve a trade-off between modelable data and uncertainty with minimal prior knowledge? This involves two key components: 1) assuming possible model structures and trying to fit these to the data and 2) characterizing uncertainty. Finally, it is important to note that the additive model is only one of many available for extracting "true" data and error values.

5.2 UNCERTAINTY CHARACTERIZATION

To begin with, one needs to understand how to characterize events that are uncertain, or in other words, unpredictable. Let us start with the examination of the most unassuming of experiments, the toss of a coin. Suppose one were to flip a coin and try to predict the outcome, while the coin is still in midair. It should be quite clear that the answer to this question is uncertain, *i.e.*, under normal circumstances, one cannot definitively say what the coin will show upon landing. This is quite different from an experiment in which a coin having heads on both sides is tossed. The key conceptual insight is the appreciation of the fact that although the exact outcome of a given coin toss is not certain, the total set of possibilities certainly is. Thereby, instead of focusing on determining the exact outcome of a coin toss (which is inherently uncertain), one introduces the notion of "probability" of an outcome, which is the likelihood that an experiment will yield that outcome [19, 38].

5.2.1 A Simple Probability Model

Probabilities can be understood within a simple framework. Consider an experiment. The set of all possible outcomes of an experiment is referred to as the sample space of the experiment. The probability of any outcome can be computed using the following assumptions:

a) The outcomes of the experiment are mutually exclusive.
b) All outcomes are equally likely.
c) The sample space is the exhaustive set of possible outcomes of the experiment.

If these conditions are met, probability of any event related to the experiment can be computed. The probability of any event is simply the ratio of the number of outcomes in the sample space that are favorable to the event to the total number of outcomes in the sample space. If n mutually exclusive outcomes which make up an exhaustive sample space are equally likely, the probability of each outcome is $1/n$. In a coin toss, the possible outcomes (H and T) have an equal probability of 0.5. Consider the experiment involving the toss of three coins at the same time. The sample space is, $S = \{HHH, HHT, HTH, HTT, THH, TTH, THT, TTT\}$. Consider the probability of getting two heads in this case (event). This is the ratio of the number of outcomes having two heads to the total number outcomes $= 3/8$.

5.2.2 Computing Probabilities from Experimental Data

While tossing a coin, the assumption of equal probabilities is valid only if the coin is unbiased, *i.e.*, equally likely to show heads or tails. Now, suppose that the coin is biased but we don't know the extent of this bias. Since the outcomes are no longer equally likely, one cannot use the simple model to determine the probability of any event associated with the coin toss. In this case, analysis of experimental data is required. The probability of heads and tails can be computed by carrying out the coin toss a large number of times. We no longer rely on the sample space but simply say that the probability of any event is the ratio of the number of outcomes of the experiments favorable to the event to the total number of times the experiments are performed. One can use this definition to calculate the probability of getting heads/tails in the case of a biased coin, and also to calculate the probabilities of getting two heads when three identical biased coins are tossed simultaneously and so on. The calculated probabilities based on the frequencies will stabilize at the "true" probabilities as the number of experiments that are performed increases.

5.3 RANDOM VARIABLES AND PROBABILITY MASS FUNCTIONS

We know that tossing three coins at the same time has eight probable outcomes. In general, tossing n coins at the same time will have 2^n possible outcomes. It should be evident that as n increases, the number of possible outcomes also increases exponentially, and representing these outcomes can become quite cumbersome. A simple approach to representing the outcomes of an experiment is desirable. This can be achieved through the definition of random variables, which are simply functions that map the sample space of an experiment to the set $\mathbb{R}$ of real numbers. Consider a simple coin toss. Suppose we define a random variable X that is associated with this experiment, and assign H to 1 and T to 0. X can therefore take values 0 and 1. The probability of the coin toss yielding a head $= P(X = 1) = 0.5$ and a tail $= P(X = 0) = 0.5$.

The idea of random variables is explored with another example. Consider a book club which has 16 members and meets once a week. Suppose one is interested in understanding the number of people that attend each meeting. This is similar to a coin toss in that although the outcome of any experiment (the number of people attending a given meeting) is uncertain, the total set of possibilities is fixed. A random variable X can be defined to denote the number of members present in a given meeting. X can take values $(0 \leq x \leq 16, x \in$ whole numbers). Once we know the probability of each outcome of the experiment, we can use these probabilities to calculate the probability of various events associated with the experiment. A simple probability model will lead to $P(X = 0) = 1/17$, $P(X = 1) = 1/17, \ldots, P(X = 16) = 1/17$. Intuitively, this seems like a very poor estimate of the probabilities. It is more likely that the probability of very few people attending or most people attending are both rather low, and the highest probabilities are likely for attendance in the middle range (maybe between 6 and 10).

5.4   DERIVING MODEL PROBABILITY DISTRIBUTION FUNCTIONS

The fundamental challenge is as follows—one would like to assign a probability to every value of the random variable (each value of the random variable represents a possible outcome of the experiment), however assuming each outcome is equally likely seems too simplistic. Let us define a function which maps random variable X to probabilities. Such a function, denoted by p, is called the probability mass function of the random variable X. In essence, it assigns a probability (or "weight/mass") to every possible value of X [16, 41]. We write, $P(X = x) = p(x), 0 \leq x \leq 16$.

The next question is obviously related to how one can determine this probability mass function. In some cases, it can be derived theoretically using certain assumptions. A probability model based on an understanding of the underlying random processes is required in these cases. Another approach is to estimate the probabilities using statistics. Since the book club meets once a week, that is approximately four meetings a month, this translates to about two hundred meetings a year. If the number of people attending these meetings is recorded, then the frequency of attendance can be used to determine probabilities. The estimates will improve as more measurements are taken, *i.e.*, the sample size of observations is increased. However, many underlying assumptions such as the basic behavior of individuals do not change over the experimental period need to hold.

The probability mass functions can also be theoretically derived based on different sets of underlying assumptions. For example, consider the assumption that the probability of any individual member attending a meeting, which is p, is the same for all members. Further, assume that a member attending the meeting is not dependent on which other member attends the meeting. $P(X = 1)$ is probability of one person attending the meeting. Consider a random member A. The probability of A attending a meeting is p. The probability

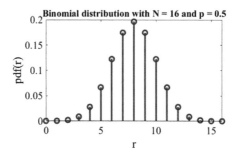

FIGURE 5.1 Binomial distribution.

that A attends the meeting and others do not is $(p) \times (1-p)^{15}$. To compute $P(X=1)$, we should consider not just A but also all the others. Therefore, probability of one person attending the meeting $= (16C_1) \times (p) \times (1-p)^{15}$. This can be generalized to, $P(X=r) = (16C_r) \times (p)^r \times (1-p)^{(16-r)}$.

This is our first theoretically derived probability mass function. These functions are also called probability distribution functions. The function we derived is the well-known binomial distribution. The graph of this distribution is shown in Figure 5.1 (for $p = 0.5$). As can be seen, the results are more intuitive when compared to the standard probability model. It has the number of people attending a given meeting on the X-axis, and the corresponding probability on the Y-axis. As speculated, the highest probabilities seem to be in the 6 to 10 attendance range, and the probabilities drop off at either very high or very low attendance.

5.4.1 Alternate Model Distributions for the Group Meeting Attendance Problem

Another possible assumption is to divide all members of the book club into say four groups of four based on likelihood of attendance probabilities. This is different from the previous model in which the probability of attendance of every member was assumed to be the same (p). Several possible density functions based on the four probabilities (p_i being probability for the ith group) are plotted in Figure 5.2. In the case where three groups have zero probability of attendance and one has probability 1, the probability of exactly four people attending the meeting is always 1 (top right in Figure 5.2). Similar analysis of the distribution plots can be made based on the assumed probabilities. The most appropriate assumptions can be identified by comparing these results with actual data.

5.4.2 Another Example – Students Clearing Exams

Consider the case of a student taking an exam that can be attempted multiple times. Let the probability of the student passing the exam be p; we also assume

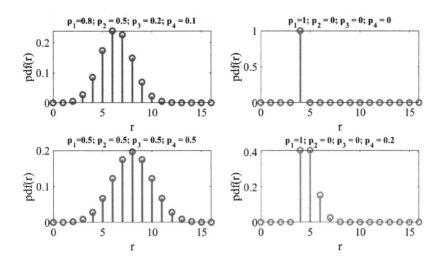

FIGURE 5.2 Alternate distributions with more groupings.

that p remains constant every time the student takes the exam. Therefore, the probability of the student failing is $(1 - p)$. Let X be a random variable representing the number of times the student must take the exam before passing. Therefore, $X = 1, 2, 3, 4, 5, \ldots$ (possible values of X). Our aim is to find the probability mass function of X. Consider the case when the student has to take the exam r times before passing. This implies that the student fails the first $(r - 1)$ times and passes the rth time. We can therefore write, $P(X = r) = (1 - p)^{(r-1)} \times (p)$, which can be recognized as the well-known geometric distribution.

5.4.3 Summary of Discussions on Theoretical Distributions

The discussions so far can be summarized into the following four points—

(a) One may use certain logical assumptions to derive theoretical distribution functions such as the binomial distribution, geometric distribution.
(b) Certain distributions appear in numerous situations (binomial distribution, normal distribution). This endows these distributions with a certain universality and hence it is useful to study and analyze them further.
(c) Experimental data can be used to derive probability distribution functions.
(d) Statistical properties of the data can be compared with those obtained from the theoretical distribution to identify the best fit and validate the assumptions made.

5.5   PROPERTIES OF PROBABILITY DISTRIBUTION FUNCTIONS

Probability distribution functions to model random variables and derivation of theoretical distributions for some examples were described. Further, some preliminary discussion of how these theoretical distributions can be bench-marked was also provided. Formal benchmarking procedures have not been described yet. In this section, fundamental properties of the distribution func-tions will be described, followed by an introduction to continuous variables and the extension of ideas of probability distribution function to continuous variables (most of what we have seen till now is for discrete variables). We start with the properties of distributions functions for discrete variables.

a) Probabilities cannot be negative, therefore, $p(X = x) \geq 0, \forall\, x$.
b) Since the sum of the probabilities for all events must add up to one,
$$\left(\sum_{\forall x} p(X = x) \right) = 1.$$

One can verify that these properties hold for the geometric and binormal distributions discussed earlier.

5.5.1   Continuous Random Variables and Their Distributions

Until now all the random variables that were discussed are discrete random variables, *i.e.,* random variables which take specific values. We now turn our attention to another class of random variables, the continuous random vari-ables. Continuous random variables can take a value in any finite or infinite range in the number line. Some examples are: the time taken to finish an activity, and the temperature of a sample etc.

Consider a random variable which can take values between 1 and 2. Con-sider the problem of determining the probability that the random variable takes the value 1.5. Applying the probability ideas described earlier, the prob-ability of any event is the ratio of the number of favorable outcomes to the total number of outcomes. In this case, the number of favorable outcomes is 1, while the total number of outcomes is infinite (the random variable can take any value from 1 to 2 along the number line). Hence, the probability that the random variable will take a value of 1.5 is zero. Consequently, the probability of the random variable taking any specific value between 1 and 2 is 0. However, this seems quite counter-intuitive; surely a value of 1.5 is not impossible. One way to rationalize this is through the following argument. Impossible events have a probability of zero, however, zero probability does not imply an impossible event.

Rather than the probability of the random variable taking a particular value (like 1.5), suppose one wants to know the probability of the random variable taking a value in a range, such as between 1 and 1.5. Intuitively, this prob-ability should be 0.5. However, this result is not so clear when the standard probability model is applied; this yields what appears to be an indeterminate answer (infinity divided by infinity). We thus turn to mathematics to provide a formalization for this idea, by introducing the concept of a "measure", which assigns a value to every subset of a set.

Very informally, this can be thought of as the "size" of the set. We then think of probability as the ratio of the measure of favorable events to the measure of all outcomes. In the current example, the measure can be thought of as the length of the set, which is 0.5 for 1 to 1.5 and 1 for 1 to 2. Hence, the probability of the random variable assuming a value between 1 and 1.5 is 0.5. This notion of a range rather than a value carries over to distribution functions as well. The probability density function (pdf) for continuous variables is [13, 38],

$$f(X = x) = f(x) = \lim_{dx \to 0} \left(\frac{P(x \leq X \leq x + dx)}{dx} \right) \tag{5.2}$$

Equation 5.2 can be rewritten as the probability that a continuous random variables takes a value in a very small range

$$P(x \leq X \leq x + dx) = f(x)dx \tag{5.3}$$

It should be noted that $f(x)$ does not provide the probability of a random variable assuming the value x. Instead, it provides a function, integration of which, can be used to compute the probability of the random variable taking values in an interval.

5.5.2 Summary of Distributions for Continuous Random Variables and Their Properties

The probability density function can also be thought of as the normalized frequency of data per unit length. Every continuous random variable can be characterized/modeled by a probability density function. We now look at some properties of the probability density functions:

(a) $f(X = x) \geq 0, \forall x$
(b) $\int_{-\infty}^{\infty} f(x)dx = 1$, since the probability for a random variable to take some value in the whole range, that is from negative infinity to positive infinity must be 1.
(c) The probability of the random variable assuming values in the range a to b is equal to $\int_a^b f(x)dx$
(d) One can also define another useful function called the cumulative distribution (cdf), which is the probability of the random variable assuming a value less than the given value x i.e., $P(X \leq x)$. It is denoted by F. $F(x) = P(X \leq x) = \int_{-\infty}^{x} f(x)dx$

To summarise, uncertainty was characterized using random variables. Random variables could be discrete or continuous. For continuous variables, one can only talk about the random variable taking values in an interval of finite non-zero width.

5.6 QUALITATIVE VALIDATION OF RANDOM VARIABLE PROBABILITY DISTRIBUTION FUNCTIONS

Previously we described random variables and how theoretical models can be built for the distribution functions based on certain assumptions. We alluded to being able to verify the hypothesized distribution using experimental data. In this section, certain plots, referred to as q-q and p-p plots that can be generated using the available data will be described. These plots can be used to qualitatively understand the strength of the hypothesized distributions.[1]

Consider two variables x_1 and x_2, where one is interested in developing a model. From Chapter 3, we know that if x_1 and x_2 are deterministic, a model would be of the form $a_1x_1 + a_2x_2 = b_1$. An alternate form is $x_2 = cx_1 + b$. When these variables are random, how does one build a model? Obviously, one could still talk about a model of the form $a_1x_1 + a_2x_2 = b_1$ between these variables; implications of such relationships will be discussed later.

There are crucial differences between models for deterministic and random variables. Consider two deterministic variables x_1 and x_2. Assuming x_1 to be the independent variable and x_2 to be the dependent variable, and that a linear model is valid, plot Figure 5.3 shows x_2 as a function of x_1. Looking at the plot, one could categorically say that the data does not fully fit the notion of a linear model by observing the discrepancies. Incidentally, the data was generated through a quadratic model with a small coefficient for the quadratic term. When it comes to random variables, such categorical statements are not possible. All statements and conclusions have a probability of being true. The higher this probability, the more likely that the model, on average, will provide more accurate answers.

Let us now consider a random variable x_1. In the case of deterministic variables, modeling a single variable is not very interesting. One may immediately be inclined to disagree and point out that the deterministic variable may be characterized as being constant, increasing, decreasing, and so on. However, on a little more reflection, it will become readily apparent that these statements make sense only since we already implicitly assume the existence of another independent variable (such as time or space) with respect to which the deterministic variables exhibit certain behaviors. As a result, meaningful models in the case of deterministic variables are generated only if one considers at least two variables. In contrast, models can be generated for a single random variable. As described before, this is the pdf of the random variable. The central questions that need to be addressed are: generation of a model from a single (random) variable data, and the validation of the model against a benchmark. It is easy to see that validation here is slightly more complicated than the deterministic case due to the apparent lack of a benchmark.

---

[1] In this chapter, whenever samples are used, the sample number is denoted by a superscript.

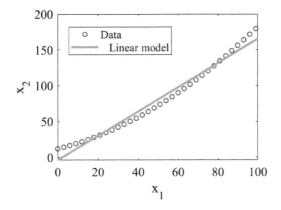

FIGURE 5.3 Model for deterministic variables.

TABLE 5.3

Data for Random Variable x_1

x_1	0.53	1.83	−2.25	0.86	0.31	−1.30	−0.43

Consider a random variable x_1, say the temperature of a surface in some arbitrary units. Assume that seven measurements of the temperature were made as documented in Table 5.3. Modeling this temperature as a random variable is distinctly different from usual modeling methods, for example modeling the temperature of an ideal gas as a function of its pressure, volume, and number of moles. In such a case, one would write $T = PV/nR$, which represents the dependence of the output variable T on several input variables (P - Pressure, n - Number of moles, R - Universal gas constant). In this case of modeling T as a random variable x_1, one is looking for a pdf that characterizes the random variable.

Mathematically, $pdf(x_1) = f(x_1, \mathbf{p})$, where $\mathbf{p}$ represents a vector of parameters. The pdf of the temperature of the surface doesn't provide the probability of the surface taking any particular temperature; instead, the probability that the temperature of the surface will fall within a given range can be computed. For example, suppose, $pdf(x_1) = (1/k)\, x_1$, where k is a constant for normalization. Now, the probability that the temperature of the surface falls in the range T_1 to T_2 would be $(1/k)\left((T_2^2 - T_1^2)/2\right)$. In other words, if you make n measurements of the surface temperature, where n is large, you would expect that $(n/k)\left((T_2^2 - T_1^2)/2\right)$ values would lie between T_1 and T_2.

In deterministic modeling of the temperature of an ideal gas, a set of measured temperature values, which are the "actual" values, will be available. Also, based on the measured pressure, volume, and number of moles, the predicted temperature values can be calculated based on the model. Consider

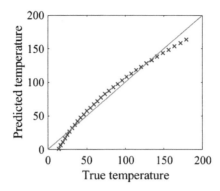

FIGURE 5.4 Parity plot to assess goodness of model.

Figure 5.4, where the true and predicted temperature values are plotted; this is referred to as a parity plot.

In a parity plot in the x-axis, the actual output value at each sample is plotted. On the y-axis, the corresponding output value estimated for the corresponding samples are plotted. For a perfect model, these datapoints at different samples will be on a line with unity slope and zero intercept. Deviations from this reference line can be used to visually assess the goodness of the fit. The key point to note here is that for every prediction there is a reference value that can be used to benchmark the quality of the model. Contrast the above example with Figure 5.5. The model is the pdf we predict. The area under the curve in any range (obtained by integrating the pdf) represents the probability that the random variable will take a value in that range.

The experimental pdf is obtained by sampling several measurements and plotting the fraction of points in each range (histogram). Since this can be done only with a broad enough range comprising significant number of datapoints, the experimental pdf is a step like function as shown in Figure 5.5. The model pdf, shown as the bold curve, can then be compared with this experimental pdf. It should then be clear that the "resolution" of a pdf constructed from data will improve as the number of obtained datapoints increases. It is interesting to note that the y-axis values in this case are generated from the data itself as there is no reference output for each sample.

An important question given the discussion above is, how does one verify if a model pdf fits the data? One approach is to calculate certain quantities based on the model pdf and compare them with the same quantities calculated from data. The assumed model pdf is appropriate if these quantities match. The first such statistical quantity of interest is the "m-quantiles". Quantiles represent the probability or fraction of points below a certain threshold value. For example, to obtain the 2-quantile for a pdf, one determines the point on the x-axis that splits the pdf into two regions each with 50% of the total area under the pdf curve. The 2-quantile for a dataset is called the median and it

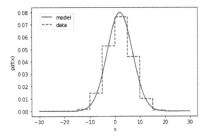

FIGURE 5.5 pdf model for a random variable.

splits the data into two groups with an equal number of points. Extending this idea, $(m-1)$ points are required to split data into m quantiles, and if the total number of datapoints is n, there will be (n/m) datapoints between any two quantiles (similarly, for a pdf, the area bounded by the region between any two quantiles will be $1/m$).

5.6.1 Computing Quantiles

For a pdf $f(x, \mathbf{p})$ identifying m quantiles involves identifying $(m-1)$ points, say $q^1, q^2 \ldots q^{(m-1)}$, which split the pdf into regions with the area under the pdf curve being $(1/m)$. Hence, the area between the first and i^{th} quantile would be (i/m). Thus, all $(m-1)$ quantile points (q^i) can be determined using the following equation (Quantile equation)

$$\int_{-\infty}^{q_i} f(x, \mathbf{p})dx = \frac{i}{m} \ \forall \ i \in [1 \quad m-1] \tag{5.4}$$

The area between quantiles q^{i+1} and q^i is $(i+1)/m - i/m = 1/m$, consistent with what was discussed earlier. For $q = 3$, one will derive two points $(q^1$ and $q^2)$ which divide the pdf into three regions of equal area. This aspect of quantiles can be linked to the fundamental property of the pdf, which is that the area bounded by any two points on the x-axis is the probability that the random variable takes values between those two points. Since the area bounded between any two quantile points is equal, the probability that the random variable takes values between those points must be equal. Hence, if we sample values from the random variable, the quantile points must divide the dataset into groups with an equal number of points.

To determine if a theoretical pdf fits a dataset, m quantiles for both the pdf (using the equation given above) and the dataset (reorder the data from the smallest to largest value, and pick every $(n \times i/m)^{th}$ datapoint (integer that is the floor of ceiling of this number) $\forall \ i \in [1 \quad m-1]$, where n is the total number of points in the dataset) and label them as x^i. Plot x^i vs q^i. The closer the obtained plot is to 45^0 line, greater confidence one may have that the theoretical pdf fits the data. This confidence will increase as m (number of

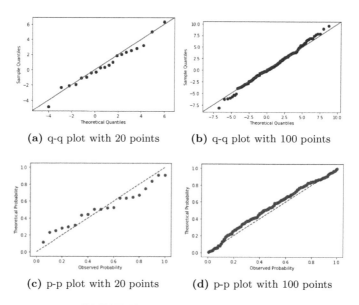

(a) q-q plot with 20 points   (b) q-q plot with 100 points

(c) p-p plot with 20 points   (d) p-p plot with 100 points

FIGURE 5.6 q-q and p-p plots.

quantiles) increases. A plot that compares the computed quantiles based on a theoretically assumed distribution and the ones identified from data is called a $q-q$ plot [33]. Such a plot for data with 20 datapoints is shown in Figure 5.6a. To generate the plot, a theoretical distribution has to be assumed; we do not describe the assumed distribution here and various options for probability distributions will be described later. From the plot, one might validate if the assumed distribution seems like a reasonable choice for the data. Data from the same distribution (but now with 100 datapoints) is shown in Figure 5.6b. It can be seen that the fit is even better with more data if the assumed distribution is accurate.

5.6.2 Computing Probabilities

Another statistical quantity that can be computed is the cumulative probability. Given a point x^i, the probability that a datapoint will lie in the region $(-\infty, x^i)$ can be computed (Probability equation)

$$F(x^i) = \int_{-\infty}^{x^i} f(x)dx \qquad (5.5)$$

$F(x^i)$ can be computed for all the chosen points—this is called as the cumulative density function (cdf). The probabilities can be compared against the ones computed from the data. To generate probabilities from data, the following process is followed. The data is sorted in an ascending order. This sorted data is partitioned into m equal bins. The $(m - 1)$ partition points are $x^1 \ldots x^{(m-1)}$. The cumulative probabilities computed at these points are $i/m \; \forall \; i \in [1 \quad m - 1]$. These values are compared against the ones computed using equation 5.5 (for x^i) and plotted. This graph is the p-p plot. If this plot is close to the 45^0 line, the greater confidence one may have that the theoretical pdf fits the data. The p-p plots corresponding to the datapoints used to generate the two q-q plots in Section 5.6.1 are shown in Figures 5.6c and 5.6d. These plots essentially provide the same information; however, q-q plots seem to be preferred in many applications.

This chapter started with the aim of looking at decomposing data matrices into true and uncertainty matrices. This led us to exploring the notion of uncertainty and ways in which uncertainty could be quantified through the definition of probability density functions. Theoretical development of density functions was explained through some examples. The idea of validating the hypothesized density functions through the use of q-q and p-p plots was described. However, the limitations of what has been seen till now is that the comparisons are visual and not quantitative. Hence, acceptance of the hypothesized distribution function is subject to individual interpretation. Further, one needs to know both the form of the hypothesized distribution and the parameters of the distribution to develop the p-p and q-q plots. Methods that address both these concerns will be discussed in what follows.

Example 5.1. *Consider the following samples of a random variable x.*

$$x = [1.15 \quad -1.01 \quad 3.24 \quad -2.92 \quad 0.3 \quad 0.02 \quad 1.14 \quad 1.13 \quad -1.23 \quad 0.44 \quad 0.17 \quad 1.76]$$

Let us say we hypothesized that "x" follows the normal distribution, $f(x) = \mathcal{N}(0.4, \; 6.25)$ (mean $= 0.4$ and variance $= 6.25$). Comment on the goodness of the proposed model using q-q and p-p plots.

Solution 5.1. *To find the observed probability, let us first sort the samples.*

$$-2.92 \quad \underset{x^1}{-1.23} \quad -1.01 \quad \underset{x^2}{0.02} \quad 0.17 \quad \underset{x^3}{0.3} \quad 0.44 \quad \underset{x^4}{1.13} \quad 1.14 \quad \underset{x^5}{1.15} \quad 1.76 \quad 3.24$$

The quantiles will be as shown above if we consider 6 quantiles. Each quantile will have a probability of 1/6. For the proposed normal distribution, theoretical quantiles are q^i such that

$$F(q^i) = \frac{i}{6} \; \text{Where F is the cdf} \Rightarrow q^i = F^{-1}\left(\frac{i}{6}\right) \qquad \forall \; i = 1:5$$

For the proposed distribution, $\mathcal{N}(0.4, \; 6.25)$,

$$q^1 = -2.02; \quad q^2 = -0.68; \quad q^3 = 0.4; \quad q^4 = 1.48; \quad q^5 = 2.82$$

x^i vs q^i as given in Figure 5.7a is the q-q plot. The points are

$$\begin{pmatrix} -2.02 \\ -1.23 \end{pmatrix}, \begin{pmatrix} -0.68 \\ 0.02 \end{pmatrix}, \begin{pmatrix} 0.4 \\ 0.3 \end{pmatrix}, \begin{pmatrix} 1.48 \\ 1.13 \end{pmatrix}, \begin{pmatrix} 2.82 \\ 1.15 \end{pmatrix}$$

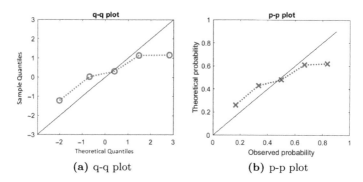

(a) q-q plot   (b) p-p plot

FIGURE 5.7 q-q and p-p plots for Example 5.1.

p-p plot will be the plot of sample and theoretical probabilities correspond-
ing to x^i as shown in Figure 5.7b. Sample cdf corresponding to x^i are
1/6, 2/6, 3/6, 4/6, and 5/6. But for the proposed distribution $\mathcal{N}(0.4, 6.25)$,

$$P^1 = F(-1.23) = 0.26; \qquad P^2 = F(0.02) = 0.44; \qquad P^3 = F(0.3) = 0.48$$
$$P^4 = F(1.13) = 0.61; \qquad P^5 = F(1.15) = 0.62$$

Since both p-p plot and q-q plot aren't close to 45^0 line (y=x line), the
proposed pdf is not a good fit for the given data

5.7   ESTIMATING PARAMETERS OF A DISTRIBUTION

Parameters of the hypothesized distributions can be estimated from data.
Parameter estimation ideas will be described through the definition of some
distributions and parameters associated with those distributions. The most
important distribution in statistics is the standard normal distribution. The
form of this distribution function is shown in equation 5.6. This is a distribution with two parameters (μ and σ). Two other distributions, the binomial
and geometric distributions, have already been introduced. These distributions were characterized by one parameter. Another example, common in me-
chanical/reliability engineering, is the Weibull distribution shown in equation
5.7 (with shape parameter, $\beta > 0$ and scale parameter $\eta > 0$)

$$f(x, \mathbf{p} = [\mu\ \sigma]) = \frac{1}{\sqrt{2\pi}\sigma} e^{-\frac{(x-\mu)^2}{2\sigma^2}} \tag{5.6}$$

$$f(x, \mathbf{p} = [\beta \; \eta]) = \begin{cases} \frac{\beta}{\eta} (\frac{x}{\eta})^{\beta-1} e^{(\frac{-x}{\eta})^{\beta}} & x \geq 0 \\ 0 & x < 0 \end{cases} \qquad (5.7)$$

Estimating the parameters of a pdf is generally the first step in the validation of hypothesized distributions. Post parameter estimation, p-p and q-q plots can be generated as described in the previous sections. It is interesting to observe that the original slightly abstract conceptualization of uncertainty has been converted to a prosaic parameter estimation problem that is similar to estimating the parameters such as slope and intercept in a linear model.

5.7.1   Mean and Variance

There are several approaches for estimating parameters for pdfs from data. We start with the definition of two very important quantities: the mean and variance (shown in equations 5.8 and 5.9).

$$\text{Mean} = \mu = \int x f(x, \mathbf{p}) dx \qquad (5.8)$$

$$\text{Variance} = \sigma^2 = \int (x - \mu)^2 f(x, \mathbf{p}) dx \qquad (5.9)$$

These quantities can be computed for any pdf. These are the most used quantities in data science (so much so that there are companies with these quantities in their names!!). The mean μ describes an average value that a random variable takes and variance describes how much the values of the random variable fluctuates around this average value. It is not surprising that these quantities are important in decision-making. Whenever an engineer makes a decision, an average will provide a target value to aim for and variance will provide the variability around this value that needs to be addressed. For example, a design could be optimized for average behavior and made resilient to deviations from average behavior. Mean and variance are functions of the parameters $\mathbf{p}$ of the distribution.

When equations 5.8 and 5.9 are used with $f(x, \mathbf{p})$ being normal, as shown in equations 5.10 and 5.11, then the computed quantities are directly the mean and variance parameters of the normal distribution. For other distributions the mean and variance will be a function of the parameters of that distribution.

$$\mu = \int x \frac{1}{\sqrt{2\pi}\sigma} e^{-\frac{(x-\mu)^2}{2\sigma^2}} dx \qquad (5.10)$$

$$\sigma^2 = \int (x - \mu)^2 \frac{1}{\sqrt{2\pi}\sigma} e^{-\frac{(x-\mu)^2}{2\sigma^2}} dx \qquad (5.11)$$

$$\text{Mode} : \; x^* = \arg\max_{x} \; f(x, \mathbf{p}) \qquad (5.12)$$

$$\text{Median} : \; x^* \text{such that} \int_{-\infty}^{x^*} f(x, \mathbf{p}) dx = 0.5 \qquad (5.13)$$

For the sake of completeness, it is worth mentioning that there are other measures for average behavior such as median (50% percentile point) and mode (point at which pdf takes the maximum value) that are described using equations 5.12 and 5.13. As with the mean, these quantities can be computed for any distribution. However, for the standard normal distribution, the mode, mean and median are all the same and equal to μ.

While theoretical definitions for mean and variance are shown in equations 5.8 and 5.9, the computation of these quantities from data is shown in equations 5.14 and 5.15.

$$\text{Mean} = \bar{x} = \frac{1}{n} \sum_{i=1}^{n} x^i \tag{5.14}$$

$$\text{Variance} = \frac{1}{n-1} \sum_{i=1}^{n} (x^i - \bar{x})^2 \tag{5.15}$$

Example 5.2. *Find the mean, mode, median, and variance of exponential distribution for which pdf is given by $f(x) = \lambda e^{-\lambda x}$ ($\forall\, x \geq 0$).*

Solution 5.2.

$$\text{Mean}\,(\mu) = \int_0^{\infty} x\,\lambda e^{-\lambda x}\,dx = \left[\lambda x \int e^{-\lambda x}\,dx - \int \lambda \left(\int e^{-\lambda x} \right) \right]_0^{\infty}$$

$$= \left[\lambda x \frac{e^{-\lambda x}}{-\lambda} + \int e^{-\lambda x}\,dx \right]_0^{\infty} = 0 + 0 - \left[-0 - \frac{1}{\lambda} \right] = \frac{1}{\lambda}$$

$$\text{Variance}\,(\sigma^2) = \int_0^{\infty} (x - \mu)^2\, f(x)\,dx = \int_0^{\infty} \left(x - \frac{1}{\lambda} \right)^2 \lambda e^{-\lambda x}\,dx$$

$$\sigma^2 = \int \lambda x^2 e^{-\lambda x}\,dx - \int 2x e^{-\lambda x}\,dx + \int \frac{1}{\lambda} e^{-\lambda x}\,dx$$

$$\int \lambda x^2 e^{-\lambda x}\,dx = \lambda x^2 \frac{e^{-\lambda x}}{-\lambda} - \int 2\lambda x \frac{e^{-\lambda x}}{-\lambda}\,dx = \left[-x^2 e^{-\lambda x} + \int 2x e^{-\lambda x}\,dx \right]_0^{\infty}$$

$$\sigma^2 = \left[-x^2 e^{-\lambda x} + \frac{e^{-\lambda x}}{-\lambda^2} \right]_0^{\infty} = \frac{1}{\lambda^2}$$

$$\text{Mode}(x^*) = \arg\max_x \; \lambda e^{-\lambda x} = 0$$

$$\text{Median}\,(m) \Rightarrow \int_{-\infty}^{m} \lambda e^{-\lambda x}\,dx = 0.5$$

$$\text{Since } f(x) = \begin{cases} \lambda e^{-\lambda x} & \forall\, x \geq 0 \\ 0 & \forall\, x < 0 \end{cases}, \int_{-\infty}^{m} \lambda e^{-\lambda x}\,dx = \int_0^{m} \lambda e^{-\lambda x}\,dx = \left[\lambda \frac{e^{-\lambda x}}{-\lambda} \right]_0^{m}.$$

$$\left[-e^{-\lambda x} \right]_0^{m} = -e^{-\lambda m} + 1 \Rightarrow 1 - e^{-\lambda m} = 0.5 \Rightarrow e^{\lambda m} = 2 \Rightarrow m = \frac{\ln 2}{\lambda}$$

Example 5.3. *Find the mean, mode, median, and variance of the 3-parameter weibull distribution given by*

$$f(x, \mathbf{p} = [\eta, \beta, \gamma]) = \frac{\beta}{\eta} \left(\frac{x - \gamma}{\eta}\right)^{\beta - 1} \exp\left(-\frac{(x - \gamma)^\beta}{\eta^\beta}\right) \quad x > \gamma$$

Solution 5.3.

$$\text{Mean}, \mu = \int_0^\infty x f(x) \, dx = \int_0^\gamma x f(x) \, dx + \int_\gamma^\infty x f(x) \, dx$$

$$= 0 + \frac{\beta}{\eta} \int_\gamma^\infty x \left(\frac{x - \gamma}{\eta}\right)^{\beta - 1} \exp\left(-\frac{(x - \gamma)^\beta}{\eta^\beta}\right) dx$$

Substitute $y = \left(\frac{x-\gamma}{\eta}\right)^\beta$, then $dy = \beta \left(\frac{x-\gamma}{\eta}\right)^{\beta - 1} \left(\frac{1}{\eta}\right) dx$.

$$\Rightarrow \mu = \int_0^\infty x e^{-y} dy = \int_0^\infty \left(\eta y^{\frac{1}{\beta}} + \gamma\right) e^{-y} dy = \eta \int_0^\infty y^{\frac{1}{\beta}} e^{-y} dy + \gamma \int_0^\infty e^{-y} dy$$

$$\mu = \eta \Gamma\left(\frac{1}{\beta} + 1\right) - \gamma(0 - 1) = \gamma + \eta \Gamma\left(\frac{1}{\beta} + 1\right)$$

Where $\Gamma(z)$ is a Gamma function $\Gamma(z) = \int_0^\infty x^{z-1} e^{-x} dx$

$$\text{Variance}, \sigma^2 = E(x^2) - (E(x))^2 = \int_\gamma^\infty x^2 f(x) \, dx - \mu^2$$

$$= \int_0^\infty \left(\eta y^{\frac{1}{\beta}} + \gamma\right)^2 e^{-y} dy - \mu^2$$

$$= \int_\gamma^\infty \eta^2 y^{\frac{2}{\beta}} e^{-y} dy + \int_\gamma^\infty \gamma^2 e^{-y} dy + 2\eta\gamma \int_\gamma^\infty y^{\frac{1}{\beta}} e^{-y} dy - \mu^2$$

$$= \eta^2 \Gamma\left(\frac{2}{\beta} + 1\right) + \gamma^2 + 2\eta\gamma \Gamma\left(\frac{1}{\beta} + 1\right) - \mu^2$$

$$\sigma^2 = \eta^2 \Gamma\left(\frac{2}{\beta} + 1\right) + \gamma^2 + 2\eta\gamma \Gamma\left(\frac{1}{\beta} + 1\right) - \gamma^2$$

$$- \left(\eta \Gamma\left(\frac{1}{\beta} + 1\right)\right)^2 - 2\eta\gamma \Gamma\left(\frac{1}{\beta} + 1\right)$$

$$\sigma^2 = \eta^2 \left(\Gamma\left(\frac{2}{\beta} + 1\right) - \left(\Gamma\left(\frac{1}{\beta} + 1\right)\right)^2\right)$$

$$\text{Mode}, x^* = \arg\max_x f(x)$$

x that maximizes $f(x)$ will also maximize $\ln f(x)$. Hence, mode $x^* = \arg\max_x \ln f(x)$. To find x that maximizes $\ln f(x)$,

$$\ln f(x) = \ln\left(\frac{\beta}{\eta}\right) + (\beta-1)\ln\left(\frac{x-\gamma}{\eta}\right) - \left(\frac{x-\gamma}{\eta}\right)^{\beta}$$

$$\left.\frac{\partial \ln f(x)}{\partial x}\right|_{x^*} = 0 \Rightarrow (\beta-1)\times\frac{\eta}{x^*-\gamma}\times\frac{1}{\eta} - \frac{\beta}{\eta}\left(\frac{x^*-\gamma}{\eta}\right)^{\beta-1} = 0$$

Rearranging,

$$\frac{\beta}{\eta}\left(\frac{x^*-\gamma}{\eta}\right)^{\beta-1} = \frac{(\beta-1)}{(x^*-\gamma)} \Rightarrow \left(\frac{x^*-\gamma}{\eta}\right)^{\beta} = \left(1-\frac{1}{\beta}\right)$$

$$x^* = \gamma + \eta\left(1-\frac{1}{\beta}\right)^{\frac{1}{\beta}}$$

$$\text{Median (m)} \Rightarrow \underbrace{\int_0^m f(x)\,dx}_{cdf,\ F(x)} = 0.5 \ \ (\text{Since } x \geq 0).$$

Using the formula for cdf of the 3-parameter Weibull distribution,

$$F(x) = 1 - e^{-\left(\frac{x-\gamma}{\eta}\right)^{\beta}} \Rightarrow F(m) = 1 - e^{-\left(\frac{m-\gamma}{\eta}\right)^{\beta}} = 0.5$$

$$\ln 0.5 = -\left(\frac{m-\gamma}{\eta}\right)^{\beta} \Rightarrow \left(\frac{m-\gamma}{\eta}\right)^{\beta} = \ln 2 \Rightarrow m = \eta\,(\ln 2)^{\frac{1}{\beta}} + \gamma$$

5.7.2 Method of Moments

One idea for estimating the parameters of a distribution is based on a linear algebraic view; generate as many equations as there are parameters. Different approaches use different ideas to derive the necessary equations. The first approach that we describe is called the method of moments. A kth order moment is defined for a pdf, which is given by equation 5.16.

$$\text{Moment(k)} = M_k = \int x^k f(x,\mathbf{p})dx \tag{5.16}$$

Given a pdf, moments of several orders, denoted by M_k can be computed. It can be easily verified that mean $\mu = M_1$, and variance $\sigma^2 = M_2 - \mu^2$.

The basic idea behind the method of moments is to compare theoretically generated moments with moments estimated from data. Each such comparison will provide one equation. A general k^{th} order moment can be computed from data using equation 5.17. Comparison of k moments as shown in equation 5.18 can be used to identify k parameters of a pdf.

$$M_{k,s} = \frac{1}{n}\sum_{i=1}^n (x^i)^k \tag{5.17}$$

$$M_{k,s} = M_k \ \forall\ k \tag{5.18}$$

Example 5.4. *For the data given below, fit as exponential distribution using method of moments.*

x	1	4	7	15	21	30

Solution 5.4. *For exponential distribution*

$$f(x, \lambda) = \lambda e^{-\lambda x}$$

$$\textit{First moment, } M_1 = \int x f(x, \lambda)\, dx = \mu = \frac{1}{\lambda}$$

$$\textit{For the given data, } M_{1,s} = \frac{1}{6} \sum_{i=1}^{n} x^i = \frac{1}{6}(78) = 13$$

$$\textit{Equating } M_{1,s} \textit{ and } M_1 \Rightarrow \hat{\lambda} = \frac{1}{13} = 0.077$$

Hence, the exponential distribution is, $f(x, \lambda) = 0.077 e^{-0.077x}$. It may be noted that one could also match multiple moments with this parameter in a least squares sense.

5.7.3 Maximum Likelihood Estimation

The second approach that we describe is the maximum likelihood approach. In contrast to method of moments, which doesn't take recourse to optimization, in maximum likelihood estimation (MLE) method, an optimization formulation is solved to identify the parameters of the distribution. However, unlike the input-output modeling problem, one doesn't have target output values to setup an objective function. In the MLE approach, this problem is addressed by proposing an objective function as given in equation 5.19, where $x_1 \ldots x_n$ are the values of the independently drawn samples of the random variable and $f(x, \mathbf{p})$ is the hypothesized distribution with unknown parameters.

$$\text{maximize } f_{mle}(\mathbf{p}) = \prod_{i=1}^{m} f(x^i, \mathbf{p}) \tag{5.19}$$

The rationale for this objective function is as follows. In MLE, one is verifying if the given data came from a hypothesized distribution. Consequently, data sampled from a distribution should have a high probability of being identified with that distribution. The objective function ensures that, when maximized, collectively, all the datapoints have a high probability of having been sampled from the hypothesized distribution. The pdf value at each of these points is identified as an outcome of the optimization solution—in some sense this can be thought of as unsupervised model building.

Of course, we know from Chapter 4 that the solution to this optimization problem is given by

$$\nabla f_{mle}(\mathbf{p}) = 0 \tag{5.20}$$

Example 5.5. *For the data given in Example 5.4, fit an exponential distribution using MLE*

Solution 5.5.

$$\lambda = \underset{\lambda}{\arg\max} \ \prod_{i=1}^{6} f(x^i, \lambda)$$

$$= \underset{\lambda}{\arg\max} \ \underbrace{\lambda^6 \, e^{-\lambda} \, e^{-4\lambda} \, e^{-7\lambda} \, e^{-15\lambda} \, e^{-21\lambda} \, e^{-30\lambda}}_{J}$$

$$J = \lambda^6 \exp\left(-\lambda\left(78\right)\right) = \lambda^6 e^{-78\lambda}$$

$$\frac{dJ}{d\lambda}\bigg|_{\hat{\lambda}} = 6\hat{\lambda}^5 e^{-78\hat{\lambda}} + \hat{\lambda}^6 e^{-78\hat{\lambda}}\left(-78\right) = 0 \implies \hat{\lambda}^5 e^{-78\hat{\lambda}}\left[6 - 78\hat{\lambda}\right] = 0$$

$$\text{Since } \hat{\lambda} \text{ cannot be zero, } 6 - 78\hat{\lambda} = 0 \Rightarrow \hat{\lambda} = \frac{6}{78} = \frac{1}{13}$$

5.7.4   Modeling Unknown Distribution — Consolidation of Ideas

In summary, for a hypothesized pdf with known parameters, p-p and q-q plots can be used to understand the quality of fit. In cases where one hypothesizes a distribution but the parameter values are not known, estimation methods such as method of moments and maximum likelihood estimation can be used. Post estimation of parameters, qualitative validation using q-q and p-p plots can be pursued. In cases where the distribution is also not known, iteration over several distributions with parameter estimation can help narrow down the distribution that best fits the data.

Example 5.6. *For the data given below, fit a suitable continuous probability distribution.*

$$x = [3.09, \ 4.95, \ 4.73, \ 1.21, \ 2.27, \ 2.71, \ 2.7, \ 3.03, \ 2.97, \ 2.1]$$

Solution 5.6. *Since we do not know the original pdf, let us first try fitting a normal pdf ($f_1(x, \mathbf{p})$).*
Trial 1: Normal distribution

$$f_1(x, \mathbf{p} = [\mu, \sigma]) = \frac{1}{\sqrt{2\pi}\sigma} \exp\left(-\frac{(x - \mu)^2}{2\sigma^2}\right)$$

Using MLE,

$$\hat{\mathbf{p}} = \underset{\mathbf{p}}{\arg\max} \ \prod_{i=1}^{10} f, \left(x^i, \mathbf{p}\right) = \underset{\mathbf{p}}{\arg\max} \ J_1$$

Since 'ln' is a monotonic function, maximum of J_1 and maximum of $\ln(J_1)$ will occur at the same value of **p**. *Hence,* $\hat{\mathbf{p}} = \arg\max_{\mathbf{p}} \ln(J_1)$. *Now,*

$$J_1 = \prod_{i=0}^{10} \frac{1}{\sqrt{2\pi}\sigma} \exp\left(-\frac{(x^i - \mu)^2}{2\sigma^2}\right) = \left(\frac{1}{\sqrt{2\pi}\sigma}\right)^{10} \exp\left(-\frac{\sum_{i=0}^{10}(x^i - \mu)^2}{2\sigma^2}\right)$$

$$\ln J_1 = \ln\left(\frac{1}{\sqrt{2\pi\sigma^2}}\right)^{10} - \frac{1}{2\sigma^2}\sum_{i=1}^{10}(x^i - \mu)^2 = -5\ln(2\pi\sigma^2) - \frac{1}{2\sigma^2}\sum_{i=1}^{10}(x^i - \mu)^2$$

To maximize $\ln J_1$, $\left.\frac{\partial \ln J_1}{\partial \mu}\right|_{\hat{\mu},\hat{\sigma}} = 0$ *and* $\left.\frac{\partial \ln J_1}{\partial \sigma}\right|_{\hat{\mu},\hat{\sigma}} = 0$.

$$\frac{\partial \ln J_1}{\partial \sigma} = 0 - \frac{1}{2\sigma^2}\sum_{i=1}^{10} -2(x^i - \mu) = \frac{1}{\sigma^2}\left(\sum_{i=1}^{10} x^i - 10\mu\right)$$

For $\mu = \hat{\mu}$, $\frac{\partial \ln J_1}{\partial \mu} = 0 \Rightarrow \hat{\mu} = \frac{1}{10}\sum_{i=1}^{10} x^i = 2.976$ *(sample mean)*

$$\frac{\partial \ln J_1}{\partial \sigma} = \frac{-5}{2\pi\sigma^2} \times 2\pi \times 2\sigma - \frac{-2\sigma^{-3}}{2}\sum_{i=1}^{10}(x^i - \mu)^2 = \frac{-10}{\sigma} + \frac{1}{\sigma^3}\sum_{i=1}^{10}(x^i - \mu)^2$$

At $\mu = \hat{\mu}$ *and* $\sigma = \hat{\sigma}$, $\frac{\partial \ln J_1}{\partial \sigma} = 0 \Rightarrow \hat{\sigma}^2 = \frac{1}{10}\sum_{i=1}^{10}(x^i - \hat{\mu})^2 = 1.152$

Note that $\hat{\sigma}^2 = \frac{n-1}{n} \times$ *sample variance for MLE. To see the goodness of fit, let us generate p-p plot with 5 quantiles. Arranging data in ascending order,*

$$\underset{x^1}{1.21} \quad 2.1 \quad \underset{x^2}{2.27} \quad 2.7 \quad \underset{x^3}{2.71} \quad 2.97 \quad 3.03 \quad \underset{x^4}{3.09} \quad 4.73 \quad 4.95$$

Each quantile will have probability 1/5. *Corresponding to* x^i, *the true cdf values are as follows:*

$$P(x^1) = 0.2072; \quad P(x^2) = 0.3985; \quad P(x^3) = 0.4978; \quad P(x^4) = 0.5423$$

These are the theoretical values. From data, observed probabilities are 0.2, 0.4, 0.6, *and* 0.8. *The corresponding p-p plot is shown in Figure 5.8a. It is clear from the p-p plot that the fit is poor. These data may not be sampled from a normal distribution. Let us try Exponential distribution.*

Trial 2: Exponential distribution

$$\hat{\lambda} = \arg\max_{\lambda} \prod_{i=1}^{n} \lambda e^{-\lambda x^i} = \arg\max_{\lambda} \ln\left(\prod_{i=1}^{n} \lambda e^{-\lambda x^i}\right) = \arg\max_{\lambda} \ln J_2$$

where $J_2 = \lambda^n e^{-\lambda \sum_{i=1}^{n} x^i}$.

$$\ln(J_2) = n\ln\lambda - \lambda \sum_{i=1}^{n} x^i$$

$$\left.\frac{d\ln J_2}{d\lambda}\right|_{\hat{\lambda}} = 0 \Rightarrow \frac{n}{\hat{\lambda}} - \sum_{i=1}^{n} x^i = 0 \Rightarrow \hat{\lambda} = \frac{n}{\sum_{i=1}^{n} x^i} = \frac{10}{29.76} = 0.336$$

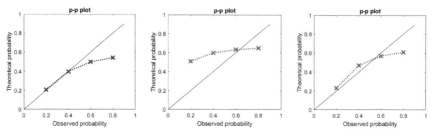

(a) p-p plot using normal dis-(b) p-p plot using exponential(c) p-p plot using log-normal
tribution   distribution   distribution

FIGURE 5.8 p-p plots for Example 5.6.

Cumulative probabilities corresponding to this exponential distribution at the 4 quantiles selected earlier are

$$P\left(x^1 = 2.1\right) = 0.5062; \quad P\left(x^2 = 2.7\right) = 0.5964$$
$$P\left(x^3 = 2.97\right) = 0.6314; \quad P\left(x^4 = 3.09\right) = 0.6459$$

Again, p-p plot (Figure 5.8b) shows that the exponential fit for the given data is poor.

Trial 3: Log-normal distribution

pdf is given by $f_3\left(x, \mathbf{p} = [\mu, \sigma]\right) = \frac{1}{x\sigma\sqrt{2\pi}} \exp\left(-\frac{(\ln x - \mu)^2}{2\sigma^2}\right)$. *If x is log-normally distributed, then* $y = \ln x$ *is normally distributed with mean* μ *and variance* σ^2. *Hence, MLE for this log-normal distribution is equivalent to MLE for normally distributed* $y = \ln x$.

$$f_4\left(y, \mathbf{p} = [\mu, \sigma]\right) = \frac{1}{\sigma\sqrt{2\pi}} \exp\left(-\frac{(y - \mu)^2}{2\sigma^2}\right)$$

Then, using MLE we have,

$$\hat{\mu} = \frac{1}{n}\sum_{i=1}^{n} y^i = \frac{1}{n}\sum_{i=1}^{n} \ln\left(x^i\right) = 1.022$$

$$\hat{\sigma}^2 = \frac{1}{n}\sum_{i=1}^{n} \left(\ln x^i - \mu\right)^2 = 0.1451$$

Hence the fitted log-normal distribution is

$$f_3\left(x, \mathbf{p} = [\mu, \sigma]\right) = \frac{1}{\sqrt{0.1451} \times x\sqrt{2\pi}} \exp\left(-\frac{(\ln x - 1.022)^2}{2 \times 0.1451}\right)$$

From cdf of this distribution,

$$P\left(x^1\right) = 0.231; \quad P\left(x^2\right) = 0.4698; \quad P\left(x^3\right) = 0.5693; \quad P\left(x^4\right) = 0.6097 \tag{5.21}$$

Ddatapoints for p-p plot: $\begin{pmatrix} 0.2 \\ 0.231 \end{pmatrix}, \begin{pmatrix} 0.4 \\ 0.4698 \end{pmatrix}, \begin{pmatrix} 0.6 \\ 0.5693 \end{pmatrix}, \begin{pmatrix} 0.8 \\ 0.6097 \end{pmatrix}.$

Comparatively, the p-p plot (refer Figure 5.8c) for log-normal distribution looks closer to 45^0 line and hence, is a reasonable fit. With more datapoints, the fit might improve.

5.8 MIXED MODELS – JOINT IDENTIFICATION OF MODEL AND DISTRIBUTION PARAMETERS

This chapter on statistics for data science was started by describing the problem of splitting data matrices (m samples) into two matrices, model (true) data and the uncertainty data. The data that is described by a deterministic model can be assumed to be the modeler's version of the truth. Mathematically, this can be represented as $X = M(or\ T) + U$. Statistical concepts were introduced towards understanding and characterizing uncertainty. Now that the basic ideas have been described, we will show how all of this comes together in model building for data science in this chapter. Two examples will help describe how these ideas come together.

Consider n samples of data on two variables $\mathbf{x} = [x_1\ x_2]$. This data can be organized as a matrix

$$X = \begin{bmatrix} x_1^1 & x_2^1 \\ \vdots & \vdots \\ x_1^n & x_2^n \end{bmatrix} \tag{5.22}$$

This matrix has to be partitioned into model and uncertainty matrices. In many cases, the uncertainty matrix also contains usable information and is of importance in data science. This partitioning can now be described based on what has been learnt in this chapter. Since this partitioning problem has infinite number of solutions, assumptions are required to resolve this. To illustrate this, consider two cases. In the first case x_1 and x_2 are related with no errors in x_1, and in the second case, x_1 and x_2 are related with errors in both variables. The first case is called the standard regression model and the second case is the error-in-variables (EIV) model.

5.8.1 Mixed Models – Error only in Dependent Variable

In the first case, let us also assume that the variables are linearly related, that is, $x_2 = \alpha x_1$ and the uncertainty associated with x_2 (e) can be captured by a pdf $f(e, \mathbf{p})$ with a parameter vector $\mathbf{p}$. One can setup a MLE problem when α and $\mathbf{p}$ are not known, as shown in equation 5.23. The objective function will simplify based on the type of pdf that is assumed to describe the error. The most common distribution that is chosen is the normal distribution and this will lead to an ordinary least squares objective function.

$$f_{mle}^{case1}(\alpha, \mathbf{p}) = \prod_{i=1}^{n} f(x_2^i - \alpha x_1^i, \mathbf{p}) \tag{5.23}$$

Solution to the optimization problem will provide one with $\hat{\alpha}$ and $\hat{\mathbf{p}}$, and the decomposition $(X = \hat{M} + \hat{U})$ can be completed with this solution as shown in equation 5.24

$$\hat{M} = \begin{bmatrix} x_1^1 & \hat{\alpha} x_1^1 \\ \vdots & \vdots \\ x_1^n & \hat{\alpha} x_1^n \end{bmatrix} \quad \hat{U} = \begin{bmatrix} 0 & \hat{e}^1 = x_2^1 - \hat{\alpha} x_1^1 \\ \vdots & \vdots \\ 0 & \hat{e}^n = x_2^n - \hat{\alpha} x_1^n \end{bmatrix} \tag{5.24}$$

5.8.2 Mixed Models – Errors in both Dependent and Independent Variables

In the second case also let us assume that the variables are linearly related, that is, $x_2 = \alpha x_1$. The uncertainty associated with error in x_1 (e_1) is assumed to be captured by a pdf $f_1(e_1, \mathbf{p}_1)$ with a parameter vector $\mathbf{p}_1$ and the uncertainty associated with the error in x_2 (e_2) is captured by a pdf $f_2(e_2, \mathbf{p}_2)$ with a parameter vector $\mathbf{p}_2$. The solution to this problem will depend on the distribution that is chosen for the two variables. The rationale for the objective function is also quite obvious. The datapoints that are assumed to have come from a distribution should have a large pdf value when substituted into the corresponding f function. This optimization problem can be solved using the techniques described in the optimization chapter. Assuming the distributions to be normal will lead to a total least squares (TLS) objective.

$$\max_{\alpha, \bar{x}_1^i \forall i, \mathbf{p}_1, \mathbf{p}_2} f_{mle}^{case2}(\alpha, \bar{x}_1^i \forall i, \mathbf{p}_1, \mathbf{p}_2) =$$

$$(\prod_{i=1}^{n} f_1(x_1^i - \bar{x}_1^i, \mathbf{p}_1))(\prod_{i=1}^{n} f_2(x_2^i - \alpha \bar{x}_1^i, \mathbf{p}_2))$$

Solution to the optimization problem will provide one with $\hat{x}_1^i \forall i, \hat{\alpha}$ and $\hat{\mathbf{p}}_1$, $\hat{\mathbf{p}}_2$ and the decomposition $(X = \hat{M} + \hat{U})$ can be completed with this solution as shown in equation 5.25, where $\hat{x}_2^i = \hat{\alpha} \hat{x}_1^i$.

$$\hat{M} = \begin{bmatrix} \hat{x}_1^1 & \hat{\alpha} \hat{x}_1^1 \\ \vdots & \vdots \\ \hat{x}_1^n & \hat{\alpha} \hat{x}_1^n \end{bmatrix} \quad \hat{U} = \begin{bmatrix} \hat{e}_1^1 = x_1^1 - \hat{x}_1^1 & \hat{e}_2^1 = x_2^1 - \hat{\alpha} \hat{x}_1^1 \\ \vdots & \vdots \\ \hat{e}_1^n = x_1^n - \hat{x}_1^n & \hat{e}_2^n = x_2^n - \hat{\alpha} \hat{x}_1^n \end{bmatrix} \tag{5.25}$$

5.9 SAMPLING DISTRIBUTIONS

Until now, validating the whole dataset against a distribution through qualitative plots was discussed. However, when one validates if data comes from a distribution by computing distribution parameters from the sample and comparing with the hypothesized distribution values, then one is validating values against one another, such as $\bar{x}$ (sample mean) $= \mu$ or $M_n = M_{n,s}$. In such comparisons, checking for exact equality will be meaningless as they are never likely to be equal due to the vagaries of sampling, finite sample size, and so on. A sensible option would be to evaluate the question, is $\bar{x} \approx \mu$ ($\bar{x}$ approximately equal to μ).

The question then is how approximate is good enough? Consider equation 5.26, where one equates sample mean to a population mean to identify if the sample came from a distribution with mean μ. The quantity calculated from samples is called a "statistic" [32].

$$\int_{-\infty}^{\infty} x f(x, \mathbf{p}) dx = \mu \approx \bar{x} = \frac{1}{n} \sum_{i=1}^{n} x^i \qquad (5.26)$$

Equation 5.26 compares a number (computed based on a known pdf and parameter values) on the left hand side (LHS) to a random variable (statistic) on the right hand side (RHS). The statistic in the RHS is a random variable because the value computed will vary every time the experiment is conducted. The n samples that are sampled from the same distribution will be different every time the experiment is conducted. This is precisely the reason why exact comparisons are meaningless. Whenever any statistic ψ is computed as a function of samples $\mathbf{x}$, that is $\psi = h(\mathbf{x})$, then ψ is a random variable with its own pdf; the numerical value that is computed for ψ based on the sample values is just one "realization" of the statistic; different realization will occur with different experiments. This pdf of the statistic is called the sampling distribution. Hence, before one derives a procedure for evaluating equation 5.26, sampling distributions need to be understood.

The key difficulty that has to be addressed then is how does one make decisions based on only one realization? Let us recall that q-q and p-p plots were constructed using the samples directly and hence one had multiple realizations of the random variable to work with. When all the sample values are consolidated into a single equation to compute a statistic, then there is only one realization for the computed statistic. Another question that needs to be answered is regarding the quality of the estimate itself. Since there are many methods for estimating parameters (MOM, MLE, and so on), each of these methods will result in different nonlinear transformations of the samples, and as a result each of these statistics will have different distribution functions. How does one then compare the quality of these estimates?

The notion of sampling distributions is critical to answering the questions posed above. In many cases, development of the sampling distribution is quite

involved; however, many of the sampling distributions have already been theoretically derived and are available for use. The task of a data scientist is to
understand the appropriate sampling distribution for their problem at hand.

The procedure for deriving sampling distributions for variables which are a
function of other random variables can now be described. We will restrict our
discussion to scalar random variables; equation 5.27 describes a statistic (ψ)
as a function of random variable x. Assuming that the function is monotonic
(for the sake of simplicity), one can find an inverse (g) for that function (h)
as shown in equation 5.28. Equation 5.29 uses the definition of pdf and the
notion of monotonicity to derive an expression that variable ψ takes a value
between $-\infty$ and a. Change of variables from x to ψ for the integral leads
to the final equation for the probability distribution of ψ shown in equation
5.30.

$$\psi = h(x) \tag{5.27}$$

$$x = f^{-1}(\psi) = g(\psi) \tag{5.28}$$

$$P(-\infty \leq \psi \leq a) = P(-\infty \leq x \leq g(a)) = \int_{-\infty}^{g(a)} f_x(x)dx \tag{5.29}$$

$$f_\psi(\psi) = f_x(g(\psi))g'(\psi) \tag{5.30}$$

The same procedure can be extended where ψ and $\mathbf{x}$ are vectors, where
the derivative will be replaced appropriately by a Jacobian matrix. Non-
monotonic transformation functions can also be handled, many of the important ones through symmetry arguments. As mentioned before, the process
of deriving sampling distributions can be quite involved, and hence, we will
simply take recourse to already derived sampling functions for solving data
science problems. Advanced topics such as moment generating functions and
the use of those in deriving sampling distributions are not covered here. Two
simple examples demonstrate how sampling distributions are derived.

Example 5.7. *Consider a random variable x which is uniformly distributed
in $[a, b]$, where a and b are positive. What is the sampling distribution of a
statistic $\psi = x^2$?*

Solution 5.7. *The distribution of x is $\frac{1}{b-a}$, since x is uniformly distributed.
We know that*

$$f_\psi(\psi) = f_x(g(\psi))g'(\psi)$$

*Since x is positive, $g(\psi) = \sqrt{\psi}$, retaining only the positive square root. Sub-
stituting into the equation above*

$$f_\psi(\psi) = \frac{1}{2(b-a)\sqrt{\psi}}$$

*It can be seen that this expression integrates to one between $-\infty$ and ∞ and
hence is a valid pdf.*

Example 5.8. *Consider a random variable x which is from a standard normal distribution with zero mean and unit variance. What is the sampling distribution of a statistic $\psi = x^2$?*

Solution 5.8. *In this case, since the standard normal distribution is symmetric, every ψ can be mapped to two x. Using equation 5.30 with only the positive root and multiplying the resulting expression by 2 will give us the desired pdf as below*

$$f_\psi(\psi) = \frac{e^{\frac{-\psi}{2}}}{\sqrt{2\pi\psi}}$$

Example 5.9. *Consider two random variables x_1 and x_2 such that x_1 follows a normal distribution with mean 10 and standard deviation 4, while x_2 follows an exponential distribution $(\lambda e^{-\lambda x})$ with $\lambda = 0.2$.*

(a) Find an approximate sampling distribution of $\psi = (x_1 - x_2)^2$ using Monte-Carlo simulations.

(b) Find a pdf that best fits the sampling distribution of ψ and estimate the parameters.

Solution 5.9. *Generate 100,000 samples of both x_1 and x_2 from the respective distributions and calculate $\psi = (x_1 - x_2)^2$. Figure 5.9a shows the histogram of the generated data, and thus, it is the sampling distribution of ψ generated using Monte-Carlo distributions.*

We make an attempt to fit the obtained sampling distribution with three different distributions: normal distribution, exponential distribution and t-distribution. MLE was used to find the parameters that best fit the data for each of the distributions. The obtained distributions with best fit parameters are plotted along with the histogram in Figure 5.9b. Although it is evident that the normal distribution gives poor fit, it is unclear whether the best fit is using exponential or t-distribution. To find the best fit, q-q plots are generated for all three distributions and shown in Figure 5.9c. Note that the q-q plot clearly shows that the t-distribution best fits the data as it is closest to the 45^0 line. The parameters of the estimated t-distribution are degrees of freedom = 2.55, location = 44.05, and scale = 42.7.

5.10   IMPORTANT SAMPLING DISTRIBUTIONS

Sampling distributions are critical for hypothesis testing. There are some very commonly used statistics in data science and sampling distributions for these statistics have already been derived. We will describe these in this section.

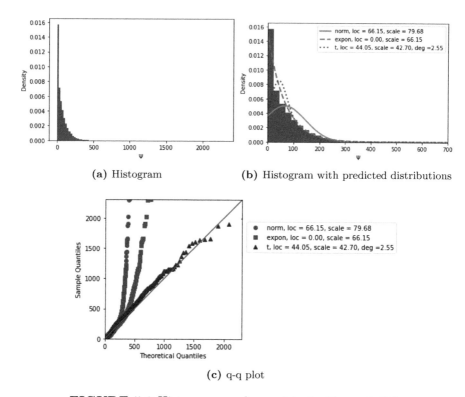

(a) Histogram

(b) Histogram with predicted distributions

(c) q-q plot

FIGURE 5.9 Histograms and q-q plots for Example 5.9.

5.10.1 z-Distribution

In many cases, a Gaussian random variable x with mean μ and variance σ^2 is transformed into a variable called Z through the following linear transformation[2]

$$Z = \frac{x - \mu}{\sigma} \sim z\text{-distribution} \qquad (5.31)$$

It can be shown that the statistic Z will have a standard normal distribution with zero mean and unity variance (z-distribution). An interesting observation about this transformation is that it converts any normal distribution $\mathcal{N}(\mu, \sigma^2)$ to $\mathcal{N}(0, 1)$. When one attempts to compute the probability that $P(x^1 \leq x \leq x^2)$ for a normal distribution, the integration required to compute this quantity doesn't have a closed form solution. As a result, a table is generated (called the z-table), where the integral is computed and listed

---

[2]Statistic is represented using Z while the distribution is represented using z. Note that Z is a random variable, not to be confused as a matrix.

for several Z^i values $\left(P(-\infty \leq Z \leq Z^i)\right)$. This table is enough to compute $P(x^1 \leq x \leq x^2)$ for any normal distribution as shown below

$$P(x^1 \leq x \leq x^2) = P(-\infty \leq x \leq x^2) - P(-\infty \leq x \leq x^1)$$

$$P(-\infty \leq x \leq x^i) = P\left(-\infty \leq Z \leq Z^i = \frac{x^i - \mu}{\sigma}\right)$$

5.10.2 Sampling Distribution of Mean of Data from Normal Distribution

Let us consider the statistic $\bar{x}$. In an experiment, x^i represents the ith sample that is being drawn. Let us assume that each sample is being drawn from the same normal distribution $\mathcal{N}(\mu, \sigma^2)$ and are also independent. Formal definition of independence will be explored later. For now, to understand sampling distributions described in this section, a layman understanding of independence will be used. Independent samples drawn from a distribution do not have any effect on any of the other samples that are drawn. In this case, the sampling distribution for the statistic $\bar{x}$ can be shown to be $\mathcal{N}(\mu, \sigma^2/n)$ [7]. This is a very important result that is often used in data science. This is because, in many cases, average properties are compared to generate insights from data. An interesting thing to notice is that the variance of the average of samples is less than the variance of the sample; one would realize that this is to be expected. As $n \to \infty$, it is clear than variance will go to zero and $\bar{x} = \mu$. The $\bar{x}$ result can be generalized to a linear combination of independent random variables x^i (equation 5.32), which are from $\mathcal{N}(\mu, \sigma^2)$.

$$x_{lc} = \sum_{i=1}^{n} c^i x^i \tag{5.32}$$

$$\text{pdf of } x_{lc} = \mathcal{N}\left(\mu \Sigma c^i, \sigma^2 \Sigma (c^i)^2\right) \tag{5.33}$$

It can be seen that when equation 5.33 is applied to $\bar{x}$, all the c^is will be $\frac{1}{n}$ and the original result for $\bar{x}$ can be obtained.

$$\bar{x} = \frac{1}{n} \sum x^i,$$

$$\mu_{\bar{x}} = E(\bar{x}) = \frac{1}{n} \sum E(x^i) = \mu$$

$$\sigma_{\bar{x}}^2 = E(\bar{x} - \mu_{\bar{x}})^2 = E\left(\frac{\sum x^i}{n} - \mu\right)^2 = E\left(\frac{\sum (x^i - \mu)}{n}\right)^2$$

$$= \frac{1}{n^2} E\left[\left(\sum_{i=1}^{n} (x^i - \mu)\right)\left(\sum_{j=1}^{n} x^j - \mu\right)\right]$$

$$= \frac{1}{n^2} E \sum_{i=1}^{n} \sum_{j=1}^{n} (x^i - \mu)(x^j - \mu) = \frac{1}{n^2}\left[n\sigma^2 + 0\right] = \frac{\sigma^2}{n}$$

Example 5.10. *Consider two random variables x_1 and x_2 such that x_1 follows a normal distribution with mean 10 and standard deviation 4, while x_2 follows an exponential distribution ($\lambda e^{-\lambda x}$) with $\lambda = 0.2$.*

(a) *Find an approximate sampling distribution of $\psi = (\bar{x}_1 - \bar{x}_2)^2$ using Monte-Carlo simulations when $\bar{x}_i$ is calculated from a population of size 20.*
(b) *Find a pdf that best fits the sampling distribution of ψ and estimate the parameters.*

Solution 5.10. *To find sampling distribution of $\psi = (\bar{x}_1 - \bar{x}_2)^2$, we need large number of samples of $\bar{x}_1$ and $\bar{x}_2$. Generate 20 samples of both x_1 and x_2 from the respective distributions and calculate their means to get a single sample for $\bar{x}_1$ and $\bar{x}_2$. Repeating this process 1,000 times, we get 1,000 samples for $\bar{x}_1$ and $\bar{x}_2$ and we use them to calculate ψ. Figure 5.10a shows the histogram of the generated 1,000 samples of ψ, and thus, it is the sampling distribution of ψ generated using Monte-Carlo distributions.*

We attempted to fit the obtained sampling distribution with four different distributions: normal distribution, exponential distribution, t-distribution, and log-normal distribution. MLE was used to find the parameters that best fit the data for each of the distributions. The obtained distributions with best fit parameters are plotted along with the histogram in Figure 5.10b. Although it is evident that the exponential distribution gives poor fit, it is unclear whether the best fit is using normal, t or log-normal distribution. To find the best fit, q-q plot has been generated for all three distributions and are given in Figure 5.10c. Note that the q-q plot clearly shows that the log-normal distribution best fits the data as it is closest to the 45^0 line. The parameters of the estimated log-normal distribution are sigma = 0.24682, location = −27.21, and scale = 51.23.

5.10.3   Central Limit Theorem

Central limit theorem (CLT) is of critical importance in the derivation of sampling distributions. There are slightly different versions of this result, we describe this result from the viewpoint of the sampling distribution of $\bar{x}$. When the sampling distribution was derived previously for $\bar{x}$, it was assumed that x^is were all independent and from $\mathcal{N}(\mu, \sigma^2)$. CLT addresses the case where the variables are from a pdf $f(x, \mathbf{p})$, which is different from a normal distribution. The result is given in equation 5.34.

$$\text{pdf of } \bar{x} \underset{n \to \infty}{\Rightarrow} \mathcal{N}(\mu, \sigma^2/n) \tag{5.34}$$

$$\mu = \int_{-\infty}^{\infty} x f(x, \mathbf{p}) dx; \qquad \sigma^2 = \int_{-\infty}^{\infty} (x - \mu)^2 f(x, \mathbf{p}) dx$$

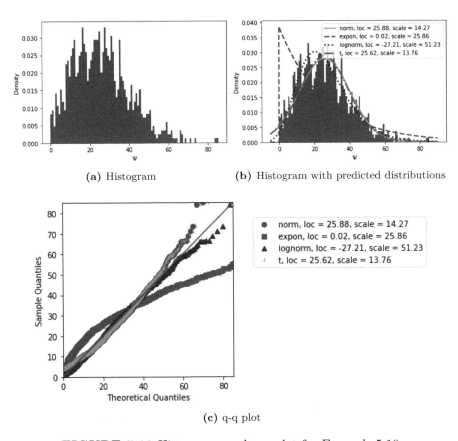

(a) Histogram

(b) Histogram with predicted distributions

(c) q-q plot

FIGURE 5.10 Histograms and q-q plot for Example 5.10.

This is indeed a remarkable result, where irrespective of the original distribution (with whatever be the values of the parameters) from which the independent samples are drawn, the mean tends to a normal distribution. While this result is valid as $n \to \infty$, in practice, with reasonably small values of n, one starts observing the convergence to a normal distribution. When we describe ML algorithms later, there will be considerable reliance on the assumption of normality for random variables. One of the justifications is that many random variables are naturally quite well modeled by normal distributions. A second and more powerful reason is that CLT guarantees that the statistic will be normal irrespective of the original distribution from which the samples were drawn if one were working with averages. Thus, CLT guarantees that averaging converts all distributions to a normal distribution.

Example 5.11. *Show that the distribution of sample mean of an uniform distribution converges to normal distribution.*

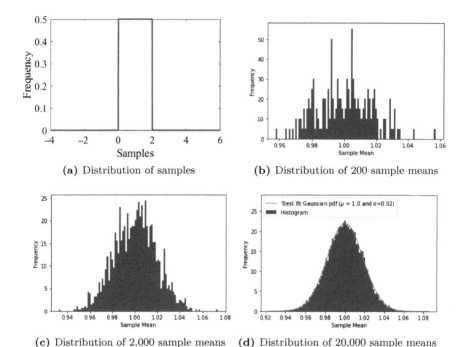

(a) Distribution of samples   (b) Distribution of 200 sample means

(c) Distribution of 2,000 sample means   (d) Distribution of 20,000 sample means

FIGURE 5.11 Distribution of sample mean of a uniform distribution.

Solution 5.11. *Consider a uniform distribution between 0 and 2 as shown in Figure 5.11a. Let us say we generate 1,000 samples from this distribution and calculate the sample mean. If we repeat the same experiment 200 times, we will obtain 200 sample means.*

The histogram (using 100 bins) of the sample means generated using the 200 values is given in Figure 5.11b. Note that we cannot clearly say that this distribution is normal. However, as we increase the number of realizations of sample means, the distribution converges to normal distribution. This is evident from Figures 5.11c and 5.11d. As we increase the number of realizations from 200 to 2,000 and then to 20,000, the distribution of sample mean gets closer to normal distribution. For the case with 20,000 sample means, the best fit Gaussian pdf is shown in the plot for comparison (Figure 5.11d).

Example 5.12. *Show that the distribution of sample mean of a Poisson distribution $\left(f(k; \eta) = \frac{\eta^k e^{-\eta}}{k!} \right)$ converges to normal distribution.*

Solution 5.12. *Consider a Poisson distribution with $\eta = 5$ as shown in Figure 5.12a. Let us say we generate 1,000 samples from this distribution and calculate the sample mean. 200 sample means can be generated by repeating the experiment 200 times.*

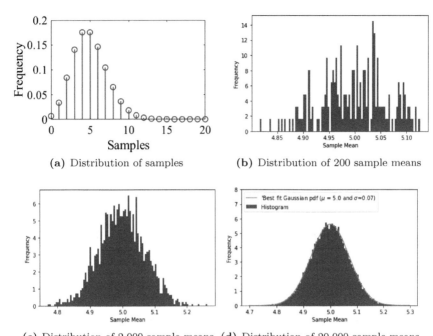

(a) Distribution of samples

(b) Distribution of 200 sample means

(c) Distribution of 2,000 sample means   (d) Distribution of 20,000 sample means

FIGURE 5.12 Distribution of sample mean of a Poisson distribution.

The histogram (using 100 bins) of the sample means generated using the 200 values is shown in Figure 5.12b. Note that we cannot clearly say that this is a normal distribution. However, as we increase the number of realizations of sample means, the distribution converges to normal distribution. This is evident from Figures 5.12c and 5.12d. As we increase the number of realizations from 200 to 2,000 and then to 40,000, the distribution of sample mean is closer to normal distribution.

A stronger version of this (with some additional constraints) generalizes CLT to sum of random variables that are independent but not identically distributed, that is each draw can come from a different distribution. This provides an even stronger basis for the assumption of normality if one were working with summation or averages of random variables.

Example 5.13. *Show that sum of samples from three distributions converge to normal distribution.*

Solution 5.13. *Consider 3 distributions:*

a Uniform distribution between 0 and 2

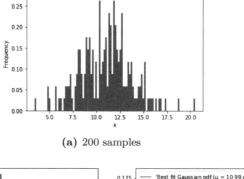

(a) 200 samples

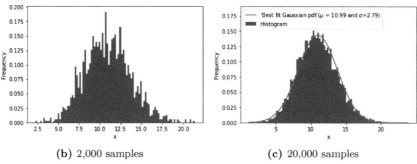

(b) 2,000 samples   **(c)** 20,000 samples

FIGURE 5.13 Sum of uniform, poisson, and binomial distribution.

b Poisson distribution with $\eta = 5$
c Binormial distribution with $N = 10$ and $p = 0.5$

If we take 200 samples from each distribution and sum them up. Let this new random variable be referred as 'x'. Now, histogram of x generated from this 200 samples is given in Figure 5.13a. The distribution of x is not evident from this plot. However, if we take more number of samples, we can see that the distribution converges to a normal distribution. This can be seen from Figures 5.13b and 5.13c. As we increase the number of samples from 200 to 20,000, the distribution looks more and more like a normal distribution.

5.10.4 *t*-Distribution

Another important distribution that is used considerably in data science is the *t*-distribution. This sampling distribution is a slight variant of the *z*-distribution. Assume that we have samples x^i from $\mathcal{N}(\mu, \sigma^2)$, then

$$T = \frac{\bar{x} - \mu}{s/\sqrt{n}} \sim t_{n-1} \tag{5.35}$$

$$\text{where } s^2 = \frac{1}{n-1} \sum_{i=1}^{n} (x^i - \bar{x})^2 \tag{5.36}$$

T is the statistic computed from samples and it can be shown to follow a t-distribution[3]. The subscript $n-1$ denotes the degrees of freedom associated with this statistic. The functional form of the t-distribution pdf is not shown here; however, much like the z-distribution, the integral between $-\infty$ and T values can be computed for any number of degrees of freedom using standard statistical packages in R, Python, and so on. A major difference between T and Z is that in calculating T, the population σ is an unknown and hence an estimate s of σ computed using equation 5.36 is used instead. Another difference is that the t-distribution will also depend on the number of samples used in the computation of the t-statistic and hence the degrees of freedom is also required for the computation of the t-distribution. Conceptually, when $n \to \infty$, t-distribution will tend to a z-distribution. This inherently means that the estimate $s \to \sigma$ as $n \to \infty$. t-statistic is useful to compare if a sample came from a distribution when the population variance is unknown.

Example 5.14. *Find t-statistic for the following samples given that the mean of the distribution is 10.*
Samples: $\begin{bmatrix} 8 & 11 & 9 & 2 & 5 & 3 & 4 & 5 & 9 & 1 \end{bmatrix}$

Solution 5.14. *Sample mean of the data, $\bar{x} = 5.7$. Sample variance of the data,*

$$s^2 = \frac{1}{n-1}\sum_{i=1}^{n}(x^i - \bar{x})^2 = \frac{102.1}{9} = 11.34 \implies s = 3.368$$

t-statistic will be as follows:

$$T = \frac{5.7 - 10}{3.368/\sqrt{10}} = -4.0372$$

5.10.5 Chi-Squared Distribution

Until now, comparing sample mean to population mean was discussed. However, in many cases, comparing the sample variance with a hypothesized population variance may be of interest. This idea will be utilized in 6.2.1. Chi-squared (χ^2) statistic (C^2) and its corresponding distribution (χ^2 distribution) allows us to perform this comparison. Equation 5.37 shows how the statistic is calculated from the samples, where s is computed using equation 5.36. Similar to t-distribution, the degrees of freedom need to be specified and the integral between $-\infty$ and C^2 values can be computed for any number of degrees of freedom using standard statistical packages. One use of a χ^2 distribution is in identifying if the error that remains in the outputs after model fitting is comparable to the errors that one would expect from the sensor used to measure

---

[3]Statistic is represented using T while the distribution is represented using t. Note that T is a random variable, not to be confused as a matrix.

the variable. This will be a critical question that needs to be answered for identifying acceptable trade-offs between model and uncertainty matrices. It can also be shown that a χ^2 distribution with n degrees of freedom is exactly equivalent to sum of n random variables following standard normal distribution, *i.e.*, $\mathcal{N}(0, 1)$ (refer equation 5.38). As a corollary, equation 5.37 can be rewritten as a sum of $n - 1$ standard normal variables.

$$C^2 = \frac{(n-1)s^2}{\sigma^2} \sim \chi^2_{n-1} \qquad (5.37)$$

$$\chi^2_n \sim z_1^2 + \ldots + z_n^2 \qquad (5.38)$$

Example 5.15. *Find χ^2-statistic for the following samples given that the mean of the distribution is 10 and variance is 9.*
Samples: $\begin{bmatrix} 8 & 11 & 9 & 2 & 5 & 3 & 4 & 5 & 9 & 1 \end{bmatrix}$

Solution 5.15. *Given that $\sigma^2 = 9$. The sample variance and T can be calculated as follows:*

$$s^2 = \frac{1}{n-1} \sum_{i=1}^{n} (x^i - \bar{x})^2 = \frac{102.1}{9} = 11.34$$

$$C^2 = \frac{(n-1)s^2}{\sigma^2} = \frac{9 \times 11.34}{9} = 11.34$$

5.10.6 F-Distribution

The last sampling distribution that is described in this section is the F-distribution. Here, the problem is to confirm if two datasets came from the same distribution in terms of their variance characteristics. A general computation of F-statistic can be seen in equation 5.39. A more specific use of the F-distribution is the comparison $s_1 \approx s_2 \approx \sigma$. Contrast this with χ^2 distribution where the comparison is $s \approx \sigma$. In this case, one doesn't need the value of σ for the computation of the F-statistic as shown in equation 5.40 (since $\sigma_1 = \sigma_2$). Again, similar to t and χ^2 distributions, the integral between $-\infty$ and f values can be computed for any number of degrees of freedom using standard statistical packages. However, since there are two datasets which are used in the computation of the statistic there are two degrees of freedom $n_1 - 1$ and $n_2 - 1$.

$$f = \frac{s_1^2/\sigma_1^2}{s_2^2/\sigma_2^2} \sim F_{n_1-1, n_2-1} \qquad (5.39)$$

$$f = s_1^2/s_2^2 \sim F_{n_1-1, n_2-1} \qquad (5.40)$$

Example 5.16. *Find F-statistic for the following sample sets generated from the same distribution.*
Sample set 1: $\begin{bmatrix} 8 & 11 & 9 & 2 & 5 & 3 & 4 & 5 & 9 & 1 \end{bmatrix}$
Sample set 2: $\begin{bmatrix} 5 & 20 & 7 & 4 & 12 & 16 & 9 & 20 \end{bmatrix}$

Solution 5.16. *Sample mean of the dataset,* $\bar{x}_1 = 5.7$ *and* $\bar{x}_2 = 11.625$. *Sample variances of the data,*

$$s^2 = \frac{1}{n-1}\sum_{i=1}^{n}(x^i - \bar{x})^2 \qquad s_1^2 = \frac{102.1}{9} = 11.34 \qquad s_2^2 = \frac{289.875}{7} = 41.41$$

F-statistic will be as follows:

$$f = \frac{11.34}{41.41} = 0.274$$

Example 5.17. *Find F-statistic for the following sample sets generated from the distributions with variance* 9 *and* 11 *respectively.*
Sample set 1: $\begin{bmatrix} 8 & 11 & 9 & 2 & 5 & 3 & 4 & 5 & 9 & 1 \end{bmatrix}$
Sample set 2: $\begin{bmatrix} 5 & 20 & 7 & 4 & 12 & 16 & 9 & 20 \end{bmatrix}$

Solution 5.17. *Sample means and variances have been computed in Example 5.16. Using these values, F-statistic can be computed as follows:*

$$f = \frac{s_1^2\,\sigma_2^2}{s_2^2\,\sigma_1^2} = \frac{11.34}{41.41}\frac{11}{9} = 0.335$$

5.11 DETERMINING QUALITY OF ESTIMATES

Since any computed statistic is a random variable, the quality of the statistic has to be established through the analysis of the sampling distribution that describes the statistic. Unbiasedness and consistency are two metrics that are used to understand the quality of estimates. These two metrics answer two different questions. Consistency relates to understanding how good the estimate can be if the number of samples are increased. Unbiasedness relates to how close the averaged estimate—from repeated experiments with the same number of samples—will be to the population parameter value. It might seem a little peculiar as to how one would establish these metrics when experiments are not repeated multiple times or repeated with a large number of samples. Theoretical notions that underlie these two metrics will be described in what follows.

5.11.1 Unbiasedness

Unbiasedness refers to the ability of the estimator to accurately estimate the parameter of interest. It is important to understand how accuracy is established. If estimates (from an experiment with n samples) of a population parameter are obtained from a large number (infinite) of experiments and these estimates are averaged, and if the averaged estimate is equal to the true population parameter then the estimator is unbiased. Mathematically,

$$E(\hat{\Theta}_n) = \Theta$$

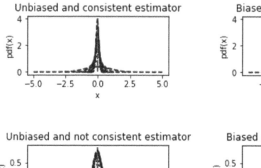
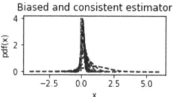
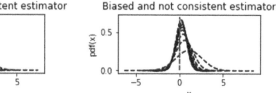

FIGURE 5.14 Pictorial representation of various estimator possibilities.

where $\hat{\Theta}_n$ is an estimate of Θ with n samples. The expectation is applied on the statistic formula and the value computed through the formula for one specific experiment has to be viewed as one realization from the sampling distribution of the statistic.

5.11.2 Consistency

Consistency of an estimate refers to whether the estimate will equal the true population value if the number of samples in a single experiment is increased. Consistency refers to the estimate converging to the true parameter when the number of datapoints are increased. Mathematically,

$$\lim_{n \to \infty} P|\hat{\Theta}_n - \Theta| > \epsilon \to 0$$

Since $\hat{\Theta}_n$ is a distribution, the equation above, if satisfied, will ensure that the deviation from the true mean of any estimate that is picked from the distribution tends to zero as the number of samples tend to infinity.

   While consistency and unbiasedness provide insights about the accuracy of the estimate when using small and large samples respectively, they are not related in the sense that all four possibilities exist when it comes to estimators. An estimator may be: (i) consistent and unbiased—$\bar{x}$ for μ, (ii) consistent but biased—s^2 (with n in the denominator instead of $(n-1)$) for σ^2, (iii) not consistent but unbiased—pick n^{th} sample as an estimate for μ, (iv) both biased and not consistent $-\sum_{i=1}^{n} x^i/k$, $k \neq n$ for μ. Visualization of distributions for estimators with the four combinations of unbiasedness and consistency can be found in Figure 5.14.

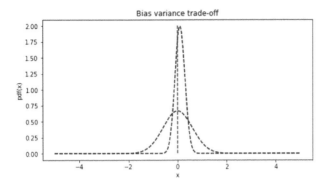

FIGURE 5.15 Pictorial representation of bias variance trade-off.

5.11.3   Bias-Variance Trade-off

We will now describe a very important concept that is of relevance in DS/ML, popularly referred to as the Bias-Variance trade-off. The variance of an estimator is computed using $E(\hat{\Theta}_n - E(\hat{\Theta}_n))^2$. Another quantity referred to as the mean-squared error (MSE) of an estimator is computed as $E(\hat{\Theta}_n - \Theta)^2$. Notice that the difference between the two formulae is in the use of the global θ value in the expectation calculation of MSE, whereas $E(\hat{\Theta}_n)$ is used in the calculation of the variance. It is easy to show that MSE = Variance + Bias^2. What this means is that in some cases, the variance may be significantly decreased by increasing the bias slightly, and an estimator with a much better MSE can be derived. This is called the bias-variance trade-off. This might be of use in some applications where a slight error can be tolerated if it can be guaranteed that the estimates will not vary considerably from a mean. This notion is captured in Figure 5.15. Of all unbiased estimators, the one with the minimum variance is called the minimum variance unbiased estimator (MVUE). An estimator which has an MSE less than any other estimator for a parameter is called the optimal estimator for that parameter.

Example 5.18.
Consider a set of datapoints $x^1, x^2, ..., x^n$ and the real values y^i associated with each datapoint x^i. Let us assume there exists a true function $f(x)$ that captures the behavior of y and is corrupted by some noise ϵ

$$y = f(x) + \epsilon$$

Let's say that we are trying to find a function $\hat{f}(x)$ that approximates the function $f(x)$ as closely as possible using data.
Define:

1. *$Bias[\hat{f}(x)] = E[\hat{f}(x) - f(x)]$*

2. $Var[\hat{f}(x)] = E[\hat{f}(x)^2] - E[\hat{f}(x)]^2$
3. $\sigma^2 = Var(\epsilon)$

Prove the following relation that establishes the bias-variance trade-off (which is one of the cornerstones of ML):

$$E\left[(y - \hat{f}(x))^2\right] = Bias[\hat{f}(x)]^2 + Var[\hat{f}(x)] + \sigma^2$$

Solution 5.18.

$$E\left[(y - \hat{f}(x))^2\right] = E\left[(f(x) + \epsilon - \hat{f}(x))^2\right]$$

$$= E\left[(f(x) - \hat{f}(x))^2\right] + E\left[\epsilon^2\right] + 2E\left[(f(x) - \hat{f}(x))\epsilon\right]$$

$$= E\left[(f(x) - \hat{f}(x))^2\right] + E\left[\epsilon^2\right] + 2E\left[(f(x) - \hat{f}(x))\right]E\left[\epsilon\right]$$

Since $E\left[\epsilon^2\right] = \sigma^2$ *and* $E\left[\epsilon\right] = 0$,

$$E\left[(y - \hat{f}(x))^2\right] = E\left[(f - \hat{f})^2\right] + \sigma^2$$

$$E\left[(f - \hat{f})^2\right] = E\left[(f - E[\hat{f}]) - (\hat{f} - E[\hat{f}])^2\right]$$

$$= E\left[(E[\hat{f}] - f)^2\right] + E\left[(E[\hat{f}] - \hat{f})^2\right]$$

$$- 2E\left[(f - E[\hat{f}])\right]E\left[(\hat{f} - E[\hat{f}])\right]$$

$$= Bias[\hat{f}(x)]^2 + Var[\hat{f}(x)] - 2(f - E[\hat{f}])\left(E\left[\hat{f}\right] - E\left[\hat{f}\right]\right)$$

$$= Bias[\hat{f}(x)]^2 + Var[\hat{f}(x)]$$

Hence,

$$E\left[(y - \hat{f}(x))^2\right] = Bias[\hat{f}(x)]^2 + Var[\hat{f}(x)] + \sigma^2$$

5.12 HYPOTHESIS TESTING

At this point we have understood that: parameters of the distribution are estimated from samples; these estimates are random variables and are characterized by a sampling distribution; the quality of the estimates are understood through an analysis of the unbiasedness and consistency properties of the estimate.

Our original interest was in checking if a random variable (for which data was available) came from a particular distribution with known distribution parameters. To do this, we devised methods to generate estimates from the data that can be compared to their "true" values. Since samples are random variables, the estimates are also random variables with their derived distributions. The numerical values for the estimates are "point" estimates from only a single realization. The "point" estimate cannot be expected to be exactly equal to the population parameter. To address this problem, a notion of risk is introduced in rejecting the sample as not being from the hypothesized distribution. Consider Figure 5.16a. Assume the distribution depicted in the picture is the sampling distribution. The Null Hypothesis (H_o) is that the "point" estimate is from this distribution. If the null hypothesis is accepted, then equation 5.26 (as an example) is accepted, else it is rejected. From Figure 5.16a, we can see that there are accept and reject regions. When a point estimate falls in a reject region, there is a probability that we are making a mistake in rejecting H_o. This risk or probability is given by the area of the shaded region. This total probability is denoted by α.

Example 5.19. *A controller is setup to control the outlet flow rate of a stirred-tank reactor. The desired flow rate is $10m^3/s$. The actual flow was then measured multiple times using a venturimeter to cross-check the performance of the controller. Following measurements were obtained from various trials.*

$$9.44,\ 11.08,\ 9.8,\ 8.55,\ 9.68,\ 10.39\ m^3/s$$

With the help of hypothesis testing comment whether the controller performance is good. Assume that the true variance is $\sigma^2 = 1$.

Solution 5.19. *Let the estimator for μ be sample mean ($\bar{x}$).*
Null hypothesis: $H_0 : \mu = 10m^3/s$ (Performance is good)
Alternate hypothesis: $H_a : \mu \neq 10m^3/s$

$$Z = \frac{\bar{x} - \mu_{\bar{x}}}{\sigma_{\bar{x}}} = \frac{\bar{x} - \mu}{\sigma/\sqrt{n}} \sim \mathcal{N}(0,1)$$

For the data given, $\bar{x} = \frac{58.94}{6} = 9.8233$ and $Z = \frac{9.8233-10}{1/\sqrt{n}} = -0.4328$. If we are interested in significance level of 0.05, then we need to find the z values of standard normal distribution corresponding to probabilities 0.025 and 0.975. $z_{\frac{\alpha}{2}} = -1.96$; $z_{1-\frac{\alpha}{2}} = 1.96$. Since $z_{\frac{\alpha}{2}} < -0.4328 < z_{1-\frac{\alpha}{2}}$ we accept the null hypothesis. Thus, the performance of the controller is good.

Example 5.20. *For the problem discussed in Example 5.19, will the conclusion change if true σ was unknown?*

Solution 5.20. *If σ is unknown, then we have to use t-statistic*

$$T = \frac{\bar{x} - \mu}{s/\sqrt{n}}$$

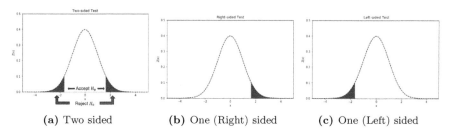

(a) Two sided   (b) One (Right) sided   (c) One (Left) sided

FIGURE 5.16 Hypothesis test.

Null hypothesis (H_0): $\mu = \mu_0$
Alternate hypothesis (H_a): $\mu \neq \mu_0$

$$\bar{x} = 9.8233; \quad s^2 = \frac{1}{n-1}\sum(x_i - \bar{x})^2 = 0.7379;$$

$$T = \frac{9.8233 - 10}{\sqrt{0.7379/6}} = -0.5038 \sim t_{n-1}$$

If we still use $\alpha = 0.05$ then the interval of accepting H_0 will be

$$\left[t_{\frac{\alpha}{2},n-1} \quad t_{1-\frac{\alpha}{2},n-1}\right] = \left[-2.5706 \quad 2.5706\right]$$

Since -0.5038 is within the interval, we accept H_0 and thus, controller performance is satisfactory.

In some cases, keeping the interest of the application in mind, one might ask questions such as what if the "point" estimate that is calculated from the data came from a distribution that has a mean greater (or lesser) than the mean of the original hypothesized distribution? In this case, the hypothesis test becomes one-sided as shown in Figures 5.16b and 5.16c. Notice that the shaded area is only to one side of the distribution as a result "point" estimates to only one side of the distribution have the possibility of being rejected.

Example 5.21. *For an industrial use, at least 97% pure ethanol was required. Distillation was carried out to purify the dilute ethanol feed. Purity of vapor outlet was measured 100 times and the average purity was found to be 96.7%.*

(a) If the true variance of the process is unknown, but the sample variance is given as 1, comment whether the true purity is acceptable with a significance of 0.1?
(b) If the variance of the process is known to be 0.64, will the conclusion change?

Solution 5.21.

$$H_0 : \mu \geq 97\% \qquad\qquad H_a : \mu < 97\%$$

(a) $s^2 = 1 \implies s = 1$

$$T = \frac{\bar{x} - \mu}{\frac{s}{\sqrt{n}}} = \frac{-0.3 \times 10}{1} = -3$$

$$t_{\alpha,99} = -1.29 \Rightarrow T < t_{\alpha,99}$$

We reject H_0. We will conclude that the required purity has not been achieved.

(b) $\sigma^2 = 0.64 \implies \sigma = 0.8$

$$Z = \frac{\bar{x} - \mu}{\frac{\sigma}{\sqrt{n}}} = \frac{96.7 - 97}{\frac{0.8}{\sqrt{100}}} = \frac{-0.3 \times 10}{0.8} = -3.75$$

For $\alpha = 0.1$, z_α (value of Z such that $\mathcal{N}(0,1)$ has a cdf of 0.1) $= -1.28$. Since $Z < z_\alpha$, we reject H_0. Thus, the required purity has not been achieved.

Example 5.22. *In the contact process for the production of sulphuric acid, temperature of not more than $450°C$ is desired. The reactor temperature was measured 25 times and it was found that the mean temperature was $450.3°C$ and the variance was 0.25. Find whether the reactor is operating at the desired condition?*

Solution 5.22.

$$H_0 : \mu \le 450°C; \qquad H_a : \mu > 450°C$$
$$T = \frac{450.3 - 450}{\frac{0.5}{\sqrt{25}}} = \frac{0.3 \times 5}{0.5} = 3$$

If we choose $\alpha = 0.05$, $t_{1-\alpha,24} = 1.71$. Since $T > t_{1-\alpha,24}$, we reject H_0.

Example 5.23. *A highly temperature sensitive reaction is being conducted in a stirred tank reactor. A controller is installed to maintain the temperature at $250°C$. Since the reaction is highly temperature sensitive, variations in temperature is not desirable. Maximum allowed variance is $0.7°C$. Temperature was measured 100 times and the sample variance was found to be 0.705. Perform hypothesis tests and comment on the performance of the controller. Choose $\alpha = 0.025$.*

Solution 5.23.

$$H_0 : \sigma^2 \le 0.7; \qquad H_a : \sigma^2 > 0.7$$
$$C^2 = \frac{(n-1)s^2}{\sigma^2} = \frac{(99) \times 0.705}{0.7} = 99.71$$

Now, if $\alpha = 0.025$, then

$$\chi^2_{0.975,99} = 128.422$$

Since $C^2 < \chi^2_{1-\alpha,99}$, we do not reject H_0. The process temperature is maintained at desired levels and the controller is working well.

Example 5.24. *There are two machines for spray coating at desired thickness. Their variability should be similar if they need to be used for the same application. To test this, various samples were collected from both machines and the summary is given below.*

$$s_1^2 = 2.1; \quad s_2^2 = 2.08; \quad n_1 = 25; \quad n_2 = 20$$

Comment on whether these machines can be used for the same application if the machines were truly built to have the same variance.

Solution 5.24.

$$H_0 : \sigma_1^2 = \sigma_2^2; \qquad H_a : \sigma_1^2 \neq \sigma_2^2$$

$$f = \frac{s_1^2}{s_2^2} = \frac{2.1}{2.08} = 1.0096$$

For $\alpha = 0.05$,

$$F_{0.025,24,19} = 0.4264; \qquad F_{0.975,24,19} = 2.4523$$

Since $F_{0.025} < f < F_{0.975}$, we accept H_0.

Example 5.25. *Let $\psi = (x_1 - x_2)^2$ be an unmeasured state of a system. x_1 and x_2 are measured variables. Let x_1 follow a normal distribution with mean 10 and standard deviation 4, and x_2 follow an exponential distribution $(\lambda e^{-\lambda x})$ with $\lambda = 0.2$.*

(a) *Find the bounds of ψ corresponding to a significance level of 0.05 for a 2-sided test.*
(b) *Does ψ corresponding to the sample $x_1 = 6.2$ and $x_2 = 0.12$ belong to the fitted distributions given in Example 5.9?*
(c) *How can one understand the notion of significance level (use $\alpha = 0.05$) using Monte-Carlo simulations?*

Use the sampling distribution derived in Example 5.9 to perform the analysis.

Solution 5.25.

(a) *The ψ value corresponding to 2.5% probability for the estimated t-distribution is -106.6. Similarly, the ψ value corresponding to 97.5% probability for the estimated t-distribution is 194.7. Thus, $[-106.6, 194.7]$ is the region of acceptance corresponding to a significance level of $\alpha = 0.05$.*
(b) *Null hypothesis: ψ is from the given sampling distribution. For given values of x_1 and x_2, $\psi = (0.12 - 6.2)^2 = 36.97$ and this value is within the region of acceptance given in part (a). Hence, we accept the hypothesis that the given sample of ψ is from the given sampling distribution.*

(c) *The significance level indicates the number of times we are likely to correctly identify that a sample is drawn from the proposed distribution. To check this, generate 1,000 test samples x_1 and x_2 from the distributions given in Example 5.9 and calculate the corresponding ψ values. Hypothesis tests are performed for each sample. Although all of these samples are drawn from the original distributions, null hypothesis is accepted only for those within the bounds calculated in part a. Among the 1,000 test samples that were generated, 94% was found to be sampled from the sampling distribution and the remaining 6% were rejected. If the fitted sampling distribution was an exact representation of the true sampling distribution and if we had larger number of test samples, we would see that the number of samples correctly identified to be from the true distribution would be closer to 95%.*

Example 5.26. *Let $\psi = (\bar{x}_1 - \bar{x}_2)^2$ be an unmeasured state of a system. x_1 and x_2 are measured variables. Let x_1 follow a normal distribution with mean 10 and standard deviation 4, while x_2 follow an exponential distribution $(\lambda e^{-\lambda x})$ with $\lambda = 0.2$.*

(a) *Find the bounds of ψ corresponding to a significance level of 0.05 for a 2-sided test.*
(b) *Does ψ corresponding to the sample $\bar{x}_1 = 8.35$ and $\bar{x}_2 = 6.75$ belong to the fitted distribution in Example 5.10?*
(c) *How can one understand the notion of significance level (use $\alpha = 0.05$) using Monte-Carlo simulations?*

Use the sampling distribution derived in Example 5.10 to perform the analysis.

Solution 5.26.(a) *The ψ value corresponding to 2.5% probability for the estimated log-normal distribution is 3.1. Similarly, the ψ value corresponding to 97.5% probability for the estimated log-normal distribution is 59.39. Thus, $[3.1, 59.39]$ is the region corresponding to 95% confidence level ($\alpha = 0.05$).*
(b) *Null hypothesis: ψ is from the given sampling distribution.*
For given values of $\bar{x}_1$ and $\bar{x}_2$, $\psi = (8.35 - 6.75)^2 = 2.56$ and this value is outside the region of acceptance given in part (a). Hence, we reject the hypothesis that the given sample is from the given sampling distribution. Since, we have estimated this sampling distribution using Monte-Carlo simulations such that x_1 and x_2 are drawn from the distributions as described in Example 5.10, rejecting the null hypothesis indicates that the given sample of x_1 and x_2 are not drawn from the given distributions.
(c) *The significance level indicates number of times we correctly identify that a sample is drawn from the proposed distribution. To check this, we generate 1,000 test samples $\bar{x}_1$ and $\bar{x}_2$ from the distributions given in Example 5.9 and calculate the corresponding ψ values. We performed hypothesis tests for each sample. Although all of these samples are drawn from the original*

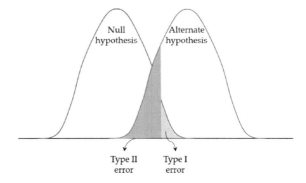

FIGURE 5.17 Type II error for a one-sided test.

distributions, null hypothesis is accepted only for those within the bounds calculated in part a. Among the 1,000 test samples that we generated, 96% was found to be sampled from the sampling distribution and the remaining 4% were rejected.

It is apparent from the foregoing discussion that the focus is on minimizing the error in rejecting a null hypothesis or minimizing the error in rejecting a sample as not coming from a distribution when it is actually from that distribution. This is visually seen from the small probability for the region of rejection. The error in wrongly rejecting the null hypothesis is called the Type I error. However, acceptance of null hypothesis when an alternate hypothesis is valid should also be considered. This is called Type II error. To calculate this error, an alternate hypothesis needs to be defined. This is shown in Figure 5.17, where Type II error is the area under the curve of the alternate hypothesis for the range of values where the null hypothesis is accepted.

Example 5.27. *For the controller problem described in Example 5.19, if the measurements actually come from a distribution with mean flow rate of 10.5 m^3/s and variance 1, what is the Type II error in declaring that the performance of the controller is good?*

Solution 5.27.

$$H_0 : \mu = 10m^3/s; \qquad H_a : \mu = 10.5m^3/s$$

For $\alpha = 0.05$, $x_{\alpha/2}$ and $x_{1-\alpha/2}$ for the null hypothesis can be calculated as:

$$z_{0.975} = 1.96 \Rightarrow \frac{x_{1-\alpha/2} - 10}{1/\sqrt{6}} = 1.96 \implies x_{1-\alpha/2} = 10.8$$

$$z_{0.025} = -1.96 \Rightarrow \frac{x_{\alpha/2} - 10}{1/\sqrt{6}} = -1.96 \implies x_{\alpha/2} = 9.2$$

The cumulative probabilities at $x_{\alpha/2}$ and $x_{1-\alpha/2}$ for the alternate distribution (with mean 10.5 and variance $1/\sqrt{6}$) are 0.0007 and 0.7689, respectively. Hence, probability of Type II error would be $0.7689 - 0.0007 = 0.7682$.

5.12.1 Confidence Intervals

In the previous discussion, a procedure for testing hypotheses on the population parameters was described. However, we might want to identify a confidence interval within which the true population parameters will lie given a "point" estimate. For example, we could seek to identify a range within which a true population mean will lie given a "point" estimate (average) for population mean. Let us assume that we know the true variance of the hypothesized distribution. From the definition of the probability density function and Figure 5.16a it can be seen that

$$P\left(-z_{\alpha/2} \le \frac{\bar{x} - \mu}{\sigma/\sqrt{n}} \le z_{\alpha/2}\right) = 1 - \alpha$$

From this, we can derive a $100(1 - \alpha)\%$ confidence interval on μ

$$\bar{x} - z_{\alpha/2}\frac{\sigma}{\sqrt{n}} \le \mu \le \bar{x} + z_{\alpha/2}\frac{\sigma}{\sqrt{n}}$$

Interestingly, we find that the confidence interval can be used to decide on the hypothesis test also. If the hypothesized true μ is in the confidence interval, we then accept the null hypothesis. These are completely equivalent statements. However, if we want to understand if the true μ will actually be in the range, then we realize that this has to be a categorical yes or no. Either the true mean μ is in the range or it is not in the given range. What then is the interpretation of this $100(1 - \alpha)\%$ confidence interval? The confidence interval only means that if we adopt the same procedure of drawing n samples from the distribution and generate the confidence intervals (both $\bar{x}$ and the confidence intervals will be different each time), then $100(1 - \alpha)\%$ of times, the true μ will be in the range that is decided from that experiment.

Example 5.28. *For the controller problem described in Example 5.19, find the 95% confidence interval and comment on controller performance.*

Solution 5.28.

$$H_0: \mu = 10; \qquad H_a: \mu \ne 10$$

Given $\bar{x} = 9.8233$, $\sigma^2 = 1$, $\alpha = 0.05$, *and* $n = 6$. *Since this is a two-tailed* *test, we check $z_{\frac{\alpha}{2}} \leq \frac{\bar{x}-\mu}{\frac{\sigma}{\sqrt{n}}} \leq z_{1-\frac{\alpha}{2}}$.*

$$\Rightarrow \mu \leq \bar{x} - z_{\frac{\alpha}{2}} \frac{\sigma}{\sqrt{n}} \text{ and } \bar{x} - z_{1-\frac{\alpha}{2}} \frac{\sigma}{\sqrt{2}} \leq \mu$$

$$\Rightarrow \bar{x} - z_{1-\frac{\alpha}{2}} \frac{\sigma}{\sqrt{n}} \leq \mu \leq \bar{x} - z_{\frac{\alpha}{2}} \frac{\sigma}{\sqrt{n}}$$

$$\Rightarrow 9.8233 - \left(+1.96 \times \frac{1}{\sqrt{6}}\right) \leq \mu \leq 9.8233 + \frac{1.96}{\sqrt{6}}$$

$$\Rightarrow 9.023 \leq \mu \leq 10.623 \Rightarrow \text{ This is the confidence interval.}$$

Since the postulated μ (or desired μ) is within the confidence interval, we can say that the controller performance is good and we accept H_0.

Example 5.29. *Find confidence interval for the problem described in Example 5.22 and decide whether the null hypothesis should be accepted or rejected.*

Solution 5.29.

$$H_0: \mu \leq 450°C; \qquad H_a: \mu > 450°C$$

$$\left. -\infty \leq \frac{\bar{x} - \mu}{\frac{s}{\sqrt{n}}} \leq t_{1-\alpha,n} \right\} \text{ Confidence region}$$

$$\bar{x} - t_{1-\alpha,n} \frac{s}{\sqrt{n}} \leq \mu \leq \infty$$

$$450.3 - 1.71 \times \sqrt{\frac{0.25}{25}} = 450.13 \implies 450.13 \leq \mu \leq \infty \Rightarrow CI$$

Since the confidence interval does not include $450°C$, we reject H_0.

Example 5.30. *A survey was conducted to find the distribution of height of males in a state. Height of a random subset of 100 males were measured and a variance of 10cm was obtained. Find the confidence interval for population variance.*

Solution 5.30. *Given, $s^2 = 10, n = 100$. Let $\alpha = 0.05$.*

$$\chi^2_{\frac{\alpha}{2},n} \leq \frac{(n-1)s^2}{\sigma^2} \leq \chi^2_{1-\frac{\alpha}{2},n} \implies \sigma^2 \leq \frac{(n-1)s^2}{\chi^2_{\frac{\alpha}{2},n}} \text{ and } \frac{(n-1)s^2}{\chi^2_{1-\frac{\alpha}{2},n}} \leq \sigma^2$$

Thus, the confidence interval σ^2 is

$$\frac{(n-1)s^2}{\chi^2_{1-\frac{\alpha}{2},n}} \leq \sigma^2 \leq \frac{(n-1)s^2}{\chi^2_{\frac{\alpha}{2},n}}$$

For the given data,

$$\frac{99 \times 10}{128.422} \le \sigma^2 \le \frac{99 \times 10}{73.36}$$
$$7.71 \le \sigma^2 \le 13.49 \Rightarrow CI$$

Example 5.31. *For the problem described in Example 5.24, find the confidence interval and comment on whether one should accept/reject the null hypothesis*

Solution 5.31.

$$H_0 : \sigma_1^2 = \sigma_2^2; \qquad H_a : \sigma_1^2 \ne \sigma_2^2$$

$$F_{0.025}(24, 19) \le \frac{s_1^2/\sigma_1^2}{s_2^2/\sigma_2^2} \le F_{0.975}(24, 19)$$

$$0.4264 \times \frac{s_2^2}{s_1^2} \le \frac{\sigma_2^2}{\sigma_1^2} \le 2.4523 \times \frac{s_2^2}{s_1^2} \implies 0.4224 \le \frac{\sigma_2^2}{\sigma_1^2} \le 2.429$$

Since $\frac{\sigma_2^2}{\sigma_1^2} = 1$, falls in this range, we accept H_0.

5.13 DISTRIBUTIONS OF MULTIPLE RELATED RANDOM VARIABLES

Until now single random variables have been discussed without consideration of how these random variables might be related to each other. Study of multivariable random variables is quite involved. For example, relationships between multiple variables could be direct or through joint distributions and so on. However, understanding multiple random variables together, at least at an introductory level, is critical for understanding several ML concepts. In this section, multivariable concepts are introduced through simple examples.

Probability density functions (pdf), $f(x)$, define random variables in the univariate continuous case. In the multivariable case, for example, considering two variables, the pdf becomes $f_{x,y}(x, y)$—a function of both variables. In the single variable case the pdf can be used to compute the probability that the random variable takes a value in a certain range. In a similar manner, the multivariable pdf is a means to calculate the probability that the variables take values in a specified area (two-dimensions), volume (three-dimensions) and hyper-volume (multiple-dimensions). As a result, the function $f_{x,y}(x, y)$ is called a joint pdf. Mathematically

$$P(a_1 \le x \le b_1 \text{ and } a_2 \le y \le b_2) = \int_{a_2}^{b_2} \int_{a_1}^{b_1} f_{x,y}(x, y) dx dy$$

It is also possible to find the probability for a range of values for a single variable (in the multivariable case) and this is achieved through the definition of marginal distributions. In a two variable case, two marginal distributions $f_x(x)$ and $f_y(y)$ can be derived as shown below

$$f_x(x) = \int_{-\infty}^{\infty} f_{x,y}(x,y)dy$$

$$f_y(y) = \int_{-\infty}^{\infty} f_{x,y}(x,y)dx$$

Now it is possible to find the probability that x will take values in a range as shown below (a completely equivalent approach will work for y)

$$P(a_1 \leq x \leq b_1) = \int_{a_1}^{b_1} f_x(x)dx$$

Example 5.32. *Consider two biased coins being tossed 100 times. Out of these 100 trials, 20 trials had both heads and 25 trials had both tails. Among the rest, 25 trials had head in the first coin while tail in the second coin and 30 trials with tail in the first coin and head in the second coin. Find the joint probability distribution for this example.*

Solution 5.32. *Joint probability distribution is calculated and shown in the following table.*

		Coin 2 (y)	
		H	T
Coin 1 (x)	H	$f(H,H) = 0.2$	$f(H,T) = 0.25$
	T	$f(T,H) = 0.3$	$f(T,T) = 0.25$

Example 5.33. *Calculate the marginal probabilities for the data given in Example 5.32.*

Solution 5.33. *Marginal probabilities of coin 1 are as follows:*

$$f_x(H) = 0.2 + 0.25 = 0.45; \qquad f_x(T) = 0.3 + 0.25 = 0.55$$

Marginal probabilities of coin 2 are as follows:

$$f_y(H) = 0.2 + 0.3 = 0.5; \qquad f_y(T) = 0.25 + 0.25 = 0.5$$

Example 5.34. *Let x and y be two random variables having the following joint distribution*

$$f_{x,y}(x,y) = \begin{cases} cxy & if\ 0 < x < 1\ and\ 2 < y < 4 \\ 0 & otherwise \end{cases}$$

Find the constant "c". Find the marginal distribution $f_x(x)$ and $f_y(y)$.

Solution 5.34.

$$\int_2^4 \int_0^1 f_{x,y}dxdy = 1 \Rightarrow \int_2^4 \left[0.5cx^2y\right]_0^1 dy = 0.25c\left[y^2\right]_2^4 = 1 \Rightarrow c = 1/3$$

$$f_{x(x)} = \int_2^4 f_{x,y}(x,y)\,dy = \int_2^4 \frac{xy}{3}dy = \frac{x}{6}\left[y^2\right]_2^4 = 2x$$

$$f_y(y) = \int_0^1 \frac{xy}{3}dx = \frac{y}{6}\left[x^2\right]_0^1 = \frac{y}{6}$$

Another distribution that is of major interest in DS and ML is the conditional distribution. Conditional distributions can be used to answer questions regarding the probability that a variable will take certain values given that the other variables have already taken certain values. In other words, can one find the probability of occurrence of certain events conditioned on the fact that certain other events have already occurred. The conditional probability can be derived as a function of joint and marginal probabilities as below

$$f_{x|y} = \frac{f_{x,y}(x,y)}{f_y(y)}; \qquad f_{y|x} = \frac{f_{x,y}(x,y)}{f_x(x)}$$

Example 5.35. *For the data given in Example 5.32, find the probability that the coin 2 gives a tail given that the first coin gave a head.*

Solution 5.35. *The probability that the coin 2 gives a tail given that the first coin gave a head would be the conditional probability $f(y = T|x = H)$.*

$$f(y = T|x = H) = \frac{f(H,T)}{f_x(H)} = \frac{0.25}{0.45} = 0.56$$

Example 5.36. *For the system described in Example 5.34, find the conditional probability distribution given y*

Solution 5.36.
$$f_{x|y} = \frac{f_{x,y}(x,y)}{f_y(y)} = \frac{6xy}{3y} = 2x$$

Example 5.37. *Consider the continuous random variables x and y with the joint pdf*

$$f_{x,y}(x,y) = \left(\frac{1+xy}{c}\right) \forall \begin{array}{l} -1 < x < 1 \\ -1 < y < 1 \end{array}$$

(a) Find the constant "c".
(b) Find $f_y(y)$ and $f_x(x)$.
(c) Find $P\left(0 < x < \frac{1}{2},\ 0 < y < \frac{1}{4}\right)$.
(d) Find $P\left(0 < x < \frac{1}{2}\middle|0 < y < \frac{1}{4}\right)$.
(e) Find $P\left(0 < x < \frac{1}{2},\ 0.5 - \Delta y < y < 0.5 + \Delta y\right)$. Also comment on the probability of $\Delta y = 0$.
(f) Find $P\left(0 < x < \frac{1}{2}\middle|\ y = 0.5\right)$.

Solution 5.37.

(a) For a pdf, $\int_{-1}^{1}\int_{-1}^{1} f_{x,y}(x,y)\,dx\,dy = 1$

$$\Rightarrow \int_{-1}^{1}\left[\frac{x}{c} + \frac{x^2 y}{2c}\right]_{-1}^{1} dy = \int_{-1}^{1}\left(\frac{2}{c} + 0\right) dy = 1$$

$$\Rightarrow \frac{2}{c}[y]_{-1}^{1} = \frac{4}{c} = 1 \Rightarrow c = 4 \qquad \Rightarrow f(x,y) = \left(\frac{1+xy}{4}\right)$$

(b)

$$f_y(y) = \int_{-1}^{1}\left(\frac{1+xy}{4}\right) dx = \left[\frac{x}{4} + \frac{x^2 y}{8}\right]_{-1}^{1} = \frac{1}{2}$$

$$f_x(x) = \int_{-1}^{1}\left(\frac{1+xy}{4}\right) dy = \left[\frac{y}{4} + \frac{xy^2}{8}\right]_{-1}^{1} = \frac{1}{2}$$

(c) $P\left(0 < x < \frac{1}{2},\ 0 < y < \frac{1}{4}\right)$

$$= \int_{0}^{\frac{1}{4}}\int_{0}^{\frac{1}{2}}\left(\frac{1+xy}{4}\right) dx\,dy = \int_{0}^{\frac{1}{4}}\left[\frac{x}{4} + \frac{x^2 y}{8}\right]_{0}^{\frac{1}{2}} dy$$

$$= \int_{0}^{\frac{1}{4}}\left(\frac{1}{8} + \frac{y}{32}\right) dy = \left[\frac{y}{8} + \frac{y^2}{64}\right]_{0}^{\frac{1}{4}} = \frac{1}{32} + \frac{1}{16 \times 64} = \frac{1}{32}\left(1 + \frac{1}{32}\right) = \frac{33}{1024}$$

(d)

$$P\left(0 < x < \frac{1}{2}\middle|0 < y < \frac{1}{4}\right) = \int_{0}^{\frac{1}{2}} f_{x|0<y<\frac{1}{4}}dx = \int_{0}^{\frac{1}{2}}\frac{f(x, 0 < y < \frac{1}{4})}{\int_{0}^{\frac{1}{4}} f_y dy}dx$$

$$= \int_{0}^{\frac{1}{2}}\left[\frac{\int_{0}^{\frac{1}{4}}(\frac{1+xy}{4})dy}{\frac{1}{2} + \frac{1}{4}}\right] dx = 8\int_{0}^{\frac{1}{2}}\left[\frac{y}{4} + \frac{xy^2}{8}\right]_{0}^{\frac{1}{4}}$$

$$= 8\int_{0}^{\frac{1}{2}}\frac{1}{16} + \frac{x}{16 \times 8}dx = 8\left[\frac{1}{16} + \frac{\frac{1}{2}}{16 \times 8}\right]_{0}^{\frac{1}{2}}$$

$$= \frac{8}{32}\left(1 + \frac{1}{32}\right) = \frac{33}{128}$$

Or

$$P\left(0 < x < \frac{1}{2} \middle| 0 < y < \frac{1}{4}\right) = \frac{P\left(0 < x < \frac{1}{2},\ 0 < y < \frac{1}{4}\right)}{P\left(0 < y < \frac{1}{4}\right)}$$

From part c, we know, $P\left(0 < x < \frac{1}{2},\ 0 < y < \frac{1}{4}\right) = \frac{33}{1024}$. *Now,*

$$P\left(0 < y < \frac{1}{4}\right) = \int_0^{\frac{1}{4}} f_y\, dy = \int_0^{\frac{1}{4}} \frac{1}{2} dy = \frac{1}{2} \times \frac{1}{4} = \frac{1}{8}$$

$$\Rightarrow P\left(0 < x < \frac{1}{2} \middle| 0 < y < \frac{1}{4}\right) = \frac{\frac{33}{1024}}{\frac{1}{8}} = \frac{33 \times 8}{1024} = \frac{33}{128}$$

(e) $P\left(0 < x < \frac{1}{2},\ 0.5 - \Delta y < y < 0.5 + \Delta y\right)$

$$= \int_{0.5-\Delta y}^{0.5+\Delta y} \int_0^{\frac{1}{2}} \left(\frac{1+xy}{4}\right) dx\, dy = \int_{0.5-\Delta y}^{0.5+\Delta y} \left[\frac{x}{4} + \frac{x^2 y}{8}\right]_0^{\frac{1}{2}}$$

$$= \int_{0.5-\Delta y}^{0.5+\Delta y} \frac{1}{8} + \frac{y}{32} dy = \left[\frac{y}{8} + \frac{y^2}{64}\right]_{0.5-\Delta y}^{0.5+\Delta y}$$

$$= \frac{\Delta y}{4} + \frac{(0.5 + \Delta y)^2 - (0.5 - \Delta y)^2}{64} = \frac{\Delta y}{4} + \frac{2\Delta y}{64} = \frac{9}{32}\Delta y$$

Now, if we want $y = 0.5, \Delta y = 0,\ P\left(0 < x < \frac{1}{2},\ y = 0.5\right) = 0$. *This can be directly concluded as y is a continuous random variable and the question is to find the probability that the variable will take a value equal to 0.5. Since the probability of any continuous random variable taking a discrete value is zero,* $P\left(0 < x < \frac{1}{2},\ y = 0.5\right) = 0$.

(f) $P\left(0 < x < \frac{1}{2},\ y = 0.5\right)$

$$= \int_0^{\frac{1}{2}} f_{x|y=0.5}\, dx = \int_0^{\frac{1}{2}} \frac{f_{(x, y=0.5)}}{f_{y=0.5}}\, dx = \int_0^{\frac{1}{2}} \left(\frac{1 + 0.5x}{4}\right) \times \frac{1}{1/2} dx$$

$$= \frac{1}{2} \left[x + \frac{x^2}{4}\right]_0^{\frac{1}{2}} = \frac{1}{2} \left(\frac{1}{2} + \frac{1}{16}\right) = \frac{9}{32}$$

Or,

$$P\left(0 < x < \frac{1}{2},\ y = 0.5\right) = \lim_{\Delta y \to 0} \frac{P\left(0 < x < \frac{1}{2},\ 0.5 - \Delta y < y < 0.5 + \Delta y\right)}{P\left(0.5 - \Delta y < y < 0.5 + \Delta y\right)}$$

From part e, we have, $P\left(0 < x < \frac{1}{2},\ 0.5 - \Delta y < y < 0.5 + \Delta y\right) = \frac{9}{32}\Delta y$.

$$P\left(0.5 - \Delta y < y < 0.5 + \Delta y\right) = \int_{0.5+\Delta y}^{0.5-\Delta y} f_y\, dy = \frac{1}{2} [y]_{0.5+\Delta y}^{0.5-\Delta y} = \Delta y$$

$$\Rightarrow P\left(0 < x < \frac{1}{2} \middle| y = 0.5\right) = \frac{9}{32}$$

Similar to the univariate case, one can also take expectations of functions of single or multiple variables. The general expectation formula in the two variable case is

$$E(h(x,y)) = \int_{-\infty}^{\infty} \int_{-\infty}^{\infty} h(x,y) f_{x,y}(x,y) dx dy \qquad (5.41)$$

This expectation operator can be used to define many other other quantities of interest.

$$\mu_x = \int_{-\infty}^{\infty} \int_{-\infty}^{\infty} x f_{x,y}(x,y) dx dy; \qquad \mu_y = \int_{-\infty}^{\infty} \int_{-\infty}^{\infty} y f_{x,y}(x,y) dx dy \quad (5.42)$$

$$\sigma_{xx} = \int_{-\infty}^{\infty} \int_{-\infty}^{\infty} (x - \mu_x)^2 f_{x,y}(x,y) dx dy; \qquad (5.43)$$

$$\sigma_{yy} = \int_{-\infty}^{\infty} \int_{-\infty}^{\infty} (y - \mu_y)^2 f_{x,y}(x,y) dx dy \qquad (5.44)$$

$$\sigma_{xy} = \int_{-\infty}^{\infty} \int_{-\infty}^{\infty} (x - \mu_x)(y - \mu_y) f_{x,y}(x,y) dx dy \qquad (5.45)$$

As one can see, similar to the univariate case, an x-mean (μ_x) and a y-mean (μ_y) can be calculated. Variances in the x and y variables can also be calculated [4]. One another interesting quantity is the cross-variance (or covariance) between the x and y variables and this is shown as σ_{xy}. This quantity provides us information about possible linear relationships between random variables. This quantity is also reported in a normalized version as a correlation coefficient ρ_{xy}.

$$\rho_{xy} = \frac{\sigma_{xy}}{\sqrt{\sigma_{xx}\sigma_{yy}}}$$

If there is a perfect linear relationship between the two variables (with a positive slope) then $\rho_{xy} = 1$ and if the slope is negative, $\rho_{xy} = -1$. This is easy to verify from the equation for ρ_{xy}.

When there are multiple random variables in a problem, a natural question of interest is if these are related or independent. Independence of random variables can be defined through either the relationship between their joint and marginal distributions

$$f_{x,y}(x,y) = f_x(x) f_y(y)$$

or through a relationship between the conditional and marginal distributions.

---

[4]Until now, σ^2 was used to represent variance while σ was used to represent standard deviations. However, for multiple related random variables, notations σ_{xx}, σ_{yy}, and σ_{xy} are used to represent variance in x, variance in y, and covariance, respectively. This notation is used as the covariance (σ_{xy}) can also be negative. Similarly, sample variances and covariance for multivariable case are represented using s_{xx}, s_{yy} and s_{xy}.

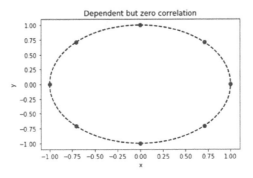

FIGURE 5.18 Example of zero correlation but dependent variables.

$$f_{x|y}(x) = f_x(x); \qquad f_{y|x}(y) = f_y(y)$$

The key notion of independence is that one variable can vary completely independent of the other variable and hence conditional and the marginal distributions are the same, that is conditioning on one variable doesn't have any effect on the other.

Correlation and independence are related but slightly different ideas. If two variables are independent, then the correlation between these two variables will be zero. However, if two variables have a zero correlation, it doesn't mean that they are independent. This is because correlation tries to capture linear relationships, while the notion of independence is much broader. Figure 5.18 shows an example case where the correlation computed from the data will be zero, whereas the variables are clearly related through a quadratic equation.

Example 5.38. *Consider the following dataset for two random variables x and y.*

x	3.1	1.5	2.6	4.2
y	10	12	11	14

Determine

(a) sample variance of x, s_{xx}
(b) sample variance of y, s_{yy}
(c) sample cross-variance between x and y, s_{xy}
(d) sample correlation coefficient

Solution 5.38.

(a) $s_{xx} = \frac{1}{n-1}\sum_{i=1}^{n}(x^i - \bar{x})^2 = \frac{3.77}{3} = 1.26$

(b) $s_{yy} = \frac{1}{n-1}\sum_{i=1}^{n}(y^i - \bar{y})^2 = \frac{8.75}{3} = 2.9167$

(c) $s_{xy} = \frac{1}{n-1}\sum_{i=1}^{n}(x^i - \bar{x})(y^i - \bar{y}) = \frac{2.45}{3} = 0.8167$

(d) Sample correlation coefficient $\rho_{xy} = \frac{s_{xy}}{\sqrt{s_{yy}s_{xx}}} = 0.426$

Example 5.39. *If the joint probability distribution between variables x and y is*

$$\begin{cases} \lambda^2 e^{-\lambda(x+y)} & \forall\ 0 < x < \infty \\ & 0 < y < \infty \\ 0 & otherwise \end{cases}$$

Find whether x and y are independent through

(a) the relation between joint and marginal pdf
(b) the relation between marginal and conditional pdf

Solution 5.39.

$$f_{x,y} = \lambda^2 e^{-\lambda(x+y)}$$

(a)

$$f_x(x) = \int_0^\infty f_{x,y}dy = \lambda^2 e^{-\lambda x}\int_0^\infty e^{-\lambda y}dy = \frac{\lambda^2 e^{-\lambda x}}{-\lambda}\left[e^{-\lambda y}\right]_0^\infty = \lambda e^{-\lambda x}$$

$$f_y(y) = \int_0^\infty \lambda^2 e^{-\lambda(x+y)}dx = \lambda^2 e^{-\lambda y}\frac{\left[e^{-\lambda x}\right]_0^\infty}{-\lambda} = \lambda e^{-\lambda y}$$

$$f_x f_y = \lambda^2 e^{-\lambda(x+y)} = f_{x,y}$$

x and y are independent.

(b)

$$f_{x|y}(x) = \frac{f_{x,y}(x,y)}{f_y(y)} = \frac{\lambda^2 e^{-\lambda(x+y)}}{\lambda e^{-\lambda y}} = \lambda e^{-\lambda x} = f_x(x)$$

$$f_{y|x}(x) = \frac{f_{x,y}(x,y)}{f_x(x)} = \frac{\lambda^2 e^{-\lambda(x+y)}}{\lambda e^{-\lambda x}} = \lambda e^{-\lambda y} = f_y(y)$$

x and y are independent.

Example 5.40. *Consider the joint pdf $f_{x,y} = |x|$ where $x \in \begin{bmatrix} -1 & 1 \end{bmatrix}$ and $y \in \begin{bmatrix} 0 & 1 \end{bmatrix}$. Find σ_{xx}, σ_{yy}, σ_{xy}. Find whether x and y are uncorrelated and independent?*

Solution 5.40.

$$\mu_x = \int_0^1 \int_{-1}^1 x\,|x|\,dx\,dy = \int_0^1 \int_{-1}^0 -x^2 dx\,dy + \int_0^1 \int_0^1 x^2 dx\,dy$$

$$= \int_0^1 -\left[\frac{x^3}{3}\right]_{-1}^0 dy + \int_0^1 \left[\frac{x^3}{3}\right]_0^1 dy = \frac{-1}{3}(-1) + \frac{1}{3} = \frac{2}{3}$$

$$\mu_y = \int_{-1}^1 \int_0^1 y\,|x|\,dy\,dx = \int_{-1}^1 \frac{|x|}{2} dx = \int_{-1}^0 \frac{-x}{2} dx + \int_0^1 \frac{x}{2} dx$$

$$= \left[\frac{-x^2}{4}\right]_{-1}^0 + \left[\frac{x^2}{4}\right]_0^1 = -\left(\frac{-1}{4}\right) + \frac{1}{4} = \frac{1}{2}$$

$$\sigma_{xx} = \int_{-1}^1 \int_0^1 \left(x - \frac{2}{3}\right)^2 |x|\,dy\,dx = \int_{-1}^1 \left(x - \frac{2}{3}\right)^2 |x|\,dx$$

$$= \int_{-1}^0 \left(x - \frac{2}{3}\right)^2 (-x)\,dx + \int_0^1 \left(x - \frac{2}{3}\right)^2 x\,dx$$

$$= \int_{-1}^0 \left(-x^3 + \frac{4}{3}x^2 - \frac{4}{9}x\right) dx + \int_0^1 \left(x^3 - \frac{4}{3}x^2 + \frac{4}{9}x\right) dx$$

$$= \left[\frac{-x^4}{4} + \frac{4}{9}x^3 - \frac{2}{9}x^2\right]_{-1}^0 + \left[\frac{x^4}{4} - \frac{4}{9}x^3 + \frac{2}{9}x^2\right]_0^1$$

$$= -\left(\frac{-1}{4} - \frac{4}{9} - \frac{2}{9}\right) + \frac{1}{4} - \frac{4}{9} + \frac{2}{9} = \frac{1}{2} + \frac{4}{9} = \frac{13}{18}$$

$$\sigma_{yy} = \int_{-1}^1 \int_0^1 \left(y - \frac{1}{2}\right)^2 |x|\,dy\,dx = \int_{-1}^1 \int_0^1 \left(y^2 - y + \frac{1}{4}\right) |x|\,dy\,dx$$

$$= \int_{-1}^1 \left(\frac{1}{3} - \frac{1}{2} + \frac{1}{4}\right) |x|\,dx = \frac{1}{12}\left[\int_{-1}^0 -x\,dx + \int_0^1 x\,dx\right]$$

$$= \frac{1}{12}\left(\frac{[-x^2]_{-1}^0}{2} + \frac{[x^2]_0^1}{2}\right) = \frac{1}{12}\left(\frac{1}{2} + \frac{1}{2}\right) = \frac{1}{12}$$

$$\sigma_{xy} = \int_{-1}^1 \int_0^1 \left(y - \frac{1}{2}\right)\left(x - \frac{2}{3}\right) |x|\,dy\,dx$$

$$= \int_{-1}^1 |x|\left(x - \frac{2}{3}\right)\left(\frac{1}{2} - \frac{1}{2}\right) dx = 0$$

Since $\sigma_{xy} = 0$, x and y are uncorrelated. To find whether x and y are independent,

$$f_x(x) = \int_0^1 |x|\, dy = |x|$$

$$f_y(y) = \int_{-1}^1 |x|\, dx = \int_0^1 x\, dx + \int_{-1}^0 -x\, dx = \frac{1}{2} + \frac{1}{2} = 1$$

Since $f_{x,y} = f_x \times f_y$, we can say that x and y are independent.

Finally, one important joint distribution that is used in ML is the multivariate Gaussian distribution. Similar to the univariate case, multivariate Gaussian is parameterized by a mean and variance, however, in the multivariable case ($x_1 \ldots x_n$ are the variables), these become a vector and matrix respectively. Let $\boldsymbol{\mu}$ be the vector of means and V is called the variance-covariance matrix as define below

$$\boldsymbol{\mu} = \begin{bmatrix} E(x_1) \\ \vdots \\ E(x_n) \end{bmatrix}; \qquad V_{ij} = E((x_i - E(x_i))(x_j - E(x_j))^T)$$

then the multivariate Gaussian is the following equation

$$f(x_1 \ldots x_n) = \frac{1}{2\pi^{n/2}|V|^{1/2}} exp^{(-1/2(\mathbf{x}-\boldsymbol{\mu})^T V^{-1}(\mathbf{x}-\boldsymbol{\mu}))}$$

EXERCISE PROBLEMS

Exercise 5.1. *Find the mean, mode, median and variance of Poisson distribution for which probability mass function is $f(k; \eta) = \Pr(X = k) = \frac{\eta^k e^{-\eta}}{k!}$.*

Exercise 5.2. *Find the mean, mode, median and variance of Pareto distribution for which probability density function is given by*

$$f_x(x) = \begin{cases} \frac{\alpha x_m^\alpha}{x^{\alpha+1}} & x \geq x_m, \\ 0 & x < x_m. \end{cases}$$

Exercise 5.3. *Find the mean, mode, median and variance of logistic distribution for which probability density function is given by $\frac{e^{-(x-\mu)/s}}{s\left(1+e^{-(x-\mu)/s}\right)^2}$*

Exercise 5.4. *For the data given below, fit a Poisson distribution using method of moments*
$x = \begin{bmatrix} 6 & 3 & 2 & 3 & 2 & 5 & 2 & 6 & 2 & 3 \end{bmatrix}$

Exercise 5.5. *For the data given in the question above, fit a poisson distribution using MLE*

Exercise 5.6. *For the data given below, fit a Logistic distribution using method of moments*
$$x = \begin{bmatrix} -0.284 & 5.702 & 1.411 & 3.229 & -0.234 & 3.828 & 0.939 & 4.274 & 4.593 \end{bmatrix}$$

Exercise 5.7. *For the data given in the question above, fit a Logistic distribution using MLE*

Exercise 5.8. *For the data given below, fit a suitable continuous probability distribution.*
$$x = \begin{bmatrix} 2.603 & -1.783 & 0.5718 & 7.7102 & -0.4321 & 6.1125 & 3.3073 & 14.1037 \end{bmatrix}$$

Exercise 5.9. *A controller is set up to control the temperature inside a stirred-tank reactor. The desired temperature is 25^0C . The actual temperature was then measured multiple times to cross-check the performance of the controller. Following measurements were obtained from various trails.*
$26.4^0C \quad 26.13^0C \quad 24.96^0C \quad 23.55^0C \quad 22.73^0C \quad 25.59^0C$
With the help of hypothesis testing, comment whether the controller performance is good. Assume that the true variance is $\sigma^2 = 1.4$

Exercise 5.10. *For the problem discussed above, will the conclusion change if true σ was unknown?*

Exercise 5.11. *Consider the following samples of a random variable x.*
$$x = \begin{bmatrix} 1.15 & -1.01 & 3.24 & -2.92 & 0.3 & 0.02 & 1.14 & 1.13 & -1.23 & 0.44 & 0.17 & 1.76 \end{bmatrix}$$

Let us say we hypothesized that x follows the log-normal distribution with mean 0.5 and standard deviation 2. Comment on the goodness of the proposed model using q-q and p-p plots.

Exercise 5.12. *For an industrial use, anthracite coal is required (at least 86% carbon content). There were three batches of coal and the carbon content for each batch of coal was measured 10 times and the average content was found to be 85.7, 86.7 and 84.8 in each beach.*

(a) *If the sample variance of the batches were 1.2, 0.9 and 1.32 respectively, comment whether these coal batches can be used for the required application.*
(b) *If the variance in the measurement is known to be 0.75, will the conclusions change?*
(c) *Find the 95% confidence interval and use it to comment whether these coal batches can be used for the required application.*

Exercise 5.13. *The level of chemical mixture in a stirred tank reactor has to be maintained below 90% of the tank height. A controller is installed to maintain the level at desired conditions. The reactor level was measured 20 times and it was found that the mean level was 2.65m while the reactor height was 3m. The variance of the level sensor is 0.63. Find whether the reactor is operating at the desired conditions based on both hypothesis testing and confidence interval? If the actual level is 2.65 m, calculate Type II error.*

Exercise 5.14. *A pressure sensitive reaction is being carried out in a reactor and the maximum allowed variations in pressure is $0.07\,Pa$. Pressure inside the reactor was measured 8 times and the sample variance was found to be 0.09. Comment on the true variance of the process using both confidence intervals and by performing hypothesis testing.*

Exercise 5.15. *There are two machines for spray coating at desired thickness. Their variability should be similar if they need to be used for the same application. To test this, various samples were collected from both machines and the summary is given below*

$$s_1^2 = 1.71 \qquad s_2^2 = 0.95$$
$$n_1 = 25 \qquad n_2 = 15$$

Comment on whether these machines can be used for the same application if the machines were truly built to have the same variance.

Exercise 5.16. *Let x and y be two random variables having the following joint distribution*

$$f_{x,y}(x,y) = \begin{cases} cx^2y & if\ 0 < x < 1\ and\ 2 < y < 4 \\ 0 & otherwise \end{cases}$$

(a) Find the constant 'c'.
(b) Find the marginal distribution $f_x(x)$ and $f_y(y)$.
(c) Find whether x & y are independent through the relation between joint and marginal pdf.
(d) Find whether x & y are independent through the relation between marginal and conditional pdf.
(e) Find whether x & y are correlated.

Exercise 5.17. *Consider the continuous random variables x and y with the joint pdf*

$$f_{x,y}(x,y) = \left(\frac{1+xy^2}{c}\right) \forall \begin{array}{l} -1 < x < 1 \\ -1 < y < 1 \end{array}$$

(a) Find the constant 'c'
(b) Find $f_y(y)$ and $f_x(x)$
(c) Find $P\left(0 < x < \frac{1}{2}, 0 < y < \frac{1}{4}\right)$
(d) Find $P\left(0 < x < \frac{1}{2} | 0 < y < \frac{1}{4}\right)$
(e) Find $P\left(0 < x < \frac{1}{2}, 0.5 - \Delta y < y < 0.5 + \Delta y\right)$. Also comment on the probability of $\Delta y = 0$.
(f) Find $P\left(0 < x < \frac{1}{2} | y = 0.5\right)$

Exercise 5.18.
An algorithm was setup in a sensor manufacturing company to detect faulty sensors. The sensor will be rejected if the test is positive and the sensor will be accepted if the test is negative. 93% of the non-defective sensors were accepted and 4% of the defective sensors also were accepted. In addition, 94% of the sensors are non-defective and 6% of the sensors are defective.

1. *What is the probability that a sensor is accepted?*
2. *If a sensor is accepted, what is the probability that it is non-defective?*
3. *If a sensor is rejected, what is the probability that it is non-defective?*

Exercise 5.19. *Consider a random sample $X_1, X_2, X_3, ...X_n$ with a distribution $\mathcal{N}(\mu, \sigma^2)$. Find the Cramer Rao lower bound for the following*

1. *μ when σ^2 is known*
2. *μ when σ^2 is unknown*
3. *σ^2 when μ is known*
4. *σ^2 when μ is unknown*

6 Function Approximation Methods

In function approximation methods, the goal is to predict the output variables (dependent) as a function of input variables (independent). Both the input and output variables can be vectors. Models can be linear or nonlinear.

6.1 SETTING UP THE PROBLEM

It would be worthwhile to describe the problem setup for all the function approximation methods right at the beginning of this chapter. Mathematically, given a data matrix X (of size $m \times n$) with m samples and n variables, function approximation methods derive models to predict $\mathbf{y}$ for each of the datapoints. The most important aspect of a model is its predictive capability. If all the datapoints are used in building (training) the model, then one will be left with no data to test the model. As a result, the generalization capability of the model will become hard to establish.

This problem is addressed by splitting the data into training and test sets. That is, the m samples are split into two sets of m_1 and m_2 samples: X^{train} ($m_1 \times n$) and X^{test} ($m_2 \times n$) . The model is built using the dataset X^{train} and tested on the dataset X^{test}. There are several approaches to splitting the data (80-20 split, randomization, and so on), and the metrics for evaluation of the test data will also differ. These ideas will be described as a part of the example problems.

The same idea is relevant also for classification problems where the generalization capability of the classifier has to be tested. This is done on the test data using metrics that are relevant for classifiers, such as the confusion matrix and other performance measures (recall, precision, accuracy, and so on). These will be described through examples in the classification chapter.

6.2 PARAMETRIC METHODS

In parametric methods, the model $f(\mathbf{x}, \mathbf{p})$ is characterized by both the form of the function and also the parameters ($\mathbf{p}$). Once the model parameters are derived from data, then these method do not need training data to make predictions. Various parametric function approximation methods will be described in this section. The most important of these is the linear regression method.

DOI: 10.1201/b23276-6

6.2.1 Linear Regression

Before simple linear regression is described, different correlation metrics between pairs of variables will be introduced.

6.2.1.1 Quantities that Indicate Relationships between Variables

The goal of function approximation methods is to identify if there are relationships between variables. While identifying relationships, it might also be interesting to understand the relationship between a binary pair of variables. This might provide a more qualitative understanding of the strength of the relationship between variables. In Chapter 5, we have already discussed correlation as an approach to understanding if variables vary together. We will explore these techniques in a little more detail here. Two important observations to make here are: a zero value for any of these correlation numbers described here does not imply independence of variables, and correlation doesn't imply causation. The first observation stems from the fact that the correlation definitions attempt to capture if there is a linear relationship (Pearson), or if the variables vary together monotonically (Spearman), or if there is an ordinal association between the variables (Kendall). Independence is a broader statistical definition. In some cases, independence may be inferred from correlation, but not always. The second observation stems from the fact that these binary measures are symmetric and hence, while a relationship may be inferred, causation cannot be.

The first correlation that is described is the Pearson correlation [9]. Consider two scalar variables, x and y, with n samples. Pearson correlation ρ^P is calculated as follows:

$$\rho^P = \frac{s_{xy}}{\sqrt{s_{xx}}\sqrt{s_{yy}}} = \frac{\sum x^i y^i - n\bar{x}\bar{y}}{\sqrt{\sum x^{i2} - n\bar{x}^2}\sqrt{\sum y^{i2} - n\bar{y}^2}} \tag{6.1}$$

The second quantity is the Spearman correlation [40, 10]. In this case, the variables x and y are sorted in ascending order independently. A rank from 1 to n is assigned to x and y independently - let the rank variables be r_x and r_y. The lowest x value will get a rank 1 and so on. If a value is repeated multiple times, then an average position rank is given. For example, after sorting, if a value repeats in the third, fourth, and fifth positions, then all these values in x will be converted to a value of 4 (average of 3,4,5) in r_x. The x value in the sixth position will be converted to a value of 6 in r_x and this will continue. A similar rank conversion is performed for y. Now Spearman correlation is

$$\rho^S = \frac{s_{r_x r_y}}{\sqrt{s_{r_x r_x}}\sqrt{s_{r_y r_y}}} \tag{6.2}$$

The last correlation that is described is the Kendall correlation [21]. Given n datapoints, $^nC_2$ binary pairs are chosen. Each of the binary pairs is labeled as a concordant or discordant pair. A pair is concordant when either $(x^i > x^j$ and $y^i > y^j)$ or $(x^i < x^j$ and $y^i < y^j)$ holds. Otherwise, it is a discordant pair. Data with repeats in x and y can be ignored for simplicity. Let n_c be the number of concordant pairs and n_d the number of discordant pairs. Then,

$$\rho^K = \frac{n_c - n_d}{{}^nC_2} \tag{6.3}$$

Example 6.1. *For the following data find the Pearson correlation for the pairs (x, y_1) and (x, y_2) and comment on the relationships between variables in each pair.*

x	1	3	10	2	9	6	10	1	5	2
y_1	3	9	30	6	27	18	30	3	15	6
y_2	4	12	103	7	84	39	103	4	28	7

Solution 6.1. *Looking at the data, it is clear that $y_1 = 3x$ and $y_2 = x^2 + 3$. As a result, the pair (x, y_1) has a linear relationship, whereas (x, y_2) has a nonlinear relationship.*

$$\rho^P_{xy_1} = \frac{s_{xy_1}}{\sqrt{s_{y_1y_1}}\sqrt{s_{xx}}}; \qquad \rho^P_{xy_2} = \frac{s_{xy_2}}{\sqrt{s_{xx}}\sqrt{s_{y_2y_2}}}$$

$$\bar{x} = 4.9 \qquad \bar{y}_1 = 14.7 \qquad \bar{y}_2 = 39.1$$

$$s_{xy_1} = \frac{1}{n-1}\sum \left(x^i - \bar{x}\right)\left(y^i_1 - \bar{y}_1\right) = \frac{1}{9} \times 362.7 = 40.3$$

$$s_{y_1y_1} = 120.9; \quad S_{x,x} = 13.433; \quad s_{y_2y_2} = 1729.4; \quad s_{xy_2} = 149.57$$

$$\Rightarrow \rho^P_{xy_1} = \frac{40.3}{\sqrt{13.433} \times \sqrt{120.9}} = 1; \quad \rho^P_{xy_2} = \frac{149.57}{\sqrt{13.433 \times 1729.4}} = 0.9813$$

Note that we get $\rho = 1$ for the case where variables are linearly related, and a value less than 1 when the relationship is nonlinear.

Example 6.2. *For the data below, find the Pearson correlation.*

x	1	3	10	2	9	6	10	1	5	2
y	10	1	8	9	9	1	4	3	9	5

Solution 6.2. *Unlike Example 6.1, in this case data has been randomly generated.*

$$\bar{x} = 4.9 \qquad \bar{y} = 5.9$$

$$s_{xy} = 0.7667; \quad s_{xx} = 13.433; \quad s_{yy} = 12.32$$

$$\Rightarrow \rho^P = \frac{0.7667}{\sqrt{12.32 \times 13.433}} = 0.0596 \sim 0$$

Note that we get a value close to 0 when x and y are randomly sampled.

Example 6.3. *For the following data, find the Spearman correlation and comment on the relationship between variables x and y.*

x	2	3	-1	2	2	-1	-2	1	4	0
y	6	9	-3	6	6	-3	-6	3	12	0

Solution 6.3. *From the data, it is evident that $y = 3x$ which is a monotonic relation.*

x	-2	-1	-1	0	1	2	2	2	3	4
r_x	1	2.5		4	5	7			9	10

y	-6	-3	-3	0	3	6	6	6	9	12
r_y	1	2.5		4	5	7			9	10

So the data matrix of r_x and r_y is

x	2	3	-1	2	2	-1	-2	1	4	0
r_x	7	9	2.5	7	7	2.5	1	5	10	4
y	6	9	-3	6	6	-3	-6	3	12	0
r_y	7	9	2.5	7	7	2.5	1	5	10	4

$$s_{r_x r_x} = \frac{1}{n-1} \sum \left(r_x^i - \bar{r}_x\right)^2 = 8.89; \qquad s_{r_y r_y} = \frac{1}{n-1} \sum \left(r_y^i - \bar{r}_y\right)^2 = 8.89$$

$$s_{r_x r_y} = \frac{1}{n-1} \sum \left(r_x^i - \bar{r}_x\right)\left(r_y^i - \bar{r}_y\right) = 8.89$$

$$\rho^S = \frac{8.89}{\sqrt{8.89 \times 8.89}} = 1 \Rightarrow \text{Monotonic relation}$$

Example 6.4. *For the data given below, find the Spearman correlation and comment on the relation between x and y.*

x	2	3	-1	2	2	-1	-2	1	4	0
y	4	9	1	4	4	1	4	1	16	0

Solution 6.4.

r_x	7	9	2.5	7	7	2.5	1	5	10	4
r_y	6.5	9	3	6.5	6.5	3	6.5	3	10	1

$$s_{r_x r_y} = 6.1667; \qquad s_{r_x r_x} = 8.8889; \qquad s_{r_y r_y} = 8.3889$$

$$\rho^S = \frac{6.1667}{\sqrt{8.8889 \times 8.3889}} = 0.7141$$

As $\rho^S \ll 1$, it shows that x and y have a non-monotonic relation. If we look at the data, we can see that $y = x^2$, which is a non-monotonic relation.

Example 6.5. *If x and y are the ranks of six candidates in an interview as marked by two interviewers. Find the Kendall correlation between x and y.*

x	5	2	1	3	4	6
y	6	1	2	3	4	5

TABLE 6.1

Calculations for Example 6.5

x_i	x_j	y_i	y_j	Concordant (c) / Discordant (d)
5	2	6	1	c
5	1	6	2	c
5	3	6	3	c
5	4	6	4	c
5	6	6	5	d
2	1	1	2	d
2	3	1	3	c
2	4	1	4	c
2	6	1	5	c
1	3	2	3	c
1	4	2	4	c
1	6	2	5	c
3	4	3	4	c
3	6	3	5	c
4	6	4	5	c

Solution 6.5. *Concordant and discordant pairs are calculated and reported in Table 6.1. Note that there are 13 concordant and 2 discordant pairs.*

$$\rho^K = \frac{13 - 2}{15} = 0.73$$

Example 6.6. *In Example 6.5, if there was a third interviewer (z) who ranked the candidates the same way as that of interviewer (x), then final the Kendall correlation between their ranks, x and z.*

Solution 6.6. *Since x and z are exactly the same, for any pair, if $x_i > x_j$, then $z_i > z_j$, Thus all pairs are concordant.*
Hence,

$$\rho^K = \frac{15 - 0}{15} = 1$$

6.2.1.2   Univariate Linear Regression

In univariate linear regression [35, 1], a model between a dependent scalar variable y and an independent scalar variable x is identified. The form of the

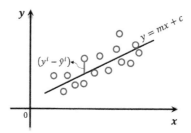

FIGURE 6.1 Univariate linear regression.

model is $y = \beta_1 x + \beta_0$, with the assumption that the measurements of x (x^i, i^{th} sample) are error free and the measurements of y (y^i) have an additive error. The equation below consolidate these ideas, where e^i is a realization of random error (e) at every sampling instant

$$y = \beta_1 x + \beta_0; \qquad y^i = \beta_1 x^i + \beta_0 + e^i$$

$$\min_{\beta_0, \beta_1} E = \sum_{i=1}^{n} e^{i^2} = \sum_{i=1}^{n} (y^i - \beta_1 x^i - \beta_0)^2 \qquad (6.4)$$

Figure 6.1 depicts the problem setup pictorially. The errors are distances from the sample points to the regression line. Clearly, one would like to collectively minimize this error. Minimizing the sum of all errors is not appropriate as this will only bias errors to be negative and not necessarily zero. However, minimizing the sum of squared errors is suitable as all squared quantities are positive (refer equation 6.4). Since sum of squares of errors are minimized, this is also called the total least squares approach. If the errors are assumed to be realized from a Gaussian distribution, then the total least squares objective can be derived from a maximum likelihood formulation. Extending this idea, any p-norm can be used as an objective function. If one were to use 1-norm, then the objective function will become the sum of absolute values of the errors.

Notice that y^i and x^i are sample values and the objective function is written only in terms of the unknowns β_0 and β_1. Equation 6.4 is an unconstrained nonlinear programming problem and the solution to this optimization problem is obtained by $\frac{\partial E}{\partial \beta_0} = 0$ and $\frac{\partial E}{\partial \beta_1} = 0$. Using these equations and after some algebraic manipulations, we will find that

$$\hat{\beta}_1 = \frac{\sum_{i=1}^{n} (x^i - \bar{x})(y^i - \bar{y})}{\sum_{i=1}^{n} (x^i - \bar{x})^2} \qquad (6.5)$$

$$\hat{\beta}_0 = \bar{y} - \hat{\beta}_1 \bar{x} \qquad (6.6)$$

Hats for $\hat{\beta}_0$ and $\hat{\beta}_1$ denote that these are estimates for the true β_0 and β_1 respectively. The material from Chapter 5 can now be used in analyzing the results of this linear regression solution. Since we have estimates for β_0 and β_1, noise realizations $\hat{e}^i \ \forall \ i$ can also be computed. If the errors are hypothesized to be from a Gaussian distribution and samples $\hat{e}^i$ are available, one can verify if these are from the hypothesized distribution using a q-q plot. A quantitative approach to verifying if the univariate linear model is a good approximation is through equation 6.7 (for R^2) and equation 6.8 (for adjusted R^2, R^2_{adj}). p is the number of predictor variables in the model. It is quite easily seen that R^2 value is bounded between 0 and 1. One would get a value of zero when $\hat{y}^i = \bar{y} \ \forall \ i$ (under-parameterized, only one parameter β_0 in the model) and $R^2 = 1$ when $\hat{y}^i = y^i \ \forall \ i$ (over-parameterized with n parameters). Obviously larger R^2 values without over-parameterization are better. R^2_{adj} value penalizes the metric when more parameters are used and, in some cases, might provide a better apple-to-apple comparisons between models. The R^2_{adj} is always less than R^2 as seen from equation 6.9, which is a rearrangement of equations 6.7 and 6.8. As a result, R^2_{adj} can take negative values.

$$R^2 = 1 - \frac{\sum(y^i - \hat{y}^i)^2}{\sum(y^i - \bar{y})^2} \tag{6.7}$$

$$R^2_{adj} = 1 - \frac{\sum(y^i - \hat{y}^i)^2/(n-p-1)}{\sum(y^i - \bar{y})^2/(n-1)}, \ n > p+1 \tag{6.8}$$

$$R^2_{adj} = R^2 - (1 - R^2)\frac{p}{n-p-1} \tag{6.9}$$

From equations 6.5 and 6.6, the estimates are functions of random variables (e^i through y^i) and hence they themselves are random variables. As a result, $\hat{\beta}_1$ and $\hat{\beta}_0$ values are interpreted as realizations from the distributions for these estimates. These are sampling distributions and can be derived based on the assumptions made about the true e^i distributions. Using these sampling distributions, hypothesis tests and confidence intervals can also be established for the estimated parameters. The examples that follow illustrate these ideas for univariate linear regression.

Example 6.7. *Consider the following data where x is the number of units sold and y is the profit.*

x	1	2	3	4	5	6	7	8	9	10
y	4.04	6.56	10.67	13.73	16.14	19.14	21.94	25.67	27.98	30.64

If the true relation between x and y is linear, $y = \beta_0 + \beta_1 x$, find the parameters β_0 and β_1, that best fits the data. Comment on the validity of assumptions made on the noise/error using q-q plot.

Solution 6.7.

$$\hat{\beta}_1 = \frac{\sum_{i=1}^n (x^i - \bar{x})(y^i - \bar{y})}{\sum_{i=1}^n (x^i - \bar{x})^2}; \qquad \hat{\beta}_0 = \bar{y} - \hat{\beta}_1 \bar{x}$$

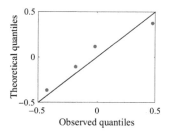

FIGURE 6.2 q-q plot for Example 6.7.

From data, $\bar{x} = 5.5,\ \bar{y} = 17.65,\ \sum_{i=1}^{n}\left(x^i - \bar{x}\right)^2 = 82.5$

$$\hat{\beta}_1 = \frac{245.985}{82.5} = 2.9816; \quad \hat{\beta}_0 = 17.65 - (2.9816 \times 5.5) = 1.25$$

$$\Rightarrow \hat{y} = 1.25 + 2.9816x; \quad Error,\ \hat{e} = y - 1.25 - 2.98x$$

The samples for $e^i (= y^i - \hat{y}^i)$ *are,*

$$\begin{bmatrix} -0.1916 & -0.6532 & 0.4752 & 0.5536 & -0.018 & 0.0004 \\ & -0.1812 & 0.5672 & -0.1044 & -0.426 & \end{bmatrix}$$

Sorted rank: $r_e = \begin{bmatrix} 3 & 1 & 8 & 9 & 6 & 7 & 4 & 10 & 5 & 2 \end{bmatrix}$. *Error samples after sorting will be as follows:*

$$\begin{bmatrix} -0.6532 & -0.426 & -0.1916 & -0.1812 & -0.1044 & -0.018 \\ & 0.0004 & 0.4752 & 0.5536 & 0.5672 & \end{bmatrix}$$

Fitting a Gaussian distribution to the error samples

$$\mu_e = \bar{e} = 0.0022; \quad \sigma_e^2 = \frac{1}{8}\sum_{i=1}^{10}\left(e^i - \mu_e\right)^2 = 0.19285; \quad \sigma_e = 0.439$$

For the distribution $\mathcal{N}(0.0022, 0.439)$, *5-quantile positions are*

$$q^i = F^{-1}\left(\frac{i}{5}\right) \Rightarrow q^1 = -0.367; \quad q^2 = -0.109; \quad q^3 = 0.114; \quad q^4 = 0.372$$

From the data, $e^1 = -0.426,\ e^2 = -0.1812,\ e^3 = -0.018,$ *and* $e^4 = 0.4752$ *(sample 2, 4, 6 and 8 in the sorted list). The corresponding q-q plot is shown in Figure 6.2. Note that the samples are close to 45° line and thus, the assumption of Gaussian noise is reasonable.*

Example 6.8. *Calculate* R^2 *and* R^2_{adj} *for Example 6.7.*

Solution 6.8.

$$R^2 = 1 - \frac{\sum\left(y^i - \hat{y}^i\right)^2}{\sum\left(y^i - \bar{y}\right)^2} = 0.9979$$

$$R^2_{adj} = R^2 - \frac{\left(1 - R^2\right)p}{n - p - 1} = 0.9979 - \frac{0.0021 \times 2}{10 - 1 - 1} = 0.9974$$

Example 6.9. *If the following test data set was given for Example 6.7, comment on the performance of the linear model developed in Example 6.7 in predicting the output for the test set.*

x	11.72	8.83	5.97	6.05	6.8
y	36.83	27.98	19.46	20.14	22.22

Solution 6.9.

$$\hat{y} = 2.9816x + 1.25$$

x	11.72	8.83	5.97	6.05	6.8
y	36.83	27.98	19.46	20.14	22.22
$\hat{y}$	36.19	27.58	19.05	19.29	21.52

$$\bar{y} = 25.326$$

$$R^2 = 1 - \frac{\sum \left(y^i - \hat{y}^i\right)^2}{\sum \left(y^i - \bar{y}\right)^2} = 1 - \frac{1.9419}{210.3375} = 0.9908$$

Since R^2 is close to 1, the estimated linear model works well for the given test data. (Note: R^2_{adj} is mainly used with training data as '$n - p - 1$' represents the degrees of freedom of $\sum \left(y^i - \hat{y}^i\right)^2$ for the training set).

Example 6.10. *If the test set is as shown below, calculate R^2 and comment on the performance of the linear model estimated in Example 6.7.*

x	25	26	27	28	29	30
y	101.75	105.75	109.75	113.75	117.75	121.75

Solution 6.10. *Using $\hat{y} = 2.9816x + 1.25$, we have*

$$\hat{y} = \begin{bmatrix} 75.79 & 78.77 & 81.75 & 84.73 & 87.72 & 90.70 \end{bmatrix}$$

$$R^2 = 1 - \frac{\sum \left(y^i - \hat{y}^i\right)^2}{\sum \left(y^i - \bar{y}\right)^2} = -16.48$$

Note that a negative value is obtained. This is because the ratio $\frac{\sum \left(y^i - \hat{y}^i\right)^2}{\sum \left(y^i - \bar{y}\right)^2}$ is greater than 1 indicating that the sum squared error (SSE) of the model prediction is larger than the SSE if the model were the average of the given data. Thus, the performance of the model is very poor in predicting the given test set. Notice that R^2 can be negative if computed on test data.

Example 6.11. *Consider the following data*

x	1	2	3	4	5	6	7	8	9	10
y	1.65	2.72	4.48	7.39	12.18	20.09	33.12	54.6	90.02	148.41

Fit polynomial models of order varying from 1 to 6 and calculate R^2 and R^2_{adj} for each model. Comment on the best model choice based on R^2 and R^2_{adj} values.

Solution 6.11. *Model equation for a polynomial model of order k is given by*

$$y = \beta_k x^k + \beta_{k-1}x^{k-1} + \dots + \beta_1 x + \beta_0 = \begin{bmatrix} x^k & x^{k-1} & \dots & x & 1 \end{bmatrix} \begin{bmatrix} \beta_k \\ \vdots \\ \beta_0 \end{bmatrix}$$

For multiple samples, one can write the above equation as $\mathbf{y} = X\boldsymbol{\beta}$. Including measurement noise, $\mathbf{y_{meas}} = X\boldsymbol{\beta} + \mathbf{e}$. Polynomial regression is essentially a multivariate regression (refer Section 6.2.1.3) with various powers of x as the variables/features used for regression.

$$\hat{\boldsymbol{\beta}} = \left(X^T X\right)^{-1} X^T \mathbf{y}; \qquad \hat{\mathbf{y}} = X\hat{\boldsymbol{\beta}}$$

Using $k = 1$ to 9, we get polynomial models of different orders and $\hat{\mathbf{y}}$ can be obtained for each model. Model parameters are given in Table 6.2.

TABLE 6.2

Model Parameters for Example 6.11

Model Order	Model Parameters
1	$[13.74, -38.12]$
2	$[2.94, -18.6, 26.56]$
3	$[0.45, -4.42, 15.33, -11.69]$
4	$[0.052, -0.704, 4.001, -7.673, 6.249]$
5	$[0.005, -0.085, 0.68, -2.18, 4.10, -0.895]$
6	$[0.0004, -0.008, 0.084, -0.38, 1.167, -0.803, 1.59]$

Once $\hat{\mathbf{y}}$ for each model is calculated, R^2 and R^2_{adj} can be computed. The values obtained are tabulated in Table 6.3. Note that both R^2 and R^2_{adj} are

---

TABLE 6.3

R^2 and R^2_{adj} for Example 6.11

Model Orders	1	2	3	4	5	6
R^2	0.75	0.97	0.998	0.999	0.999	0.999
R^2_{adj}	0.69	0.956	0.996	0.999	0.999	0.999

monotonically increasing with model order for this example. Also, R_{adj}^2 is either less than or equal to R^2. Although R^2 and R_{adj}^2 values are highest for the higher orders (4 to 6), we can see that there is no significant improvement in R^2 and R_{adj}^2 after model order 3 and only a very small improvement from quadratic to cubic model. Hence, we could say that the best model order is either 2 or 3.

Example 6.12. *For the problem discussed in Example 6.7, if the manager expects that the slope β_1 is 3 using prior knowledge. Comment whether the data can be thought as coming from the expected model given that the error variance is 1.2.*

Solution 6.12.

$$H_0 : \beta_1 = 3 \ and \ H_a : \beta_1 \neq 3$$

From data, $\beta_1 = 2.9816$, $\bar{x} = 5.5$, $s_{xx} = 82.5$ and $\sigma_e^2 = 1.2$.

$$\Rightarrow \sigma_{\hat{\beta}_1}^2 = \frac{\sigma_e^2}{s_{xx}}$$

$$Z = \frac{\hat{\beta}_1 - \beta_1}{\sqrt{\frac{\sigma^2}{s_{xx}}}} \sim N(0,1); \quad Z = \frac{2.9816 - 3}{\sqrt{\frac{1.2}{82.5}}} = -0.1526$$

For $\alpha = 0.05$, $z_{\frac{\alpha}{2}} = -1.96$ and $z_{1-\frac{\alpha}{2}} = 1.96$. Since $-1.96 < -0.1526 < 1.96$, we fail to reject the null hypothesis. Hence the data can be considered to come from the expected model.

Univariate regression:

$$E\left(\hat{\beta}_0\right) = \beta_0, \ E\left(\hat{\beta}_1\right) = \beta_1$$

If σ_e^2 is error variance,

$$\sigma_{\hat{\beta}_1}^2 = \frac{\sigma_e^2}{s_{xx}}; \quad \sigma_{\hat{\beta}_0}^2 = \sigma_e^2 \frac{\sum \left(x^i\right)^2}{n \, s_{xx}}$$

$$s_{xx} = \sum_{i=1}^{n} \left(x^i - \bar{x}\right)^2; \quad s_{xy} = \sum_{i=1}^{n} \left(x^i - \bar{x}\right)\left(y^i - \bar{y}\right)$$

$$\text{Estimate of } \sigma_e^2 : \hat{\sigma}_e^2 = \sum \frac{\left(y^i - \hat{y}^i\right)^2}{n-2} = \frac{SSE}{n-2}$$

$$\text{Distribution of slope estimate} : \hat{\beta}_1 \sim N\left(\beta_1, \frac{\sigma_e^2}{s_{xx}}\right)$$

Example 6.13. *For the problem discussed in Example 6.12, if the error variance was unknown, will the conclusion change?*

Solution 6.13. *If σ_e^2 is unknown, we have to use $\hat{\sigma}_e^2$.*

$$\hat{\sigma}_e^2 = \frac{\sum (y^i - \hat{y}^i)^2}{n-2} = \frac{1.5429}{8} = 0.1929; \quad T = \frac{2.9816 - 3}{\sqrt{\frac{0.1929}{82.5}}} = -0.3805$$

Degrees of freedom is $n-2$. For $\alpha = 0.05$, $t_{\frac{\alpha}{2},8} = -2.306$ and $t_{1-\frac{\alpha}{2},8} = 2.306$. Since $t_{\frac{\alpha}{2},8} < -0.3805 < t_{1-\frac{\alpha}{2},8}$, we fail to reject null hypothesis. The data belong to the expected model.

Example 6.14. *For the problem discussed in Example 6.7, comment whether the slope term is significant or not using hypothesis testing.*

Solution 6.14. *We need to check whether $\beta_1 = 0$ or $\beta_1 \neq 0$. We could use two approaches: t-test and F-test.*
(a) *Using t-test:*

$$H_0: \quad \beta_1 = 0 \quad \text{(Slope is not significant)}$$
$$H_a: \quad \beta_1 \neq 0 \quad \text{(Slope is significant)}$$

From Example 6.13, we have $\hat{\sigma}^2 = 0.1929$ and $s_{xx} = 82.5$.

$$T = \frac{\hat{\beta}_1 - 0}{\sqrt{\frac{\hat{\sigma}^2}{s_{xx}}}} = \frac{2.9816 \times \sqrt{82.5}}{\sqrt{0.1929}} = 61.661 \qquad \left(\because \hat{\sigma}_\beta^2 = \frac{\hat{\sigma}^2}{s_{xx}} \right)$$

For $\alpha = 0.05$, $t_{\frac{\alpha}{2},8} = -2.306$ and $t_{1-\frac{\alpha}{2},8} = 2.306$. Since $61.661 > t_{1-\frac{\alpha}{2},8}$, we reject the null hypothesis. Thus, we conclude that the slope is significant.
(b) *Using F-test:*
If the slope term is insignificant, then the model is $\hat{y}_{RM} = \beta_0$ (reduced model, RM). If we find β_0 that best fits the data, we will see that $\beta_0 = \bar{y}$ (can be easily shown using MLE or optimization). Hence, $\hat{y}_{RM} = \bar{y}$ is the reduced model predictions corresponding to $\beta_1 = 0$. Predictions corresponding to $\beta_1 \neq 0$ (full model, FM) is,

$$\hat{y}_{FM} = \beta_0 + \beta_1 x = 1.25 + 2.9816x$$

Error for each sample corresponding to reduced model is $y^i - \bar{y}$ while that for the full model is $y^i - 1.25 - 2.9816x^i$.

$$SSE \text{ of reduced model} = SSE_{RM} = \sum_{i=1}^{10} (y^i - \bar{y})^2 = 734.9807$$

$$SSE \text{ of full model} = SSE_{FM} = \sum_{i=1}^{10} (y^i - 1.25 - 2.9816x^i)^2 = 1.5429$$

The slope term is significant only if SSE_{FM} is significantly less than SSE_{RM}. To test this, we could use the following hypothesis test.

$$H_0 : \sigma^2_{e,FM} = \sigma^2_{e,RM}; \qquad H_a : \sigma^2_{e,FM} < \sigma^2_{e,RM}$$

$$\sigma^2_{e,RM} = \frac{SSE_{RM}}{n-1}; \qquad \sigma^2_{e,FM} = \frac{SSE_{FM}}{n-2}$$

$$f = \frac{SSE_{FM}/(n-2)}{SSE_{RM}/(n-1)} = \frac{1.5429}{734.9807} \times \frac{9}{8} = 0.0024$$

Note that degrees of freedom are $(n-2, n-1)$. Now if we choose a significance level of $\alpha = 0.05$, $F_{0.05,8,9} = 0.2951$. Since $0.0024 \ll F_{0.05,8,9}$, we reject null hypothesis. Thus, $SSE_{FM} < SSE_{RM}$ and shows that the slope term is significant.

Another approach to perform a F-test is to check the difference between SSE_{RM} and SSE_{FM}. We want $SSE_{RM} - SSE_{FM}$ to be large.

$$f = \frac{\frac{SSE_{RM}-SSE_{FM}}{n-1-n+2}}{\frac{SSE_{FM}}{n-2}} = \frac{SSE_{RM} - SSE_{FM}}{SSE_{FM}} \times n - 2$$

$$= \frac{(734.9807 - 1.5429)}{1.5429} \times 8 = 3802.9$$

Since $f \gg F_{1-\alpha,1,8}$, we reject the null hypothesis. Thus, the slope term is significant. We can also look at p-value, the smallest value of α that would have resulted in rejection of null hypothesis.

$$p\text{-value} \,(\alpha \text{ such that } F_{1-\alpha,1,8} = 3802.9) = 5.3146 \times 10^{-12}$$

Example 6.15. *Find the confidence interval for the slope in Example 6.12.*

Solution 6.15.

$$-1.96 < \frac{\hat{\beta}_1 - \beta_1}{\sqrt{\frac{\sigma^2}{s_{xx}}}} < 1.96 \; \text{(for } \alpha = 0.05)$$

$$-1.96 \times \sqrt{\frac{\sigma^2}{s_{xx}}} < \hat{\beta}_1 - \beta_1 < 1.96\sqrt{\frac{\sigma^2}{s_{xx}}}$$

$$\Rightarrow \beta_1 < \hat{\beta}_1 + 1.96\sqrt{\frac{\sigma^2}{s_{xx}}} \; \text{and } \beta_1 > \hat{\beta}_1 - 1.96\sqrt{\frac{\sigma^2}{s_{xx}}}$$

$$\Rightarrow \hat{\beta}_1 - 1.96\sqrt{\frac{\sigma^2}{s_{xx}}} < \beta_1 < \hat{\beta}_1 + 1.96\sqrt{\frac{\sigma^2}{s_{xx}}}$$

$$\Rightarrow 2.9816 - 1.96\sqrt{\frac{1.2}{82.5}} < \beta_1 < 2.9816 + 1.96\sqrt{\frac{1.2}{82.5}}$$

$$\Rightarrow 2.7452 < \beta_1 < 3.218$$

This is the confidence interval.

Example 6.16. *Find the confidence interval for the slope in Example 6.13.*

Solution 6.16.

$$\hat{\beta}_1 - 2.306\sqrt{\frac{\hat{\sigma}^2}{s_{xx}}} < \beta_1 < \hat{\beta}_1 + 2.306\sqrt{\frac{\hat{\sigma}^2}{s_{xx}}}$$

$$2.9816 - 2.306\sqrt{\frac{0.1929}{82.5}} < \beta_1 < 2.9816 + 2.306\sqrt{\frac{0.1929}{82.5}}$$

$$2.87 < \beta_1 < 3.093$$

This is the confidence interval.

6.2.1.3 Multivariate Regression

Multivariate regression is perhaps one of the most used modeling techniques in all disciplines. Hence, from a modeling perspective, it is important to get a good understanding of this technique. The most general multivariate techniques address multiple outputs and multiple inputs (MIMOs) problems. This is the case where there are multiple dependent and independent variables. If there are no relationships between the outputs, then the MIMO problem can be decomposed to several multiple inputs, single output (MISO) problems. This is the most popular multivariate regression problem and will be considered in this section. The model form is

$$y = \beta_0 + \beta_1 x_1 + \ldots + \beta_p x_p \tag{6.10}$$

$$y_{meas} = y + e \tag{6.11}$$

where $x_1, \ldots, x_p$ are the predictor or independent variables in the model, y is the dependent or output variable (true value), y_{meas} is the measurement of the dependent variable or output, e is the measurement error. The true y can never be exactly found when there are measurement errors. To keep notation simple, we will not distinguish between y and y_{meas} (unless explicitly needed to make a point). Relationships between dependent variables have to be handled differently.

Multivariate regression problems can arise in two contexts, both of which are important in data science and ML. The first context is where there are a number of predictor variables $(x_1 \ldots x_p)$ that have an impact on the output. Another context is where there is only one predictor variable (x_1) and new predictor variables that are functions of x_1 such as $x_1^2, x_1^3 \ldots$, are generated. While both of these cases can be solved by the multivariate linear regression formulation, there is still a subtle difference. In the first case, the model is still linear while in the second case, the model is nonlinear.

The second nonlinear case is called the linear-in-parameters model, which allows us to use the tools of multivariate regression. New predictors like the ones described above can be derived in cases where there are multiple predictors also, such as the inclusion of terms like $x_1^2 x_2, x_2^3 x_3, \ldots$ and so on. This

idea of increasing the number of predictor variables is called feature generation or lifting in DS and ML, and is an important tool for solving nonlinear
problems through the same techniques used to solve linear problems.

When there are multiple (m) samples of data, we can write equation 6.11 for
these multiple samples and consolidate all the equations as shown in equation
6.12, where $\mathbf{y}$ is a vector of all output values, $\boldsymbol{\beta}$ is the vector of parameters, $\mathbf{e}$
is the vector of error realizations (at each sample point), and X is a matrix
with samples in rows and predictor variables as columns.

$$y^i = \beta_0 + \beta_1 x_1^i + \ldots + \beta_1 x_n^i + e^i$$
$$\mathbf{y} = X\boldsymbol{\beta} + \mathbf{e} \tag{6.12}$$

Similar to the univariate case, one then minimizes the total sum of squared
errors:

$$\min_{\boldsymbol{\beta}} \ \mathbf{e}^T \mathbf{e} \tag{6.13}$$

It is seen that the problem above is an unconstrained optimization problem
in decision variables $\boldsymbol{\beta}$, and a solution to this problem is achieved by solving $\partial e^T e / \partial \beta = 0$. Simple algebra using vector differentiation, will lead to the
solution

$$\hat{\boldsymbol{\beta}} = (X^T X)^{-1} X^T \mathbf{y} \tag{6.14}$$
$$\hat{y} = X\hat{\boldsymbol{\beta}} \tag{6.15}$$

From Chapter 3, we know that the inverse $(X^T X)^{-1}$ will exist if all the
columns (variables) are independent assuming that we have more samples
than variables ($m > n$). From the statistics chapter we also know that if
we consider equation 6.14 as an expression, then it is clear that $\hat{\boldsymbol{\beta}}$ itself is a
random variable and when we substitute the values for X and $\mathbf{y}$ we will get
one realization of $\hat{\boldsymbol{\beta}}$. Similarly $\hat{y}$ is also a random variable and the predicted
value is one realization of this random variable.

Once could now derive the sampling distributions for $\hat{\boldsymbol{\beta}}$ and $\hat{y}$. For this, we
need to make assumptions about the pdf of the error variable e. If we assume
that every e^i is identically and independently distributed (iid) as a normal
distribution with a mean value of zero and variance σ_e^2, then $E(\mathbf{e}^T \mathbf{e}) = \sigma_e^2 I$.

$$E(\hat{\boldsymbol{\beta}}) = E((X^T X)^{-1} X^T \mathbf{y}) = E((X^T X)^{-1} X^T (\mathbf{y}_t + \mathbf{e}))$$
$$= E((X^T X)^{-1} X^T X\boldsymbol{\beta}) + E((X^T X)^{-1} X^T \mathbf{e}) = \boldsymbol{\beta}$$
$$Cov(\hat{\boldsymbol{\beta}}) = E((\hat{\boldsymbol{\beta}} - \boldsymbol{\beta})(\hat{\boldsymbol{\beta}} - \boldsymbol{\beta})^T = (X^T X)^{-1} X^T E(\mathbf{e}\mathbf{e}^T) X (X^T X)^{-1}$$
$$= (X^T X)^{-1} X^T (\sigma^2 I) X (X^T X)^{-1}$$
$$= \sigma^2 (X^T X)^{-1} X^T X (X^T X)^{-1} = \sigma^2 (X^T X)^{-1}$$

The mean of the sampling distribution for $\hat{\beta}$ is β and the covariance matrix is $\sigma_e^2 (X^T X)^{-1}$. Now this distribution can be used to construct hypothesis tests and confidence intervals on the true model parameters β. The examples that follow describe how quantities computed from the results of multivariate regression can be used with statistical tests for detecting "goodness-of-fit", important predictor variables, and so on. In summary, the most important questions are: does the identified model characterize the data to a sufficient level of accuracy (sufficiency of the model) and, (ii) is it necessary to retain all the predictor variables in the model (necessity of all terms in the model). The first question is addressed by checking if the error that is computed follows the hypothesized distribution for the error variables. This can be done using a q-q plot or some form of a hypothesis test. The second question is addressed by multiple techniques, all of which use hypothesis tests appropriately.

Multivariate regression:

$$\hat{\beta} = (X^T X)^{-1} X^T \mathbf{y}; \quad E\left(\hat{\beta}\right) = \beta$$

$$Var\left(\hat{\beta}\right) = \sigma_e^2 (X^T X)^{-1} \text{ (Diagonal values)}$$

$$\hat{\sigma}_e^2 = \frac{\sum \left(y^i - \hat{y}^i\right)^2}{n - p - 1}$$

Example 6.17. *A food delivery chain was analyzing the time required to deliver the food and recorded the following information*

Distance between hotel and destination (d)	8.2	10.7	9	8.2	6.4	9.7	6.6	13.4	14.5	5.8
Number of items in the order (n)	7	5	5	7	1	2	1	7	6	7
Time for delivery (t)	32.35	38.08	33.03	32.11	21.74	32.77	22.69	47.73	50.53	24.85

(a) *Identify a linear model relating time for delivery (t) to independent variables distance d and the number of items (n).*
(b) *Find R^2 and R_{adj}^2 for the model built in part (a).*
(c) *Fit a normal distribution to the residual/error between true times and predicted times. Draw a q-q plot and comment on the validity of assumption.*

Solution 6.17.(a)

$$\mathbf{y} = X\beta + \mathbf{e}$$

$$\underbrace{t}_{\substack{\mathbf{y} \\ 10 \times 1}} = \underbrace{\begin{bmatrix} 1 & d & n \end{bmatrix}}_{\substack{X \\ 10 \times 3}} \underbrace{\begin{bmatrix} \beta_0 \\ \beta_1 \\ \beta_2 \end{bmatrix}}_{\substack{\beta \\ 3 \times 1}} \implies \hat{\beta} = (X^T X)^{-1} X^T \mathbf{y} = \begin{bmatrix} 1.779 \\ 3.026 \\ 0.796 \end{bmatrix}$$

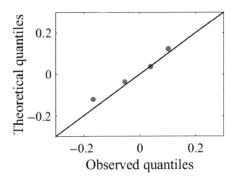

FIGURE 6.3 q-q plot for Example 6.17.

Hence the linear model is, $\hat{t} = 1.779 + 3.026d + 0.796n$.

(b)

$$R^2 = 1 - \frac{\sum \left(y^i - \hat{y^i}\right)^2}{\sum (y^i - \bar{y})^2} = 0.9998$$

$$R^2_{adj} = R^2 - \frac{\left(1 - R^2\right) \times 2}{10 - 2 - 1} = 0.9998$$

(c) Using error, $e^i = t^i - \hat{t^i}$, the error values are

$$0.187, \ -0.054, \ 0.039, \ -0.05, \ -0.199, \ 0.05, \ 0.146, \ -0.166, \ 0.102, \ -0.052$$

For fitting a normal distribution, $\hat{e} = \mathcal{N}\left(\hat{\mu}_e, \hat{\sigma}_e^2\right)$.

$$\hat{\mu}_e = \frac{1}{10} \sum_i e^i = 1.065 \times 10^{-15}; \quad \hat{\sigma}_e^2 = \frac{\sum \left(e^i - \hat{\mu}_e\right)^2}{n - 3} = 0.0209$$

$$\hat{\sigma}_e = 0.1446$$

Sorted error values are:

$$-0.199, \ -0.166, \ -0.054, \ -0.053, \ -0.052, \ 0.039, \ 0.05, \ 0.102, \ 0.146, \ 0.187$$

Observed quantiles are -0.166, -0.053, 0.039, 0.102. Theoretical quantiles using $\mathcal{N}\left(1.065 \times 10^{15}, 0.0209\right)$ are -0.122, -0.0366, 0.0366, 0.122. The corresponding q-q plot is provided in Figure 6.3. Note that the samples are close to $45°$ line and thus, the assumption of Gaussian noise is reasonable.

Example 6.18. *Perform F-test and comment whether the number of items ordered is significant in predicting the time for delivery in Example 6.17.*

Solution 6.18.
$$\text{Full model: } \hat{t}_{FM} = \beta_0 + \beta_1 d + \beta_2 n$$

If the number of items ordered was insignificant, then the reduced model is

$$\text{Reduced model: } \hat{t}_{RM} = \beta_{0_{RM}} + \beta_{1_{RM}} d$$

Fitting the full model and reduced model from given data

$$\hat{t}_{FM} = 1.779 + 3.026d + 0.796n; \quad \hat{t}_{RM} = 3.374 + 3.266d$$

Sum squared error of full model: $SSE_{FM} = 0.146$ $(dof = 10 - 3)$
Sum squared error of reduced model: $SSE_{RM} = 32.239$ $(dof = 10 - 2)$
Null hypothesis, $H_0 : \beta_2$ *is insignificant*
Alternative hypothesis, $H_a : \beta_2$ *is significant* $(SSE_{RM} - SSE_{FM} >> SSE_{FM})$

$$f = \frac{(SSE_{RM} - SSE_{FM})/1}{SSE_{FM}/(10-3)} = \frac{(32.239 - 0.146)\,7}{0.146} = 1538.71$$

$$F_{1-\alpha,1,7} = 5.59$$

Since $f >> F_{1-\alpha,1,7}$, *we reject null hypothesis. Thus, hypothesis shows that* β_2 *is significant and thus, the full model is needed. This is also evident from the p-value.*

$$p\text{-value} = 1.822 \times 10^{-9}$$

Example 6.19. *Perform t-test and comment on the significance of* β_2 *in predicting time for delivery in Example 6.17. Find confidence interval for* β_2.

Solution 6.19.

$$H_0 : \beta_2 = 0; \quad H_1 : \beta_2 \neq 0$$

$$t = \frac{\hat{\beta}_2 - 0}{\hat{\sigma}_{\beta_2}}; \quad \hat{\sigma}_\beta^2 = \hat{\sigma}_e^2 \left(X^T X\right)^{-1}$$

$$\hat{\sigma}_e^2 = \frac{1}{n-3} \sum \left(y^i - \hat{y}^i\right)^2 = \frac{1}{7} \times 0.146 = 0.0209$$

$$\left(X^T X\right)^{-1} = \begin{bmatrix} 1.3 & -0.11 & -0.0395 \\ -0.11 & 0.015 & -0.006 \\ -0.0395 & -0.006 & 0.0198 \end{bmatrix}$$

$$\hat{\sigma}_{\beta_2}^2 = 0.0209 \times 0.0198 = 0.000413; \quad \hat{\sigma}_{\beta_2} = 0.0203$$

$$T = \frac{0.796 - 0}{0.0203} = 39.2; \quad t_{\frac{\alpha}{2},n-p-1} = t_{0.025,7} = -2.365; \quad t_{0.975,7} = 2.365$$

Since $T >> t_{0.975,7}$, *we reject null hypothesis. Thus,* β_2 *is significant.*

Confidence interval:

$$-2.365 \leq \frac{\hat{\beta}_2 - \beta_2}{0.0204} \leq 2.365$$

$$\hat{\beta}_2 - 2.365 \times 0.0203 \leq \beta \leq \hat{\beta}_2 + 2.366 \times 0.0203$$

$$0.7482 \leq \beta_2 \leq 0.844 \; \left(\because \hat{\beta}_2 = 0.796 \right)$$

Example 6.20. *For the data provided in Example 6.17, split the given data into 70% training set and 30% test set. Build the linear model relating "d" and "n" using the training set and report the performance of the developed model in predicting "t" for the test set.*

Solution 6.20. *Samples given are shuffled and seven of them are chosen as the training set. The rest of the samples are in the test set.*
    Training set:

d_{tr}	10.7	6.4	8.2	14.5	13.4	9.7	8.2
n_{tr}	5	1	7	6	7	2	7
t_{tr}	38.8	21.74	32.35	50.53	47.73	32.77	32.11

    Test test:

d_{te}	6.6	9	5.8
n_{te}	1	5	7
t_{te}	22.69	33.03	24.85

Following the same steps as described in Example 6.17, using the training set $(X_{tr} = \begin{bmatrix} d_{tr} & n_{tr} \end{bmatrix})$,

$$\hat{\beta} = \left(X_{tr}^T X_{tr} \right)^{-1} X_{tr}^T t_{tr} = \begin{bmatrix} 1.672 & 3.0253 & 0.8144 \end{bmatrix}^T$$

For this training set,

$$R^2 = 0.999; \quad R^2_{adj} = 0.999$$

If we use this model to predict time for the test set, we get

$$\hat{t}_{te} = [24.92, \; 32.9721, \; 22.4536]$$

Note that the predicted values of time are quite close to the true values. R^2 value corresponding to this test set is 0.998. This high R^2 value for the test set shows that the model is very good for the given data.

Akaike Information Criteria and Bayesian Information Criteria (BIC) are two metrics that help in model selection. They are defined as follows:

$$AIC = 2k - 2\ln(L)$$
$$BIC = k\ln(n) - 2\ln(L)$$

where n is the number of samples, k is the number of parameters in the model and L is the likelihood function. If we assume that the error is Gaussian,

$$L = \prod_{i=1}^{n} \frac{1}{\sqrt{2\pi\sigma^2}} \exp\left(\frac{-e_i^2}{2\sigma^2}\right)$$

$$\ln L = \frac{-n}{2}\ln(2\pi) - \frac{n}{2}\ln(\sigma^2) - \frac{\sum_{i=1}^{n} e_i^2}{2\sigma^2}$$

If σ^2 is unknown, we use $\sigma^2 = \frac{SSE}{n} = \frac{\sum e_i^2}{n}$ as the estimate (MLE),

$$\Rightarrow -2\ln L = n\ln(2\pi) + n\ln\left(\frac{SSE}{n}\right) + n$$

$$\Rightarrow AIC = 2k + n\ln(2\pi) + n\ln\left(\frac{SSE}{n}\right) + n$$

$$BIC = k\ln(n) + n\ln(2\pi) + n\ln\left(\frac{SSE}{n}\right) + n$$

Example 6.21. *Consider the following data that relates the average ambient temperature to the average electricity bill in a household.*

Average Temperature (x)	6.6	26.1	6.3	27.6	14.7	18.3	23.1	15.6	9	5.7
Electricity Bill (y)	118.2	5607	105.3	6616.8	1043.7	1971.6	3907.8	1239	264	83.4

Find the best polynomial model that predicts the electricity bill given the average temperature using Akaike Information Criteria (AIC) and Bayesian Information Criteria (BIC). Use the entire data for building the model.

Solution 6.21. *Polynomial models of orders ranging from 1 to 6 were built using the given data. Table 6.4 provides a summary of the models. It is evident from the table that SSE is continuously decreasing. However, if we look at polynomial orders higher than a quadratic model, there is only a slight reduction in SSE values. If we pick a higher order model, the performance may be poor on test data. Hence, a lower order polynomial may be preferred. AIC and BIC metrics are useful in deciding the best model. AIC and BIC metrics penalize additional parameters used in reducing SSE.*

AIC and BIC values for each of the models are provided in Table 6.4. Both AIC and BIC values decrease first and then increase. Note that the model with lowest AIC value is model 4 (Fourth order polynomial). The same model has the lowest BIC value. Note that it is not necessary that AIC and BIC values are minimum for the same model. Based on AIC and BIC values, a 4th order polynomial is chosen as the best model for the given data.

Example 6.22. *For Example 6.21, find the best polynomial model using F-test.*

Solution 6.22. *F-statistic for each model was found by comparing the error variance of model with the previous model. Let $\hat{\sigma}_p^2$ and $\hat{\sigma}_{p-1}^2$ be the error variance of pth and $(p-1)^{th}$ order polynomial models respectively.*

$$\hat{\sigma}_p^2 = \frac{SSE_p}{n-(p+1)} = \frac{SSE_p}{n-p-1}; \quad \hat{\sigma}_{p-1}^2 = \frac{SSE_{p-1}}{n-(p-1)-1} = \frac{SSE_{p-1}}{n-p}$$

$$H_0 : \sigma_p^2 = \sigma_{p-1}^2; \quad H_a : \sigma_p^2 < \sigma_{p-1}^2$$

$$f_p = \frac{(SSE_{p-1}-SSE_p)/1}{SSE_p/n-p-1}$$

If $f_p \gg F_{1-\alpha,1,n-p-1}$, we reject null hypothesis which implies that it is important to add the additional term of pth order polynomial (the term $\beta_p x^p$).

TABLE 6.4

AIC and BIC Values for Example 6.21

Model order	Optimal model parameters	SSE	AIC	BIC
1	[−2143.18, 277.05]	5078172	163.76	164.36
2	[897.23, −219.13, 15.34]	39949.94	117.31	118.21
3	[5.07, 3.52, 0.03, 0.31]	0.2135	−2.08	−0.87
4	[7.86, 2.63, 0.12, 0.3 5.9×10^{-5}]	0.151851	−3.5	−1.98
5	[7.88, 2.62, 0.12, 0.3, 6.25×10^{-5}, -3.95×10^{-5}]	0.15185	−1.5	0.32
6	[7.74, 2.68, 0.11, 0.31, 7.3×10^{-6}, 1.33×10^{-6}, -1.35×10^{-8}]	0.151848	0.5	2.62

Table 6.4 lists optimum parameters for polynomial order ranging from 1 to 6. The corresponding f_p and $F_{1-\alpha,1,n-p-1}$ values are provided in Table 6.5. Note that, for the first three models, the null hypothesis is rejected implying that the additional model complexity significantly improved the SSE and hence should be included in the model. (For the linear model $(p = 1)$, the $(p-1)^{th}$ model would be equivalent to $\hat{y} = \bar{y}$).

For model orders greater than 3, null hypothesis was accepted and hence the additional model complexities were insignificant. Based on F-test, the best model order is 3 (Cubic polynomial model).

Example 6.23. *For the problem discussed in Example 6.21, perform k-fold validation for each polynomial order model and comment on the model that fits the data best.*

TABLE 6.5
F-Test Results for Example 6.22

Model order, p	SSE	f_p	$F_{0.95,1,n-p-1}$	Reject/Accept H_0
1	5078172	75.79	5.32	Reject
2	39949.94	882.79	5.59	Reject
3	0.2135	1122231.4	5.99	Reject
4	0.151851	2.033	6.61	Accept
5	0.15185	1.61×10^{-5}	7.71	Accept
6	0.151848	4.42×10^{-5}	10.13	Accept

Solution 6.23. *k-fold validation helps us to validate the model on unseen data. We first split the data randomly to k folds. Here $k = 5$ is chosen. Since we have 10 samples, each group will have 2 samples. The 5 groups are as given below:*

Group	1		2		3		4		5	
x	14.7	23.1	6.6	27.6	26.1	15.6	18.3	9	6.3	23.1
y	1043.7	3907.8	118.2	6616.8	5607	1239	1971.6	264	105.3	3907.8

For the first iteration, samples from groups 1 to 4 are used together as the training data for building a linear model. The linear model built using these samples is tested on the 2 samples in group 5. SSE for this test data is calculated. The model is discarded and a new linear model is built using samples in groups 2 to 5 as training data. This model is tested on 2 samples in group 1. This process is repeated until all groups have been tested once. The total SSE of the linear model is the sum of SSE calculated in each iteration for the respective test data. The results of each iteration can be seen in Table 6.6. The same process can be repeated for all model orders. SSE for all the model orders are provided in Table 6.7.

Note that unlike the SSE reported in Table 6.5 for the training data, SSE is computed on test data and hence, SSE initially decreases and then increases. The initial decrease can be attributed to improved prediction with higher model orders. The increase in SSE after third order polynomial is due to overfitting of the data by higher order polynomials. Model of order 4 can be considered as the best model since it has the lowest SSE. However, there is very minimal improvement in SSE from model order 3 to 4. This is more evident from the

TABLE 6.6

K-fold Validation for Linear Model (Example 6.23)

Iteration 1	Training set	Test set	Linear Model Parameters	SSE
1	1:4	5	−2863.27 309.75	2429654.73
2	2:5	1	−2073.87 282.95	1392982
3	3:5,1	2	−1957.23 267.46	2877204.438
4	4,5,1,2	3	−1957.23 267.46	1293361.50
5	5,1:3	4	−2064.55 280.33	1234433.89
			Total SSE	9227637.25

TABLE 6.7

SSE (k-fold) and F-test Values for Polynomial Models (Example 6.23)

Model order(p)	Total SSE (Test set)	f_p	$F_{\alpha,n,n}$	Accept/ Reject H_0
1	9227637.25	0.174	0.336	Reject
2	100458.43	0.011	0.336	Reject
3	0.53	5.28×10^{-6}	0.336	Reject
4	0.51	0.964	0.336	Accept
5	1.22	2.39	0.336	Accept
6	18.38	15.038	0.336	Accept

F-test performed as described below.

$$H_0 : \sigma_p^2 = \sigma_{p-1}^2; \qquad H_a : \sigma_p^2 < \sigma_{p-1}^2$$
$$f_p = \frac{(SSE_p)/n}{(SSE_{p-1})/n}$$

Note that SSE_p and SSE_{p-1} are the sum of squared error values for the test sets of all k folds using p^{th} order polynomial model and $(p-1)^{th}$ order polynomial, respectively. Since SSE is calculated on the test sets, degrees of freedom is the same as the number of samples as the data isn't used for estimating the model parameters. f_k and $F_{\alpha,n,n}$ values ($\alpha = 0.5$) for all the models are provided in Table 6.7. Note that the null hypothesis is rejected for the first three models, while it is accepted for the remaining 3 models. This indicates that there is no significant gain in choosing models higher than the cubic polynomial model.

Example 6.24. *A company produces three different items A,B, and C. The data below shows the sale of these items in one day and the profit made by the company on that day.*

x_1 (Sales of item A)	8	11	9	8	6	10	7	13	14	6
x_2 (Sales of item B)	6	4	5	7	1	1	0	7	6	7
x_3 (Sales of item C)	49	40	23	39	6	32	7	47	26	21
Profit (y)	93.26	89.76	60.78	79.34	28.23	75.83	32.74	105.59	79.68	48.86

Find the best multilinear model that represents the relation between sales of A, B and C and the profit using forward selection method.

Solution 6.24. *In forward selection, features that improve prediction of y are iteratively added. This is repeated until additional features do not improve the prediction of y. Various metrics can be chosen to decide the order in which the features are evaluated. Correlation coefficient $(\rho_{x_i,y})$ is used here. Similarly, there are many metrics that can be used to decide the desired performance of the model. R^2 value is used here. We break the search when a R^2 value of 0.99 is achieved or when there is no significant improvement in R^2 value.*

TABLE 6.8

Correlation Coefficient Values for Example 6.24

Quantity	Value	Rank
$\rho_{x_1,y}$	0.72	II
$\rho_{x_2,y}$	0.60	III
$\rho_{x_3,y}$	0.96	I

As a first step, correlation coefficient was found between each of the independent variables x_i and the dependent variable (y). Based on the correlation coefficient values (refer Table 6.8), independent features were ordered from I to III as shown in Table 6.8. x_3 has the highest correlation coefficient and hence, is the 1st feature added to the model. We fit the model $\hat{y} = \beta_0 + \beta_3 x_3$ using the OLS method. Optimum parameters were found to be $\beta_0 = 21.7735$ and $\beta_3 = 1.6425$. R^2 value of 0.92 was obtained. As this is less than the desired value (0.99), we add the next feature, which is x_1 here.

The new model, $\hat{y} = \beta_0 + \beta_3 x_3 + \beta_1 x_1$, was identified using OLS and the optimum parameters were found to be

$$\beta = \begin{bmatrix} \beta_0 & \beta_3 & \beta_1 \end{bmatrix}^T = \begin{bmatrix} 2.0531 & 1.3688 & 3.0062 \end{bmatrix}^T$$

The new model has an R^2 value of 0.999. Hence the iterations are stopped. The best model found by forward selection is

$$\hat{y} = \beta_0 + \beta_3 x_3 + \beta_1 x_1$$

This method shows that the sales of item B doesn't contribute much to the profit. Further, the coefficient of x_1 is higher than x_3 implying that in order to improve profit, the company should focus more on the sale of item A.

Example 6.25. *For the data provided in Example 6.24, find the best multi-linear model using best subset method.*

Solution 6.25. *In the best subset method, we create models using all possible subsets of given features. For example, given 2 features x_1 and x_2, the possible subsets are $(x_1), (x_2)$, and (x_1, x_2); correspondingly, we build 3 models. Similarly, given 3 features x_1, x_2 and x_3, the subsets are $(x_1), (x_2), (x_3), (x_1, x_2), (x_1, x_3), (x_2, x_3)$ and (x_1, x_2, x_3). For this case, 7 models are built corresponding to each subset. Once all the models are identified, some metric has to be used to compare their performances and pick the best out of them. Here, we will show the uses of both AIC and BIC values. The models and their corresponding AIC and BIC values are provided in Table 6.9. Note that the model with features x_1 and x_3 has the lowest AIC and BIC values. Hence the best model is*

$$y = \begin{bmatrix} 1 & x_1 & x_3 \end{bmatrix} \begin{bmatrix} 2.053 \\ 3.006 \\ 1.37 \end{bmatrix}$$

Example 6.26. *For the data provided in Example 6.24, find the best multi-linear model using backward elimination.*

Solution 6.26. *In backward elimination, we start with the full model (including all features x_1, x_2, and x_3).*

Variables that contribute the least in predicting y are removed. One can find the variable to be removed based on any performance metric such as p-value, AIC, BIC, etc. The p-value for each feature can be easily calculated from the full model. However, when using AIC and BIC, the importance of features can be computed only by building variable removed models. The independent variable whose removal leads to the minimum AIC or BIC is identified and that variable is removed from the process.

In this example, the working of backward elimination using p-value is demonstrated. The variable with the highest p-value is identified and if the p-value is greater than 0.05, then the variable is removed. Once a variable is removed, the same procedure is repeated until there are no p values greater than the threshold.

First, the full model is identified and the parameters were found to be

$$\boldsymbol{\beta^I} = \begin{bmatrix} \beta_0^I & \beta_1^I & \beta_2^I & \beta_3^I \end{bmatrix}^T = \begin{bmatrix} 2.04 & 3.006 & 0.012 & 1.367 \end{bmatrix}^T$$
$$y = \begin{bmatrix} 1 & x_1 & x_2 & x_3 \end{bmatrix} \boldsymbol{\beta^I}$$

TABLE 6.9

AIC and BIC Values of All Models Built for Example 6.25

Subset	$\mathbf{y} = X\boldsymbol{\beta}$ Model		Optimal Parameters $(\boldsymbol{\beta})$	AIC	BIC
x_1	$y = \begin{bmatrix} 1 & x_1 \end{bmatrix}$	$\begin{bmatrix} \beta_0 \\ \beta_1 \end{bmatrix}$	$\begin{bmatrix} 7.55 \\ 6.72 \end{bmatrix}$	89.17	89.78
x_2	$y = \begin{bmatrix} 1 & x_2 \end{bmatrix}$	$\begin{bmatrix} \beta_0 \\ \beta_2 \end{bmatrix}$	$\begin{bmatrix} 44.495 \\ 5.66 \end{bmatrix}$	91.99	92.60
x_3	$y = \begin{bmatrix} 1 & x_3 \end{bmatrix}$	$\begin{bmatrix} \beta_0 \\ \beta_3 \end{bmatrix}$	$\begin{bmatrix} 21.77 \\ 1.64 \end{bmatrix}$	70.96	71.56
(x_1, x_2)	$y = \begin{bmatrix} 1 & x_1 & x_2 \end{bmatrix}$	$\begin{bmatrix} \beta_0 \\ \beta_1 \\ \beta_2 \end{bmatrix}$	$\begin{bmatrix} 1.24 \\ 5.52 \\ 3.95 \end{bmatrix}$	87.17	88.075
(x_1, x_3)	$y = \begin{bmatrix} 1 & x_1 & x_3 \end{bmatrix}$	$\begin{bmatrix} \beta_0 \\ \beta_1 \\ \beta_3 \end{bmatrix}$	$\begin{bmatrix} 2.053 \\ 3.006 \\ 1.37 \end{bmatrix}$	-10.55	-9.645
(x_2, x_3)	$y = \begin{bmatrix} 1 & x_2 & x_3 \end{bmatrix}$	$\begin{bmatrix} \beta_0 \\ \beta_2 \\ \beta_3 \end{bmatrix}$	$\begin{bmatrix} 21.78 \\ -0.009 \\ 1.644 \end{bmatrix}$	72.96	73.865
(x_1, x_2, x_3)	$y = \begin{bmatrix} 1 & x_1 & x_2 & x_3 \end{bmatrix}$	$\begin{bmatrix} \beta_0 \\ \beta_1 \\ \beta_2 \\ \beta_3 \end{bmatrix}$	$\begin{bmatrix} 2.04 \\ 3.006 \\ 0.012 \\ 1.367 \end{bmatrix}$	-9.07	-7.856

p-values for $\beta_0^I, \beta_1^I, \beta_2^I$, and β_3^I are $1.2 \times 10^{-5}, 3.5 \times 10^{-12}, 0.59$, and 4.97×10^{-14}, respectively. Note that β_2 has the highest p-value and is greater than 0.5. Hence, we remove variable x_2 from further calculations. Now, the features for the new model are x_1 and x_3. We identify a new model to obtain

$$y = \beta_0^{II} + \beta_0^{II} x_1 + \beta_0^{II} x_3 \qquad (6.16)$$

$$\boldsymbol{\beta}^{II} = \begin{bmatrix} \beta_0^{II} & \beta_1^{II} & \beta_3^{II} \end{bmatrix}^T = \begin{bmatrix} 2.053 & 3.006 & 1.37 \end{bmatrix}^T$$

The p-values of $\beta_0^{II}, \beta_1^{II}$, and β_3^{II} are $2.14 \times 10^{-6}, 5.9 \times 10^{-14}$, and 1×10^{-16}, respectively. As none of the p-values are greater than 0.05, iterations are terminated with the conclusion that equation 6.16 is the best multilinear model for the given data.

Example 6.27. *Consider the following data that relates various features like distance of a restaurant from city (x_1), customer ratings for service (x_2), ambience (x_3) and food (x_4), and the type of food served (x_5) to the average price of dinner at that restaurant for one person.*

x_1	11	14.3	12.1	10.9	8.5	12.9	8.9	17.8	19.3	7.7
x_2	4.1	2.9	3.1	4.7	0.8	0.9	0.6	4.2	4	4.4
x_3	3.1	4.5	4.5	4.2	0.7	3.6	2.2	2.8	3.5	1.4
x_4	3.4	2.2	3.5	1.2	4.8	2.1	5	1.6	3.1	2.3
x_5	5	1	2	1	2	9	1	9	7	5
y	413	496.7	584.4	542.6	248	351	330.2	147.6	206.5	305.1

Find the best multilinear model to predict y using backward elimination with AIC as the metric.

Solution 6.27. *For backward elimination using AIC as the metric, models are built with multiple feature sets and the AIC values are compared.*

Features	x_1, x_2, x_3, x_4	x_1, x_2, x_3, x_5	x_1, x_2, x_4, x_5	x_1, x_3, x_4, x_5	x_2, x_3, x_4, x_5
AIC	103.84	101.32	128.66	102.72	122.12

Note that the second model has the best AIC (lowest) indicating that the removal of x_4 from input features improved the model. Hence, x_4 is not considered for future iterations. Now, the remaining features: x_1, x_2, x_3, and x_5 are considered. Models for 3-feature subsets of these 4 features are identified. The results are provided below:

Features	x_1, x_2, x_3	x_1, x_2, x_5	x_1, x_3, x_5	x_2, x_3, x_5
AIC	104.73	129.18	100.73	120.44

The model with x_1, x_3, and x_5 has the lowest AIC and the AIC is less than the one corresponding to features x_1, x_2, x_3, and x_5. So, removal of x_2 improved the model quality.

Now, for the next iteration, we have features x_1, x_3, and x_5. Various 2-feature models are identified and AIC values are compared.

Feature	x_1, x_3	x_1, x_5	x_3, x_5
AIC	103.36	127.93	118.46

The best AIC value among 2-feature models is obtained for the one with features x_1 and x_3. However, this AIC value is higher than that of the best model in the previous iteration (model with features x_1, x_3 and x_5). It shows that removal of any feature from $\{x_1, x_3, x_5\}$ reduces the model performance. Hence, the best model is

$$y = \beta_0 + \beta_1 x_1 + \beta_3 x_3 + \beta_5 x_5$$

The optimum parameters can be identified using least squares as,

$$\beta_0 = 382.85; \quad \beta_1 = -26.16; \quad \beta_3 = 110.57; \quad \beta_5 = -8.33$$

Various performance metrics of this model are provided below:

$$R^2 = 0.967; \quad R^2_{adj} = 0.951; \quad AIC = 100.73; \quad BIC = 101.94$$

The corresponding predicted y are

$$\hat{y} = \begin{bmatrix} 396.18 & 497.94 & 547.17 & 553.72 & 221.201 & 368.44 \\ & & & 387.53 & 151.78 & 206.59 & 294.55 \end{bmatrix}$$

Example 6.28. *For the data given in Example 6.27, find the best multilinear model using forward selection method. Use AIC as the metric.*

Solution 6.28. *For forward selection, various 1 feature models and their AIC values are obtained as follows:*

Features	x_1	x_2	x_3	x_4	x_5
AIC	129.92	130.8	125.51	130.72	125.96

Note that the model with x_3 as the input feature has the lowest AIC (best AIC). Hence, we retain x_3 for all models in future iterations. In the next iteration, 2-input feature models with one of the features being x_3 are built. The AIC of these models are computed and tabulated below

Features	x_1, x_3	x_2, x_3	x_3, x_4	x_5, x_3
AIC	103.36	127.34	126.76	118.46

The model with $\{x_1, x_3\}$ as features shows the best performance among all the 4 models. Hence, we retain both x_1 and x_3 for future iterations. Now, we start with 3-feature models with 2 of the features being x_1 and x_3. The models and the corresponding AIC values are given below:

Features	x_1, x_2, x_3	x_1, x_3, x_4	x_1, x_3, x_5
AIC	104.73	104.66	100.73

It can be seen that among these three models, the one with $\{x_1, x_3, x_5\}$ has the least AIC. This AIC is also less than the best AIC in the previous iteration. So, we continue to the next iteration where we build 4-feature models with 3 of the features being x_1, x_3 and x_5. Summary of these 4-feature models are provided below:

Features	x_1, x_2, x_3, x_5	x_1, x_3, x_4, x_5
AIC	101.32	102.72

The lowest AIC among these two models is 101.32; however, this is higher than the AIC of the model with $\{x_1, x_3, x_5\}$ as features. Hence, we conclude that the best model is

$$\hat{y} = \beta_0 + \beta_1 x_1 + \beta_3 x_3 + \beta_5 x_5$$

Optimum parameters are $\beta_0 = 382.85$, $\beta_1 = -26.16$, $\beta_3 = 110.57$ and $\beta_4 = -8.33$. The forward selection method evaluated 14 models before converging.

Example 6.29. *For the data given in 6.27, find the best multilinear model using best subset method with AIC as the metric.*

Solution 6.29. *For best subset method, one has to generate all possible subsets of the given 5 features and identify models with these subsets as the input features. For 5 features, a total of 31 subsets/ models (5 1-feature models, 10 2-feature models, 10 3-feature models, 5 4-feature models, and 1 model with all 5 features) need to be evaluated. AIC for all models were calculated and the model with features $\{x_1, x_3, x_5\}$ had the lowest AIC value (100.73).*

Example 6.30. *Manufacturing cost of cylindrical tanks of various surface areas are given below:*

Curved Surface Area (x_1)	15	17	16	15	14	16	14	19	20	14
Base Area (x_2)	7	5	5	7	1	2	1	7	6	7
Total Area (x_3)	22	22	21	22	15	18	15	26	26	21
Cost (y)	797.2	793.8	758.7	792.5	536.4	651.1	543.2	937.1	943.1	755.2

If the model to predict y (cost) is given by $\hat{y} = \beta_0 + \beta_1 x_1 + \beta_2 x_2 + \beta_3 x_3$, find unknowns β_i using ordinary least squares. Comment on the performance of the identified model.

Solution 6.30.

$$\mathbf{y} = \begin{bmatrix} 1 & \mathbf{x_1} & \mathbf{x_2} & \mathbf{x_3} \end{bmatrix} \begin{bmatrix} \beta_0 \\ \beta_1 \\ \beta_2 \\ \beta_3 \end{bmatrix} = X\boldsymbol{\beta} \implies \boldsymbol{\beta} = (X^T X)^{-1} X^T \mathbf{y} \ (Using \ OLS)$$

For the given data

$$\boldsymbol{\beta} = \begin{bmatrix} 503109 & -32532 & -4080.43 & 4080.43 \end{bmatrix}^T$$

The predicted y values are given below:

Predicted $\mathbf{y}$, $\hat{\mathbf{y}} = [76335.3, 19432.1, 47883.7, 76335.3, 104787, 47883.7,$
$$104787, -37471.1, -65922.7, 104787]$$

It is evident that the predicted y is quite different from data

$$R^2 = -309975.85; \quad R^2_{adj} = -464964.28$$

The model performance is very poor. The reason for this poor performance can be attributed to multicollinearity present in the data. Data is multicollinear with

$$x_3 = x_1 + x_2$$

The rank of input matrix X is less than 4 and this makes the matrix $X^T X$ close to singular (ill conditioned). The eigenvalues of $(X^T X)$ are 7303.87, 53.97, 0.15 and -4.74×10^{-13}. Defining condition number as the ratio of largest to lowest eigenvalue,

$$\text{Condition number} = -1.54 \times 10^{16}$$

This is a very high value showing that the matrix is ill conditioned. There are multiple ways of handling such data. One such method is ridge regression.

Example 6.31. *Consider the data provided in Example 6.30. Perform ridge regression to find the optimal parameters of the model. Comment on the performance of the model in comparison to the model obtained in Example 6.30.*

Solution 6.31. *Ridge regression is linear regression with L_2 regularization, i.e., the L_2-norm of the coefficient is added as a penalty in the objective function as shown below:*

$$\min_{\beta} \|\mathbf{y} - X\boldsymbol{\beta}\|^2 + \lambda \|\boldsymbol{\beta}\|^2 = J$$

Where $\|.\|$ is the L_2-norm and λ is a tuning parameter. The solution to this unconstrained optimization problem can be found by setting $\frac{dJ}{d\beta}$ to zero.

$$\frac{dJ}{d\beta} = 0 \Rightarrow -2X^T(\mathbf{y} - X\hat{\boldsymbol{\beta}}) + 2\lambda\hat{\boldsymbol{\beta}} = 0$$

$$\Longrightarrow X^T\mathbf{y} = \left(X^T X + \lambda I\right)\hat{\boldsymbol{\beta}} \implies \hat{\boldsymbol{\beta}} = \left(X^T X + \lambda I\right)^{-1} X^T\mathbf{y}$$

It can be seen that the inverse of $X^T X + \lambda I$ is used instead of $X^T X$. Thus, by changing the value of λ, one can change the condition number of the matrix that needs to be inverted. For example, if we set $\lambda = 1$, for the given data, condition number is 7304.87. Thus, the condition number is reduced considerably from the value obtained in Example 6.30. Using this new formula

$$\boldsymbol{\beta} = \begin{bmatrix} -0.71 & 12.148 & 11.897 & 24.04 \end{bmatrix}^T$$

The predicted y values are

$$\hat{\mathbf{y}} = [793.78, \ 794.28, \ 758.09, \ 793.78, \ 541.93, \ 650.26,$$
$$541.93, \ 938.55, \ 938.80, 757.59]$$

Note that predictions are quite close to the true values. The performance metrics for this ridge regression model are

$$R^2 = 0.9996; \qquad R^2_{adj} = 0.9994$$

We can see that the model performs really well as R^2 and R^2_{adj} values are close to 1. This example demonstrates how the issue of multicollinearity can be addressed using L_2-regularization or Ridge regression.

Example 6.32. *Consider the data given in Example 6.24. Build the best multilinear model using Lasso regression.*

Solution 6.32. *Lasso (Least Absolute Shrinkage and Selection Operation) is essentially linear regression with L_1-regularization, i.e., L_1-norm of the coefficients is added as a penalty in the objective function.*

$$\min_{\beta} \; \|\mathbf{y} - X\boldsymbol{\beta}\|_2 + \lambda \|\boldsymbol{\beta}\|_1$$

where λ is a tuning parameter. Since L_1-norm is not continuously differentiable a solution is estimated numerically. Solving in a simulation tool

$$\boldsymbol{\beta} = \begin{bmatrix} 2.18 & 2.99 & 0 & 1.37 \end{bmatrix}^T$$

Note that Lasso identifies $\beta_2 = 0$ indicating that feature x_2 is not significant in predicting y. Thus, using Lasso, we are able to find the best subset of features and the corresponding optimal parameters. R^2 value of the model is 0.9999. $\lambda = 0.1$ was used. As λ is increased, more β_i would become zero.

Example 6.33. *For the data provided in Example 6.27, find the best multilinear model using Lasso regression for various values of λ. Comment on the influence of λ.*

Solution 6.33. *Table 6.10 summarizes the results.*
As λ increases, the number of significant features decreases and R^2 decreases. The best model is either $\{x_1, x_2, x_3, x_5\}$ or $\{x_1, x_3, x_5\}$ depending on the R^2 value and number of features that one wants to retain.

Example 6.34. *Consider an input variable x and an output variable y which are related using the relation $y = 3x^3 - x^2 + 5x + 3.2$ (third-order polynomial). The measurements of y contains Gaussian noise of zero mean and standard deviation 1.2. Comment on the bias-variance trade-off if one didn't know the true relation and were to build polynomial models of various orders to fit the data. Use 200 samples for fitting any model. Also, identify the model order that best fits the data.*

Solution 6.34. *To comment of bias-variance trade-off, we need to find the bias and variance in prediction while fitting data with models of various complexities. 200 samples were generated for x and y using the true model and to obtain the data corresponding to measurements, Gaussian noise of zero mean and standard deviation 1.2 was added to y data. A linear model is attempted first. The parameters of the model that best fits the data were estimated using least squares method. The same process is repeated 1000 times using 1000 different datasets. Thus, we have 1000 different linear models for relating y to x. Now, a test set of 20,000 samples is generated using the true model (x_{test} and y_{test}). Measurement noise was added to y_{test} to get y_{meas}. For each of the*

TABLE 6.10

Lasso Regression Results for Example 6.33

λ	β	Significant Features	R^2
0.1	333.98 −27.31 9.5 112.6 8.033 −7.09	x_1, x_2 x_3, x_4, x_5	0.974
10	381.04 −22.94 2.31 97.97 0 −9.86	x_1, x_2 x_3, x_5	0.9625
20	384.04 −19.39 0 86.74 0 −11.19	x_1, x_3, x_5	0.938
80	391.34 0 0 16.75 0 −19.15	x_3, x_5	0.516

1000 linear models that were built, the output variable can be predicted using x_{test} data. This data was used to compute the overall bias, variance and MSE. These three quantities were estimated for other polynomial orders using the same procedure.

Figure 6.4 and Table 6.11 show the variation of MSE, Bias² and variance with model orders. Note that Bias² decreases initially with increasing model order while variance increases with increasing model order. MSE decreases with initial increase in model orders as the effect of squared bias is dominant than the effect of variance. After model order 3, bias is almost constant while variance increases. Hence, MSE starts to increase as the effect of variance starts to dominate the effect of squared bias. This is the bias-variance trade-off. For this example, model order 3 has the lowest MSE and hence, we can conclude that the best model for the given data is a third-order polynomial.

6.2.2 Principal Component Analysis (PCA)

Until now, linear models were built with input and output variables identified *a priori* and an additional assumption that the input variables are error-free. PCA [34] is an alternate approach to building linear models without labeling variables as inputs or outputs. Further, both input and output variables are assumed to have errors (discussed earlier in Section 5.8.2). PCA is widely used

TABLE 6.11

Bias, Variance, and MSE Values for Example 6.34

Model Order	Bias^2	Variance	MSE
1	2.9e-01	1.8e-02	3.1e-01
2	2.0e-01	2.6e-02	2.3e-01
3	5.1e-06	3.0e-02	3.0e-02
4	4.6e-06	3.8e-02	3.8e-02
5	5.1e-06	4.5e-02	4.6e-02
6	7.56e-06	5.3e-02	5.3e-02
7	9e-05	6.1e-02	6.1e-02

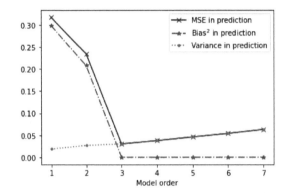

FIGURE 6.4 Bias-variance trade-off for Example 6.34.

to understand the minimum number of variables that is necessary to characterize the entire data. PCA uses SVD of the entire data matrix (including both input and output variables). As SVD can be used for model building, dimensionality reduction, noise removal, and data compression, PCA is useful for all these applications as well and as a result, has attained considerable popularity in ML. Using SVD on a data matrix Z $(= [X, \mathbf{y}]$ where X and y traditionally denoted as the input and output variables),

$$\underset{m \times n}{Z} = \underset{m \times m}{Q_1} \; \underset{m \times n}{\Sigma} \; \underset{n \times n}{Q_2^T} \tag{6.17}$$

$$\underset{m \times n}{T} = \underset{m \times n}{Z} \; \underset{n \times n}{Q_2} = \underset{m \times m}{Q_1} \; \underset{m \times n}{\Sigma} \tag{6.18}$$

where the columns of Q_1 and Q_2 are the normalized eigenvectors of ZZ^T and $Z^T Z$, respectively. T is the scores matrix. Each of the columns of matrix T is referred to as one principle component (PC). Each column in T is a linear combination of columns of Z and the eigenvectors in Q_2 act as weights. The first column represents the most significant PC (most important linear combination of variables in Z) that captures the maximum variance in the data matrix Z. There will be as many PCs as the number of variables. By storing matrices T and Q_2, one can generate the data matrix $(Z = TQ_2^T)$. Note that PCA is generally performed on mean-centered data matrix.

As discussed in SVD, denoising can be achieved by retaining only the eigenvectors corresponding to the most significant singular values. If Z contains a large number of variables, data compression can be achieved by storing matrices T_c and Q_{2c}, and regenerating denoised Z (Z_d) as needed. Dimensionality reduction is achieved using PCA by working with the most significant PCs (T_c) instead of all the variables (Z).

$$\underset{m \times n_c}{T_c} = \underset{m \times n}{Z} \underset{n \times n_c}{Q_{2c}} = \underset{m \times n_c}{Q_{1c}} \underset{n_c \times n_c}{\Sigma} \tag{6.19}$$

$$\text{Denoised data, } \underset{m \times n}{Z_d} = \underset{m \times n_c}{T_c} \underset{n_c \times n}{Q_{2c}^T} \tag{6.20}$$

The number of PCs to retain can be answered using multiple approaches. One of the popular ideas in PCA is to use the percentage of variance captured by the retained PCs. The percentage variation captured in any direction j is

$$\text{Percentage variance in data along direction } j = 100 \times \frac{\lambda_j}{\sum_{j=1}^{n} \lambda_j} \tag{6.21}$$

Sorting the eigenvalues from the highest to lowest,

$$\text{Percentage variance captured by first } k \text{ eigenvalues} = 100 \times \frac{\sum_{j=1}^{k} \lambda_j}{\sum_{j=1}^{n} \lambda_j} \tag{6.22}$$

In many applications, as a heuristic, the first k eigenvalues that capture 95% percent of variance is retained. Eigenvalues from $k+1$ to n can be considered as small, and the contribution from the terms that correspond to these eigenvalues are denoted as noise components.

Example 6.35. *Find the denoised data for the data matrix* $Z = \begin{bmatrix} 1.1 & 2.05 \\ 2 & 4.01 \end{bmatrix}$ *such that 95% variance is captured.*

Solution 6.35. *The singular values can be computed to be 5.0486 and 0.0616. The eigenvalues are 25.4884 and 0.0038.*

Contribution to percentage variance for each eigenvalue is $100 \frac{\lambda_1}{\sum_{j=1}^{n} \lambda_j}$. *Contribution of the eigenvalues 25.4884 and 0.0038 are 99.985 and 0.0149, respectively.*

Hence, denoised data that captures 95% variance should consider only one eigenvalue (or one singular value) in the compact SVD form.

$$Z_{denoised} = Q_{1t} \Sigma_t Q_{2t}^T$$

where

$$Q_{1t} = \begin{bmatrix} -0.4607 \\ -0.8876 \end{bmatrix} \text{ and } Q_{2t}^T = \begin{bmatrix} -0.4520 & -0.8920 \end{bmatrix} \text{ and } \Sigma_t = \begin{bmatrix} 5.0486 \end{bmatrix}$$

Thus, $Z_{denoised} = \begin{bmatrix} 1.0512 & 2.0747 \\ 2.0253 & 3.9972 \end{bmatrix}$.

Example 6.36. *Consider a flow process with the following flow diagram:*

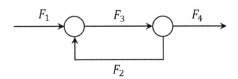

A few samples for this process are provided in Table 6.12. Perform PCA to comment on the significant variables and develop a model relating the variables.

TABLE 6.12

Flowrate Data for Example 6.36

F_1	F_2	F_3	F_4
9.59	7.92	17.83	10.06
10.33	10.32	20.52	10.1
9.04	11.41	20.66	8.84
10.98	10.17	20.9	10.58
10.18	10.17	20.5	9.98
9.82	10.81	20.61	10.11

Solution 6.36. *Mean-centering the data gives us the data matrix:*

$$Z = \begin{bmatrix} -0.40 & -2.2133 & -2.34 & 0.1150 \\ 0.34 & 0.1867 & 0.35 & 0.1550 \\ -0.95 & 1.2767 & 0.49 & -1.1050 \\ 0.99 & 0.0367 & 0.73 & 0.6350 \\ 0.19 & 0.0367 & 0.33 & 0.0350 \\ -0.17 & 0.6767 & 0.44 & 0.1650 \end{bmatrix}$$

SVD of the above data provides the following matrices:

$$Q_1 = \begin{bmatrix} -0.89 & -0.17 & 0.02 & -0.07 & 0.23 & 0.35 \\ 0.10 & 0.19 & -0.10 & -0.47 & 0.81 & -0.26 \\ 0.36 & -0.73 & -0.29 & -0.17 & 0.13 & 0.45 \\ 0.14 & 0.62 & -0.21 & -0.23 & -0.10 & 0.70 \\ 0.07 & 0.12 & -0.30 & 0.82 & 0.45 & 0.12 \\ 0.21 & -0.03 & 0.88 & 0.11 & 0.26 & 0.32 \end{bmatrix}$$

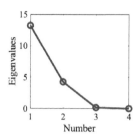

FIGURE 6.5 Eigenvalues for Example 6.36.

$$
\Sigma = \begin{bmatrix}
3.6345 & 0 & 0 & 0 \\
0 & 2.0576 & 0 & 0 \\
0 & 0 & 0.395 & 0 \\
0 & 0 & 0 & 0.1058 \\
0 & 0 & 0 & 0 \\
0 & 0 & 0 & 0
\end{bmatrix}; \quad
Q_2 = \begin{bmatrix}
0.04 & 0.71 & -0.44 & -0.54 \\
0.71 & -0.24 & 0.35 & -0.55 \\
0.69 & 0.29 & -0.22 & 0.62 \\
-0.10 & 0.59 & 0.79 & 0.12
\end{bmatrix}
$$

The eigenvalues (squares of diagonal values of Σ) are plotted in Figure 6.5. Note that the third and fourth eigenvalues are quite negligible compared to the first two eigenvalues. Hence, denoised data can be generated using the first two columns of Q_1 and Q_2.

$$
Z_{denoised} = \begin{bmatrix}
-0.4005 & -2.2200 & -2.3338 & 0.1098 \\
0.2957 & 0.1729 & 0.3725 & 0.1921 \\
-1.0112 & 1.3076 & 0.4755 & -1.0105 \\
0.9405 & 0.0526 & 0.7266 & 0.7035 \\
0.1855 & 0.1259 & 0.2497 & 0.1172 \\
-0.0100 & 0.5610 & 0.5096 & -0.1121
\end{bmatrix}
$$

Correspondingly, only the first two principal components of the data are significant.

$$PC\ 1: \begin{bmatrix} -3.2280 & 0.3751 & 1.3179 & 0.5120 & 0.2592 & 0.7637 \end{bmatrix}^T$$

$$PC\ 2: \begin{bmatrix} -0.3576 & 0.3902 & -1.4976 & 1.2837 & 0.2433 & -0.0620 \end{bmatrix}^T$$

Using PCA, we are able to capture the entire variance in data using fewer variables (PCs). Now, to build the linear model relating the original variables, least significant eigenvectors in Q_2 are selected. Thus, the linear model is

$$
\begin{bmatrix}
-0.4423 & 0.3525 & -0.2220 & 0.7942 \\
-0.5399 & -0.5525 & 0.6238 & 0.1189
\end{bmatrix}
\begin{bmatrix}
F_1 \\ F_2 \\ F_3 \\ F_4
\end{bmatrix} = 0;
$$

Converting into reduced row-echelon form, we get

$$\begin{bmatrix} 1 & 0 & -0.22 & -1.1 \\ 0 & 1 & -0.91 & 0.87 \end{bmatrix} \begin{bmatrix} F_1 \\ F_2 \\ F_3 \\ F_4 \end{bmatrix} = 0;$$

From the process flow diagram, it can be seen that the model equations relating the variables are $F_1 + F_2 - F_3 = 0$ and $-F_2 + F_3 - F_4 = 0$ which can be rearranged to $F_1 - F_4 = 0$ and $F_2 - F_3 + F_4 = 0$. These equations are almost same as the relations obtained from PCA. If the data were noise free, the last two singular values would be zero and hence, one would be able to recover the true model (not unique).

6.2.3 Neural Networks

In this section, the fundamentals of neural networks will be described. Connections to function approximation will be made throughout this section.

6.2.3.1 Neural Network Structures

Neural networks have captured the imagination of the scientific and non-scientific population. The fascination comes from the notion that neural networks are modeled based on the structure of human brain and hence might have the possibility of capturing the human cognitive processes in its entirety in the future [29, 37]. While this is quite a fanciful and aspirational notion, these structures are also quite useful in DS and ML even viewed purely as a nonlinear model form. The description that we provide in this section will take a more engineering viewpoint.

We will start with a bit of background on this field. Interestingly, the ideas that underlie neural networks have been around for several decades and this has been a fertile field of neuroscientists, psychologists, biologist, computer scientists and engineers. The basic notion is that since all cognitive processes occur in the brain, if one were able to understand the structure of the brain and then map functions to these structures, then an in-silico system that performs cognitive tasks is possible. The most fundamental units of the brain are the neurons. Neurons are connected to each other so that information in terms of electrical and chemical signals can be exchanged for collective decision making.

Computational neural networks model this fundamental component (neuron) and the interconnections through mathematical equations. A neuron receives multiple inputs from other neurons and/or exogenous inputs $(x_1 \ldots x_n)$ and generates an output (y). This is shown in Figure 6.6. The node is partitioned into two components (S and A), which can be thought of as two functions, S acting on $x_1 \ldots x_n$ to provide an intermediate output o and, A

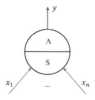

FIGURE 6.6 Representation of a node in a neural network.

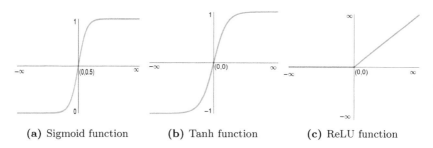

(a) Sigmoid function   **(b)** Tanh function   **(c)** ReLU function

FIGURE 6.7 Popular activation functions used in neural networks.

acting on o to give the final output. The S function is usually a summation function. Mathematically,

$$y = A(o); \quad o = S(x_1 \ldots x_n) = \sum_{i=1}^{n} x_i \tag{6.23}$$

Some of the popular activation functions used in neural networks are shown in Figure 6.7. Mathematically,

$$\text{Sigmoid: } A(o) = \frac{1}{1 + e^{-o}} \tag{6.24}$$

$$\text{Tanh: } A(o) = \frac{e^o - e^{-o}}{e^o + e^{-o}} \tag{6.25}$$

$$\text{ReLU: } A(o) = \begin{cases} o & o \geq 0 \\ 0 & o < 0 \end{cases} \tag{6.26}$$

Example 6.37. *Given 3 input features $x_1 = -3.2$, $x_2 = 4.1$ and $x_3 = 0.163$, calculate y if a single node is used with*

   a. *sigmoid activation function*
   b. *tanh activation function*
   c. *ReLU activation function*

Solution 6.37.

$$o = -3.2 + 4.1 + 0.163 = 1.063$$

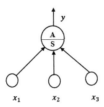

a.

$$y = \frac{1}{1+e^{-o}} \quad = \quad \frac{1}{1+e^{-1.063}} \quad = \quad 0.7433$$

b.

$$y = \frac{e^o - e^{-o}}{e^o + e^{-o}} \quad = \quad \frac{e^{1.063} - e^{-1.063}}{e^{1.063} + e^{-1.063}} \quad = \quad 0.7868$$

c. As $o > 0$, $y = 1.063$.

Assembling a neural network with neurons can now be described. Figure 6.8 depicts a fully-connected neural network structure. In this network structure one can see three layers (for the sake of simplicity) of neurons. The bottom most layer consists of nodes that are directly the input nodes. The input nodes represent input variables. Several layers follow this input layer (0th layer) leading all the way to the output layer. The single node in the output layer represents the output variable. A node in a layer has connections from all the nodes in the previous layer. The strength of the connections is represented by weights. The output of each of the nodes (except the input nodes, where the output of a node is the value of the input variable itself) is calculated using equation 6.23, except for a small change in the S function. The S function computes a weighted sum of all the inputs to the node. The weighted sum is then passed through the activation function to arrive at the final output of the node. For example, consider the output from the jth node in the kth layer($h_{k,j}$), where r_{k-1} is the number of nodes i the $(k-1)^{th}$ layer

$$h_{k,j} = A(o_{k,j} = S(h_{k-1,1}, \ldots, h_{k-1,r_{k-1}}))$$

$$S(h_{k-1,1}, \ldots, h_{k-1,r_{k-1}}) = \sum_{l=1}^{l=r_{k-1}} w_{k,lj} h_{k-1,l} \tag{6.27}$$

From this, it is clear that if values are provided for all the weights, the network computations can follow equation 6.27 starting from the known input values $x_1, \ldots, x_n$ in the forward direction all the way until $h_{m,1}$ in a m-layer network. For a MISO problem, this $h_{m,1}$ becomes the predicted value for an output $\hat{y}$ for a given set of input values. Extension of neural networks for

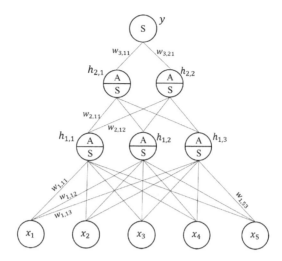

FIGURE 6.8 Complete neural network architecture.

MIMO problems follow principles similar to the extension of univariate to multivariate regression.

In summary, a neural network is a nonlinear model for the function approximation problem. Though an explicit form of the model can be written, it is more convenient to refer to the model as a weights structure W. W has as many rows as the layers in the network. Each row (r) will have as many elements as there are weights coming into the layer. The W has all the required information for predicting the output for any given input (new or from the training set), assuming that we know the form of the activation function (ReLU, sigmoid, tanh). It is possible that different nodes have different activation functions but this is not common.

Example 6.38. *Given 3 input features $x_1 = -3.2$, $x_2 = 4.1$, and $x_3 = 0.163$, calculate y if a neural network with 1 hidden layer having sigmoid activation function is used. Use two neurons in the hidden layer and the following weights*

$$w_{1,11} = 0.1; \quad w_{1,12} = 0.2; \quad w_{1,31} = 0.5; \quad w_{2,11} = 0.7;$$
$$w_{1,21} = 0.3; \quad w_{1,22} = 0.4; \quad w_{1,32} = 0.6; \quad w_{2,21} = 0.8;$$

Solution 6.38.

$$o_{1,1} = 0.1x_1 + 0.3x_2 + 0.5x_3 = 0.9915$$

$$o_{1,2} = 0.2x_1 + 0.4x_2 + 0.6x_3 = 1.0978$$

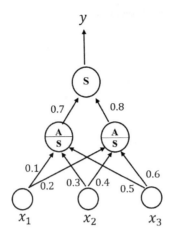

$$h_{1,1} = A(o_{1,1}) == \frac{1}{1 + e^{-1.0015}} = 0.7294$$

$$h_{1,2} = A(o_{1,2}) == \frac{1}{1 + e^{-1.1178}} = 0.7498$$

$$y = 0.7h_{1,1} + 0.8h_{1,2} = 1.1104$$

Example 6.39. *Consider the problem in Example 6.38 now with tanh activation function. Calculate* y.

Solution 6.39.

$$o_{1,1} = 0.9915; \qquad o_{1,2} = 1.0978$$

$$h_{1,1} = A(o_{1,1}) = \frac{e^{o_{1,1}} - e^{-o_{1,1}}}{e^{o_{1,1}} + e^{-o_{1,1}}} = 0.758$$

$$h_{1,2} = A(o_{1,2}) = \frac{e^{o_{1,2}} - e^{-o_{1,2}}}{e^{o_{1,2}} + e^{-o_{1,2}}} = 0.7997$$

$$y = 0.76h_{1,1} + 0.8h_{1,2} = 1.1704$$

Example 6.40. *For the problem described in Example 6.38, recalculate* y *if ReLU activation function is used in the hidden layer.*

Solution 6.40.

$$o_{1,1} = 0.9915; \qquad o_{1,2} = 1.0978$$

Both are greater than 0. Hence, $h_{1,1} = 0.9915$ *and* $h_{1,2} = 1.0978$.

$$\implies y = 1.5723$$

6.2.3.2 Training of Neural Networks

From a function approximation viewpoint, given the input matrix X and an output vector y, use of a neural network would require a choice for the structure and weights. Defining a structure would require the specification of the number layers m (the 0th layer is the input layer with as many nodes as the input variables and the mth layer is the output layer with one node), and the number of nodes in the layers $1, \ldots, m-1$. This choice is usually informed by some knowledge about the problem or decided by trial-and-error. Given a structure and data, the next task is to train the network. This is a straight-forward optimization problem (where m is the number of sample points, $\hat{y}(W) = h_{m,1}(W)$, and the superscript s represents the sample number),

$$\min_{W} \frac{1}{2} \sum_{s=1}^{m} (y^s - \hat{y}^s(W))^2 \tag{6.28}$$

It is clear that the predicted output will be a complex function of the weight structure W. The objective function represents the aim of the function approximation problem, which is to get the predicted output to match the measured output as closely as possible. It is quite easy to realize that the optimization problem posed is a NLP described in Chapter 4 and any of the methods described for solving NLPs could be used.

Example 6.41. *Given 3 input features x_1, x_2, x_3, and a neural network with 1 hidden layer having 2 neurons, show the first iteration of parameter optimization using gradient descent algorithm. Use step length as 0.5. Use sigmoid activation function in the hidden layer and the weights given in Example 6.38 as the initial guess. Given data sample: $x_1 = -3.2$, $x_2 = 4.1$, $x_3 = 0.163$, and $y = 2$.*

Solution 6.41. *Decision variable, $W = w_{k,lj}$ where $k = [1,2]$.*

$$l = \begin{cases} [1 \quad 3] & for \quad k = 1 \\ [1 \quad 2] & for \quad k = 2 \end{cases}$$

$$j = \begin{cases} [1 \quad 2] & for \quad k = 1 \\ 1 & for \quad k = 2 \end{cases}$$

Objective function:

$$\min_{W} \quad f(W) = \frac{1}{2} \sum_{s=1}^{m} (y - \hat{y}(W))^2 \quad \text{(Since only 1 sample is given)}$$

$$o_{1,1} = w_{1,11}x_1 + w_{1,21}x_2 + w_{1,31}x_3$$

$$o_{1,2} = w_{1,12}x_1 + w_{1,22}x_2 + w_{1,32}x_3$$

$$\hat{y} = w_{2,11}h_{1,1} + +w_{2,21}h_{1,2} = w_{2,11}A(o_{1,1}) + +w_{2,21}A(o_{1,2})$$

$$\frac{\partial f}{\partial w_{k,lj}} = \frac{1}{2} \times 2\,(y - \hat{y})\,\frac{-\partial \hat{y}}{\partial w_{k,lj}} = (\hat{y} - y)\,\frac{\partial \hat{y}}{\partial w_{k,lj}}$$

$$\frac{\partial f}{\partial w_{2,11}} = (\hat{y} - y)\,h_{1,1}; \qquad \frac{\partial f}{\partial w_{2,21}} = (\hat{y} - y)\,h_{1,2}$$

$$\frac{\partial f}{\partial w_{1,11}} = (\hat{y} - y)\,\frac{\partial \hat{y}}{\partial w_{1,11}} = (\hat{y} - y)\left[w_{2,11}\frac{\partial A(o_{1,1})}{\partial w_{1,11}} \right]$$

$$A(o) = \frac{1}{1 + e^{-o}} \;\Rightarrow\; \frac{dA(o)}{do} = -\left(1 + e^{-o}\right)^{-2}\left(-e^{-o}\right) = \frac{e^{-o}}{\left(1 + e^{-o}\right)^2}$$

$$\Rightarrow \frac{\partial f}{\partial w_{1,11}} = (\hat{y} - y)\left[w_{2,11}\frac{\partial A(o_{1,1})}{\partial o_{1,1}}\frac{\partial o_{1,11}}{\partial w_{1,11}} \right] = (\hat{y} - y)\,w_{2,11}\frac{e^{-o_{1,1}}x_1}{\left(1 + e^{-o_{1,1}}\right)^2}$$

Similarly,

$$\frac{\partial f}{\partial w_{1,21}} = (\hat{y} - y)\left[w_{2,11}\frac{\partial A(o_{1,1})}{\partial o_{1,1}}\frac{\partial o_{1,11}}{\partial w_{1,21}} \right] = (\hat{y} - y)\,w_{2,11}\frac{e^{-o_{1,1}}x_2}{\left(1 + e^{-o_{1,1}}\right)^2}$$

$$\frac{\partial f}{\partial w_{1,31}} = (\hat{y} - y)\,w_{2,11}\frac{e^{-o_{1,1}}x_3}{\left(1 + e^{-o_{1,1}}\right)^2}; \qquad \frac{\partial f}{\partial w_{1,12}} = (\hat{y} - y)\,w_{2,21}\frac{e^{-o_{1,2}}x_1}{\left(1 + e^{-o_{1,2}}\right)^2}$$

$$\frac{\partial f}{\partial w_{1,22}} = (\hat{y} - y)\,w_{2,21}\frac{e^{-o_{1,2}}x_2}{\left(1 + e^{-o_{1,2}}\right)^2}; \qquad \frac{\partial f}{\partial w_{1,32}} = (\hat{y} - y)\,w_{2,21}\frac{e^{-o_{1,2}}x_3}{\left(1 + e^{-o_{1,2}}\right)^2}$$

For steepest descent, $w_{k,lj}^{q+1} = w_{k,lj}^q - \alpha\left(\frac{\partial f}{\partial w_{k,lj}}\right)_{w^q}$ *where* q *is the iteration number. Given,*

$$w_{1,11}^0 = 0.1; \quad w_{1,12}^0 = 0.2; \quad w_{1,31}^0 = 0.5; \quad w_{2,11}^0 = 0.7;$$
$$w_{1,21}^0 = 0.3; \quad w_{1,22}^0 = 0.4; \quad w_{1,32}^0 = 0.6; \quad w_{2,21}^0 = 0.8; \quad \alpha = 0.5$$

From Example 6.38,

$$o_{1,2}^0 = 1.0978, \quad o_{1,1}^0 = 0.9915, \quad \hat{y}^0 = 1.1104, \quad y = 2$$
$$\text{Error, } e^0 = 1.1104 - 2 = -0.8896$$

$$w_{1,11}^1 = w_{1,11}^0 - 0.5\left(\frac{\partial f}{\partial w_{1,11}}\right)_{w^0}$$

$$= 0.1 - 0.5\left((1.1104 - 2) \times 0.7 \times \frac{e^{-0.9915}}{\left(1 + e^{-0.9915}\right)^2}\,(-3.2) \right) = -0.0967$$

$$w_{1,21}^1 = w_{1,21}^0 - 0.5\frac{\partial f}{\partial w_{1,21}}\bigg|_{w^0}$$

$$= 0.3 - 0.5\left((1.1104 - 2) \times 0.7 \times 4.1 \times \frac{e^{-0.9915}}{\left(1 + e^{-0.9915}\right)^2} \right) = 0.5520$$

$$w^1_{1,31} = w^0_{1,31} - 0.5 \frac{\partial f}{\partial w_{1,31}}\bigg|_{w^0}$$

$$= 0.5 - 0.5 \left((1.1104 - 2) \times 0.7 \times 0.163 \times \frac{e^{-0.9915}}{(1 + e^{-0.9915})^2} \right) = 0.51$$

$$w^1_{1,12} = w^0_{1,12} - 0.5 \frac{\partial f}{\partial w_{1,12}}\bigg|_{w^0}$$

$$= 0.2 - 0.5 \left((1.1104 - 2) \times 0.8 \times (-3.2) \times \frac{e^{-1.0978}}{(1 + e^{-1.0978})^2} \right) = -0.0136$$

$$w^1_{1,22} = w^0_{1,22} - 0.5 \frac{\partial f}{\partial w_{1,22}}\bigg|_{w^0}$$

$$= 0.4 - 0.5 \left((1.1104 - 2) \times 0.8 \times (4.1) \times \frac{e^{-1.0978}}{(1 + e^{-1.0978})^2} \right) = 0.6736$$

$$w^1_{1,32} = w^0_{1,32} - 0.5 \frac{\partial f}{\partial w_{1,32}}\bigg|_{w^0}$$

$$= 0.6 - 0.5 \left((1.1104 - 2) \times 0.8 \times (0.163) \times \frac{e^{-1.0978}}{(1 + e^{-1.0978})^2} \right) = 0.6109$$

$$w^1_{2,11} = w^0_{2,11} - 0.5 \frac{\partial f}{\partial w_{2,11}}\bigg|_{w^0} = 0.7 - 0.5 \times (1.1104 - 2) \times 0.7294 = 1.0244$$

$$w^1_{2,21} = w^0_{2,21} - 0.5 \frac{\partial f}{\partial w_{2,21}}\bigg|_{w^0} = 0.8 - 0.5 \times (1.1104 - 2) \times 0.7498 = 1.1335$$

$$(\because h^0_{1,1} = 0.7294 \text{ and } h^0_{1,2} = 0.7498 \quad \textit{from Example 6.38})$$

The updated weights are provided in Table 6.13.

$$o^1_{1,1} = (-0.0967 \times (-3.2)) + (0.552 \times 4.1) + (0.51 \times 0.163) = 2.6555$$

$$o^1_{1,2} = (-0.0136 \times (-3.2)) + (0.6736 \times 4.1) + (0.6109 \times 0.163) = 2.905$$

TABLE 6.13

Updated Weights for Examples 6.41 and 6.43

Decision Variable	Initial Guess	Iteration 1
$w_{1,11}$	0.1	-0.0967
$w_{1,21}$	0.3	0.552
$w_{1,31}$	0.5	0.51
$w_{1,12}$	0.2	-0.0136
$w_{1,22}$	0.4	0.6736
$w_{1,32}$	0.6	0.6109
$w_{2,11}$	0.7	1.0244
$w_{2,21}$	0.8	1.1335

$$h_{1,1}^1 = \frac{1}{1 + e^{-2.6555}} = 0.9343; \quad h_{1,1}^1 = \frac{1}{1 + e^{-2.905}} = 0.9481$$

$$\hat{y}^1 = 1.0244 \times 0.9343 + 1.1335 \times 0.9481 = 2.0318$$

$$\text{Error, } e^1 = 2.0318 - 2 = 0.0318$$

Note that magnitude of error has reduced from 0.8896 to 0.0318 after the first iteration.

Example 6.42. *Consider the following dataset.*

x_1	x_2	x_3	y
-3.2	4.1	0.163	2
-1.7	3.9	1.35	6.2
-4.16	4.33	0.78	3.5
-2.73	3.14	1.07	4.2

Assume that a neural network with 1 hidden layer having 2 neurons is used to predict y using $\mathbf{x} = [x_1, x_2, x_3]$. Hidden layer has sigmoid activation function. Show 1 iteration of parameter optimization using gradient descent algorithm if step length is 0.05. Use the following weights as initial guess. Given,

$$w_{1,11}^0 = 0.1; \; w_{1,12}^0 = 0.2; \; w_{1,31}^0 = 0.5; \; w_{2,11}^0 = 0.7;$$
$$w_{1,21}^0 = 0.3; \; w_{1,22}^0 = 0.4; \; w_{1,32}^0 = 0.6; \; w_{2,21}^0 = 0.8$$

Solution 6.42. *The neural network structure is shown in Figure 6.9.*

$$o_{1,1}^i = w_{1,11}\, x_1^i + w_{1,21}\, x_2^i + w_{1,31}\, x_3^i$$

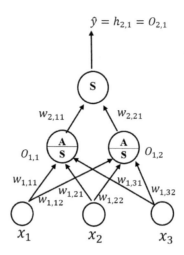

FIGURE 6.9 Neural network structure for Example 6.42.

$$o_{1,2}^i = w_{1,12}\, x_1^i + w_{1,22}\, x_2^i + w_{1,32}\, x_3^i$$

$$\hat{y}^i = h_2^i = o_2^i = w_{2,11}\, h_{1,1}^i + w_{2,21}\, h_{1,2}^i$$

$$h_{1,l}^i = A(o_{1,l}^i) = \frac{1}{1 + e^{-o_{i,l}^i}}$$

Using the above expressions and initial guesses for weights

$$\mathbf{o}_{1,1}^0 = \left[o_{1,1}^i \right]_{i=1:4}^0 = \begin{bmatrix} 0.9915 & 1.675 & 1.273 & 1.204 \end{bmatrix}^T$$

$$\mathbf{h}_{1,1}^0 = \begin{bmatrix} 0.7294 & 0.8422 & 0.7813 & 0.7692 \end{bmatrix}^T$$

$$\mathbf{o}_{1,2}^0 = \left[o_{1,2}^i \right]_{i=1:4}^0 = \begin{bmatrix} 1.0978 & 2.03 & 1.368 & 1.352 \end{bmatrix}^T$$

$$\mathbf{h}_{1,2}^0 = \begin{bmatrix} 0.7498 & 0.8839 & 0.7971 & 0.7945 \end{bmatrix}^T$$

$$\hat{\mathbf{y}}^0 = \begin{bmatrix} 1.1104 & 1.2967 & 1.1845 & 1.174 \end{bmatrix}^T; \quad SSE^0 = 39.3516$$

Now, we need to update weights such that $SSE = \sum_{i=1}^m \left(\hat{y}^i(W) - y \right)^2$ *is minimized.*

$$w_{k,lj}^{q+1} = w_{k,lj}^q - \alpha \frac{\partial f}{\partial w_{k,lj}} \bigg|_{w^q}; \qquad \frac{\partial f}{\partial w_{k,lj}} = 2\left(\hat{y}^i - y \right) \frac{\partial \hat{y}}{\partial w_{k,lj}} \bigg|_i$$

$$\frac{\partial \hat{y}}{\partial w_{2,l1}} = h_{1,l}; \qquad \frac{\partial \hat{y}}{\partial w_{1,lj}} = w_{2,ji} \frac{\partial h_{1,j}}{\partial w_{1,lj}} \qquad l = 1:3;\ j = 1:2$$

Evaluating

$$w^1_{1,11} = -0.0674, \quad w^1_{1,12} = 0.5331, \quad w^1_{1,31} = 0.5627, \quad w^1_{2,11} = 1.1458$$
$$w^1_{1,21} = 0.0281, \quad w^1_{1,22} = 0.6328, \quad w^1_{1,32} = 0.6611, \quad w^1_{2,21} = 1.2625$$

Calculating $o_{1,1}, \ o_{1,2}, \ h_{1,1}, \ h_{1,2}$ *and* $h_{2,1}$

$$\mathbf{o}^1_{1,1} = \begin{bmatrix} 2.493 & 2.953 & 3.0276 & 2.46 \end{bmatrix}^T$$
$$\mathbf{o}^1_{1,2} = \begin{bmatrix} 2.612 & 3.313 & 3.139 & 2.618 \end{bmatrix}^T$$
$$\mathbf{h}^1_{1,1} = \begin{bmatrix} 0.924 & 0.95 & 0.954 & 0.921 \end{bmatrix}^T$$
$$\mathbf{h}^1_{1,2} = \begin{bmatrix} 0.9316 & 0.965 & 0.9585 & 0.932 \end{bmatrix}^T$$

$$\hat{\mathbf{y}}^1 = \mathbf{h}^1_{2,1} = \begin{bmatrix} 2.2345 & 2.3071 & 2.3029 & 2.2322 \end{bmatrix}^T; \quad SSE^1 = 20.5146$$

Note that SSE has reduced from 39.3516 to 20.5146 using $\alpha = 0.05$.

6.2.3.3 Backpropagation Algorithm

One of the algorithms that has been specifically used in NN training is the backpropagation algorithm [47], which is a gradient descent algorithm that uses the chain differentiation rule. This algorithm will be described as a series of steps. The algorithm is an iteration between forward propagation and backward error update steps until the neural network is trained to a sufficient level of accuracy. Before the algorithm is described, some terms are described and visualized in a neural network structure (Figure 6.10). In the figure, a p layer neural network is depicted. The 0th layer is the input layer where there are as many nodes as input variables. The depiction of the node is slightly different from the one shown in Figure 6.6. In this figure, we show the output of the node within that node itself. The layers $1, \ldots, \ p - 1$ are called the hidden layers and the pth layer is called the output layer. The final output of the lth node in hidden layer k is denoted by $h_{k,l}$, and the intermediate output before the application of the activation function is denoted by $o_{k,l}$. As described before $h_{p,1}$ is the predicted output. r_k is the number of nodes in the kth layer. For every node, another variable called error, $\delta_{k,l} = \frac{\partial e}{\partial o_{k,l}}$ (for the lth node in the kth layer), is defined. e is the overall prediction error ($e = \sum_{i=1}^{m} e^i$). The weight between l^{th} node in the kth layer and j^{th} node in the $(k+1)^{th}$ layer is $w_{k+1,ij}$. These weights makeup the weights structure W. As a first step in the algorithm, W is initialized. Many strategies could be used for initialization but the most popular method is to randomize weights.

Once the weights are initialized and an activation function is chosen, it can be seen that starting from the input values, the values of all the nodes in all the hidden and output layers can be computed using equation 6.27.

In other words, given the input values, the output $(h_{p,1})$ can be predicted (forward pass). The goal of the training is to improve the weights so that the predicted output is close to the actual output for all the samples. This is achieved by iteratively updating the weights (after a combination of forward and backward pass through the network) till convergence. The forward pass has already been described. The only difference is that the forward pass will be performed for every sample $i = 1 \ldots m$, to determine the value of all the nodes in the network including the output node, which will then be the predicted output for that particular input sample. We now come to the backward pass through the backpropagation equations outlined below.

Define,

$$\hat{y}^i = h_{p,1}^i; \text{ and } \delta_{k,l}^i = \frac{\partial e^i}{\partial o_{k,l}^i}$$

where $o_{k,l}$ is the intermediate output of lth node in k^{th} layer and e^i is the prediction error for sample i defined as $e^i = 0.5\left(y^i - \hat{y}^i\right)^2 = 0.5\left(y^i - h_{p,1}^i\right)^2$.

$$h_{k,l}^i = A(o_{k,l}^i); \qquad h_{k+1,j}^i = A(o_{k+1,j}^i)$$

$$o_{k+1,j}^i = \sum_{s=1}^{r_k} w_{k+1,sj} h_{k,s}^i = \sum_{s=1}^{r_k} w_{k+1,sj} A(o_{k,s}^i) \qquad (6.29)$$

Since the effect of change in $o_{k,s}^i$ would impact $o_{k+1,j}^i \; \forall \; j = 1 : r_{k+1}$, and will in turn affect E^i, we can write $e^i = e^i(o_{k+1,j}^i) \; \forall \; j = 1 : r_{k+1}$. Using chain rule of differentiation,

$$\delta_{k,l}^i = \frac{\partial e^i}{\partial o_{k,l}^i} = \sum_{j=1}^{r_{k+1}} \frac{\partial e^i}{\partial o_{k+1,j}^i} \frac{\partial o_{k+1,j}^i}{\partial o_{k,l}^i} = \sum_{j=1}^{r_{k+1}} \delta_{k+1,j}^i \frac{\partial o_{k+1,j}^i}{\partial o_{k,l}^i}$$

From equation 6.29,

$$\frac{\partial o_{k+1,j}^i}{\partial o_{k,l}^i} = w_{k+1,lj} A'(o_{k,l}^i)$$

$$\delta_{k,l}^i = \sum_{j=1}^{r_{k+1}} \delta_{k+1,j}^i w_{k+1,lj} A'(o_{k,l}^i) = A'(o_{k,l}^i) \sum_{j=1}^{r_{k+1}} w_{k+1,lj} \delta_{k+1,j}^i \qquad (6.30)$$

Now, for m samples (m could include all samples or samples from a single batch), $e = \sum_{i=1}^{m} e^i$.

$$\delta_{k,l} = \frac{\partial e}{\partial o_{k,l}} = \sum_{i=1}^{m} \frac{\partial e^i}{\partial o_{k,l}^i} = \sum_{i=1}^{m} \delta_{k,l}^i = \sum_{i=1}^{m} \left(A'\left(o_{k,l}^i\right) \left(\sum_{j=1}^{r_{k+1}} w_{k+1,lj} \delta_{k+1,j}^i \right) \right)$$

$$(6.31)$$

For the output layer, there is only one node and $h_{p,1} = o_{p,1}$. Hence, we get,

$$\delta_{p,1}^i = \frac{\partial e^i}{\partial o_{p,1}^i} = \frac{1}{2} 2 \left(y^i - h_{p,1}^i\right) \left(\frac{-\partial h_{p,1}^i}{\partial o_{p,1}^i}\right) = -\left(y^i - h_{p,1}^i\right) \qquad (6.32)$$

$$\delta_{p,1} = \sum_{i=1}^{m} \delta_{p,1}^i = -\sum_{i=1}^{m} \left(y^i - h_{p,1}^i\right) \qquad (6.33)$$

The backpropagation starts with the computation of $\delta_{p,1}$ using equation 6.33. The $\delta_{k,l}$'s for all the other layers are computed sequentially by backpropagation through equation 6.31.

Once all the variables are computed, then the weights can be updated (including m samples) as shown below:

$$w_{k+1,lj}^{new} = w_{k+1,lj} + \eta \sum_{i=1}^{m} \left(\frac{-\partial e^i}{\partial w_{k+1,lj}}\right) = w_{k+1,lj} + \eta \sum_{i=1}^{m} \left(\frac{-\partial e^i}{\partial o_{k+1,j}}\right) \left(\frac{\partial o_{k+1,j}}{\partial w_{k+1,lj}}\right)$$

Using 6.29,

$$w_{k+1,lj}^{new} = w_{k+1,lj} - \eta \sum_{i=1}^{m} \delta_{k+1,j}^i h_{k,l}^i \qquad (6.34)$$

η is referred to as the learning rate, that is usually chosen by the user while training the network. It should be noticed (though obvious), the weights are the same for all samples, that is, a single network with the chosen weights should predict the outputs accurately for all the samples. Now using the updated weights, the algorithm continues with a forward and backward pass for the next weights update. This process is stopped when the learning achieved is satisfactory.

Example 6.43. *Consider the problem given in Example 6.41. Show 1 iteration of backpropagation to find optimal parameters. Use $\eta = 0.5$.*

Solution 6.43. *For backpropagation (given 1 sample):*

$$w_{k+1,lj}^{q+1} = w_{k+1,lj}^{q} - \eta \, \delta_{k+1,j}^{q} h_{k,l}^{q}$$

From Example 6.41,

$$o_{1,2}^0 = 1.0978; \quad w_{1,11}^0 = 0.1; \quad w_{1,12}^0 = 0.2; \quad w_{1,31}^0 = 0.5; \quad w_{2,11}^0 = 0.7$$
$$o_{1,1}^0 = 0.9915; \quad w_{1,21}^0 = 0.3; \quad w_{1,22}^0 = 0.4; \quad w_{1,32}^0 = 0.6; \quad w_{2,21}^0 = 0.8$$
$$h_{1,1}^0 = 0.7294; \quad w_{2,11}^0 = 0.7; \quad w_{2,21}^0 = 0.8$$

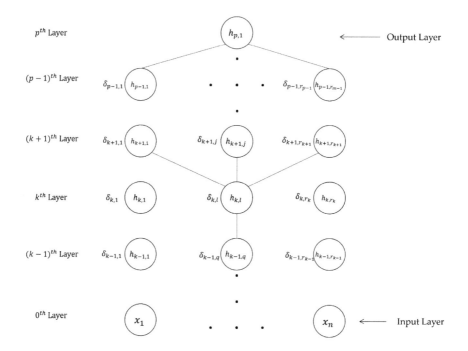

FIGURE 6.10 Backpropagation algorithm.

$$h_{1,2}^0 = 0.7498; \quad P = 2; \quad y = 2; \quad \hat{y}^0 = o_{2,1}^0 = h_{2,1}^0 = 1.1104$$
$$h_{0,11} = x_1 = -3.2; \quad h_{0,2} = x_2 = 4.1; \quad h_{0,3} = x_3 = 0.163$$

$$A(o) = \frac{1}{1 + e^{-o}}; \quad A'(o) = \frac{e^{-o}}{(1 + e^{-o})^2}$$

Iteration 1

$$\delta_{2,1}^0 = -(y - h_{2,1}) = -(2 - 1.1104) = -0.8896$$

$$\delta_{1,1}^0 = A'(o_{1,1}^0)w_{2,11}^0\delta_{2,1}^0 = \frac{e^{-0.9915}}{(1 + e^{-0.9915})^2} \times 0.7 \times (-0.8896) = -0.1229$$

$$\delta_{1,2}^0 = A'(o_{1,1}^0)w_{2,21}^0\delta_{2,1}^0 = \frac{e^{-0.1.0978}}{(1 + e^{-1.0978})^2} \times 0.8 \times (-0.8896) = -0.1335$$

Now,

$$w_{1,11}^1 = w_{1,11}^0 - \eta\delta_{1,1}^0 h_{0,1} = 0.1 - 0.5 \times (-0.1229) \times (-3.2) = -0.0967$$
$$w_{1,21}^1 = w_{1,21}^0 - \eta\delta_{1,1}^0 h_{0,2} = 0.3 - 0.5 \times (-0.1229) \times (4.1) = 0.552$$
$$w_{1,31}^1 = w_{1,31}^0 - \eta\delta_{1,1}^0 h_{0,3} = 0.5 - 0.5 \times (-0.1229) \times (0.163) = 0.51$$
$$w_{1,12}^1 = w_{1,12}^0 - \eta\delta_{1,2}^0 h_{0,1} = 0.2 - 0.5 \times (-0.1335) \times (-3.2) = -0.0136$$
$$w_{1,22}^1 = w_{1,22}^0 - \eta\delta_{1,2}^0 h_{0,2} = 0.4 - 0.5 \times (-0.1335) \times (4.1) = 0.6736$$
$$w_{1,32}^1 = w_{1,32}^0 - \eta\delta_{1,2}^0 h_{0,3} = 0.6 - 0.5 \times (-0.1335) \times (0.163) = 0.6109$$
$$w_{2,11}^1 = w_{2,11}^0 - \eta\delta_{2,1}^0 h_{1,1} = 0.7 - 0.5 \times (-0.8896) \times (0.7294) = 1.0244$$
$$w_{2,21}^1 = w_{2,21}^0 - \eta\delta_{2,1}^0 h_{1,2} = 0.8 - 0.5 \times (-0.8896) \times (0.7498) = 1.1335$$

The updated weights are same as that obtained in Example 6.41 (refer Table 6.13). Using the new weights, we can calculate new $\hat{y}$.

$$o_{1,1}^1 = (-0.0967 \times -3.2) + (0.552 \times 4.1) + (0.51 \times 0.163) = 2.6555$$
$$o_{1,2}^1 = (-0.0136 \times -3.2) + (0.6736 \times 4.1) + (0.6109 \times 0.163) = 2.905$$

$$h_{1,1}^1 = \frac{1}{1 + e^{-2.6555}} = 09343; \quad h_{1,2}^1 = \frac{1}{1 + e^{-2.905}} = 0.9481$$
$$\hat{y} = 1.02445 \times 0.9343 + 1.1335 \times 0.9481 = 2.0318; \quad y - \hat{y}^1 = 0.0318$$

The error has reduced from 0.8851 to 0.0318 after the first iteration of backpropagation. Note that $\hat{y}$ obtained after 1 iteration using backpropagation is same as that obtained using gradient descent method.

Example 6.44. *Consider the data in Example 6.42. Show 1 epoch (with whole data as a single batch) of parameter optimization using backpropagation. Use $\eta = 0.05$.*

Solution 6.44. *From Example 6.42, we have $o_{1,1}^0$, $o_{1,2}^0$, $h_{1,1}^0$, $h_{1,2}^0$, and $\hat{y}^0 \left(= h_{2,1}^0 = o_{2,1}^0\right)$.*

$$\delta_{2,1}^i = -(y^i - \hat{y}^i); \quad \delta_{1,1}^i = \delta_{2,1}^i w_{2,11} \frac{e^{-o_{1,1}^i}}{\left(1 + e^{-o_{1,1}^i}\right)^2}$$

$$\delta_{2,1} = \left[\delta_{2,1}^i\right]_{i=1:4} = \begin{bmatrix} -0.89 \\ -4.903 \\ -2.316 \\ -3.026 \end{bmatrix}; \quad \delta_{1,1} = \left[\delta_{1,1}^i\right]_{i=1:4} = \begin{bmatrix} -0.123 \\ -0.456 \\ -0.277 \\ -0.376 \end{bmatrix}$$

Similarly,

$$\delta^i_{1,2} = \delta^i_{2,1} \, w_{2,21} \frac{e^{-O^i_{1,2}}}{\left(1 + e^{-O^i_{1,2}}\right)^2}; \quad \delta_{1,2} = \left[\delta^i_{1,2}\right]_{i=1:4} = \begin{bmatrix} -0.133 \\ -0.403 - 0.3 \\ -0.395 \end{bmatrix}$$

$$w^1_{1,lj} = w^{[0]}_{1,lj} - \eta \sum_{i=1}^{4} \left(\delta^i_{1,j} \, h^i_{0,l}\right) \forall \; l = 1:3; \; j = 1:2$$

$$w^1_{2,lj} = w^{[0]}_{2,lj} - \eta \sum_{i=1}^{4} \left(\delta^i_{2,j} \, h^i_{1,l}\right) \forall \; l = 1:2; \; j = 1$$

Using above expressions, we get the updated weights as shown in the table below:

Decision Variable	Initial Guess	After Iteration 1
$w_{1,11}$	0.1	-0.0674
$w_{1,21}$	0.3	0.5331
$w_{1,31}$	0.5	0.5627
$w_{1,12}$	0.2	0.0281
$w_{1,22}$	0.4	0.6328
$w_{1,32}$	0.6	0.6611
$w_{2,11}$	0.7	1.1458
$w_{2,21}$	0.8	1.2625

Using the updated weights, we get

$$o^1_{1,1} == \begin{bmatrix} 2.493 \\ 2.953 \\ 3.0276 \\ 2.46 \end{bmatrix}; \quad O^1_{1,2} = \begin{bmatrix} 2.612 \\ 3.313 \\ 3.139 \\ 2.618 \end{bmatrix}; \quad h^1_{1,1} = \begin{bmatrix} 0.924 \\ 0.95 \\ 0.954 \\ 0.921 \end{bmatrix}$$

$$h^1_{1,2} = \begin{bmatrix} 0.9316 \\ 0.965 \\ 0.9585 \\ 0.932 \end{bmatrix}; \quad \hat{\mathbf{y}}^1 = \begin{bmatrix} 2.2345 \\ 2.3071 \\ 2.3029 \\ 2.2322 \end{bmatrix}; \quad SSE^1 = 20.5146$$

Note that SSE has reduced from 39.3516 to 20.5146. However, $\hat{\mathbf{y}}$ is still far from $\mathbf{y}$ and it takes a few iterations for it to converge to $\mathbf{y}$.

Example 6.45. *Consider the data in Example 6.42. Show 1 epoch calculation of parameter optimization using backpropagation with two batches Use $\eta = 1$.*

Solution 6.45. *Let us say our first batch has the first 2 samples and second batch has the last 2 samples.(Samples could be chosen randomly but it is shown sequentially here for the sake of simplicity.)*

The update rules are same as that used for Example 6.44, except that summation is taken for a batch rather than whole samples. Here, 1 epoch means 2 iterations (corresponding to each batch).

TABLE 6.14

Decision Variable Values for Example 6.45

Decision Variable	Initial Guess	After Iteration 1	After Iteration 2 (epoch 1)
$w_{1,11}$	0.1	0.0416	−0.0332
$w_{1,21}$	0.3	0.4141	0.4963
$w_{1,31}$	0.5	0.5318	0.554
$w_{1,12}$	0.2	0.1444	0.0656
$w_{1,22}$	0.4	0.5059	0.5925
$w_{1,32}$	0.6	0.6283	0.6515
$w_{2,11}$	0.7	0.9389	1.1219
$w_{2,21}$	0.8	1.0501	1.235

Iteration 1 (using batch 1, first 2 samples)

$$\delta_{2,1} = \begin{bmatrix} -0.8896 \\ -4.9033 \end{bmatrix} ; \quad \delta_{1,1} = \begin{bmatrix} -0.1229 \\ -0.4561 \end{bmatrix} ; \quad \delta_{1,2} = \begin{bmatrix} -0.1335 \\ -0.4025 \end{bmatrix}$$

$$w_{k+1,lj}^{1} = w_{k+1,lj}^{[0]} - \eta \sum_{i=1}^{2} \left(\delta_{k+1,j}^{i} \, h_{k,l}^{i} \right)$$

Once we calculate weights, we can get the values of $o_{1,1}$, $o_{1,2}$, $h_{1,1}$, $h_{1,2}$, and $\mathbf{y}$ similar to the approach followed in the previous examples. Calculated weights are given in Table 6.14.

Using the updated weights, we get

$$o_{1,1}^{1} = \begin{bmatrix} 1.6516 \\ 2.262 \\ 2.035 \\ 1.756 \end{bmatrix} ; \quad o_{1,2}^{1} = \begin{bmatrix} 1.714 \\ 2.575 \\ 2.0796 \\ 1.866 \end{bmatrix} ; \quad h_{1,1}^{1} = \begin{bmatrix} 0.839 \\ 0.906 \\ 0.884 \\ 0.853 \end{bmatrix}$$

$$h_{1,2}^{1} = \begin{bmatrix} 0.847 \\ 0.929 \\ 0.889 \\ 0.866 \end{bmatrix} ; \quad \hat{y}^{1} = \begin{bmatrix} 1.6777 \\ 1.8262 \\ 1.7638 \\ 1.71 \end{bmatrix} ; \quad SSE^{1} = 28.4487$$

Iteration 2 Here we use the second batch data for all the calculations.

$$\delta_{2,1} = \begin{bmatrix} -(3.5 - 1.7638) \\ -(4.2 - 1.71) \end{bmatrix} = \begin{bmatrix} -1.7362 \\ -2.49 \end{bmatrix}$$

Similarly, $\delta_{1,1} = \begin{bmatrix} -0.1666 \\ -0.2937 \end{bmatrix}$ *and* $\delta_{1,2} = \begin{bmatrix} -0.18 \\ -0.3033 \end{bmatrix}$.

$$w^2_{k+1,lj} = w^1_{k+1,lj} - \eta \sum_{i=1}^{2} \left(\delta^i_{k+1,j} \, h^i_{k,l} \right)$$

The values of weights obtained using the above equation are given in the last column of Table 6.14. Since there are only 2 batches with 2 iterations, we have used all samples exactly once. Thus we complete 1 epoch. Using the updated weights, $\hat{y}^2$ can be found out.

$$o^2_{1,1} = \begin{bmatrix} 2.231 \\ 2.74 \\ 2.72 \\ 2.24 \end{bmatrix} ; \quad o^2_{1,2} = \begin{bmatrix} 2.33 \\ 3.08 \\ 2.8 \\ 2.38 \end{bmatrix} ; \quad h^2_{1,1} = \begin{bmatrix} 0.903 \\ 0.9393 \\ 0.9381 \\ 0.9039 \end{bmatrix}$$

$$h^1_{1,2} = \begin{bmatrix} 0.911 \\ 0.956 \\ 0.9427 \\ 0.9152 \end{bmatrix} ; \quad \hat{y}^1 = \begin{bmatrix} 2.138 \\ 2.235 \\ 2.217 \\ 2.144 \end{bmatrix} ; \quad SSE^1 = 21.6165$$

Although the error has reduced from 39.3516 to 21.6165 after epoch 1, note that $\hat{y}$ is still far from y. It will take a few epochs for the error to reduce considerably.

6.3 NON-PARAMETRIC METHODS

In this section, we will describe various non-parametric methods for function approximation. While there might be some fuzzy boundaries between parametric and non-parametric models, non-parametric models will usually have one or more of these features: (i) there might not be an explicit model form that is easily identifiable, (ii) need to store training data to also perform predictions (in parametric models, only the derived model parameters are enough), and (iii) number of parameters needed might be of the same order of magnitude as the number of datapoints.

6.3.1 k-Nearest Neighbors (k-NN)

The simplest non-parametric approach for function approximation is the k-NN approach [2]. k-NN uses the philosophy that if you want to know someone, understand the neighbors. There is no training phase for k-NN, it just holds on to the x, y data. As a result k-NN is also called a lazy algorithm, where no work is done before predictions. In other words, no model is built during the training phase. When a new datapoint x_{new} is presented to the k-NN algorithm, it finds the k nearest neighbors. A distance metric is needed to identify the nearest neighbors. In most cases, an Euclidean distance is used. The parameter k is a tunable parameter of the algorithm (and not derived from data). For a new point x_{new}, the k-NN algorithm finds the k nearest neighbors and predicts the output $\hat{y} = \bar{y}$, where the average is calculated over

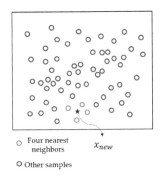

○ Four nearest
   neighbors x_{new}

○ Other samples

FIGURE 6.11 Schematic to explain k-NN.

the identified k neighbors. This is illustrated in Figure 6.11 for a case where $k = 4$.

The advantages of the k-NN algorithm are that it is very simple to setup and use and does surprisingly well in several nonlinear problems. This approach is also a simple benchmark for more complex models that one might build. The disadvantages of these methods are: distances need to be calculated to find the k-nearest neighbors (which can be a problem if the number of datapoints becomes large), susceptibility to noise, inability to capture global trends in the data, and poor predictions in sparse data regions.

Example 6.46. *Consider the following dataset.*

Height, h	1.5	1.7	1.6	1.5	1.4	1.6	1.4	1.9	2	1.4	1.8	1.5
Weight, w	71.1	103.3	26.4	27.8	21.8	94.9	90	98.3	108.1	91.9	61.5	90.2
BMI, $(y = w/h^2)$	31.6	35.74	10.31	12.36	11.12	37.07	45.92	27.23	27.23	46.89	18.98	40.09

If we use k-NN for predicting BMI from height and weight assuming the true model $\left(y = \frac{w}{h^2}\right)$ is unknown, find BMI for a person with height 1.3 m and weight 32 kg. Use k-NN with 4 neighbors and Euclidean distance as the metric for choosing neighbors.

Solution 6.46. *Using Euclidean distance, the distance of the point $\begin{bmatrix} 1.3, 32 \end{bmatrix}^T$ from all points in training dataset is evaluated. The distances are as follows:*

$$d = [39.1,\ 71.3,\ 5.61,\ 4.2,\ 10.2,\ 62.9,\ 58,\ 66.3,\ 76.1,\ 59.9,\ 29.5,\ 58.2]$$

The 4 neighbors are samples 4, 3, 5, and 11 as they have the smallest distances. Now, the predicted BMI for the given test sample is the average BMI of the 4 neighbors.

$$\hat{y} = \frac{1}{4}(12.36 + 10.31 + 11.12 + 18.98) = 13.1924$$

The true BMI for the given data is $\frac{32}{(1.3)^2} = 18.94$.

Squared error in prediction $= (13.1925 - 18.94)^2 = 33.034$

Example 6.47. *Consider the data provided in Example 6.46. If this data is used as training set for k-NN, and the following test data is given,*

Height h	1.8	1	1.6	1.7	1.5	1.2	1.2	1.8	1.2	1.1
Weight w	81.7	105.8	20.4	66.1	93.1	75.1	85.0	46.3	102.6	84.3
BMI $(y = w/h^2)$	25.22	105.8	7.97	22.87	41.38	52.15	59.03	14.29	71.25	69.67

Comment on the sensitivity of prediction quality to number of neighbors used in k-NN.

Solution 6.47. *The same procedure as described in the solution to Example 6.46 is repeated for all test samples by choosing different number of neighbors. For example, for the number of neighbors k=2,*

$$\hat{\mathbf{y}} = [43.005, \ 31.38, \ 10.715, 25.29, \ 41.98, \ 25.29, \ 43.005, \ 15.67, \ 31.485, \ 43.005]$$

$$SSE \ (k = 2) \ = \ \sum_{j=1}^{10} \left(y^j - \hat{y}^j\right)^2 \ = \ 9140.84$$

The same process is repeated for different k and the results are presented in Table 6.15.

TABLE 6.15

Sensitivity of number of neighbors used in k-NN for Example 6.47

k	2	3	4	5	6	7	8	9	10
SSE	9140.836	9169.41	8786.31	8364.83	8461.82	8273.74	8803.12	9423.20	10,220.78

Note that the SSE decreases first and increases when k increases. For the given data, k=7 shows the lowest SSE. Note that all the SSE values are reasonably high.

Example 6.48. *Comment on the sensitivity of k-NN prediction to the level of noise in the training data by using the training data given in Example 6.46 with white noise of zero mean and predefined variance added to y. Use the test data given in Example 6.47 for comparing performances. Set number of neighbors to 7.*

Solution 6.48. *For the training set, $y = \frac{w}{h^2} + \mathcal{N}(0, \sigma^2)$ is used where σ is the standard deviation of white noise added to y. Results for different σ are given in the table below.*

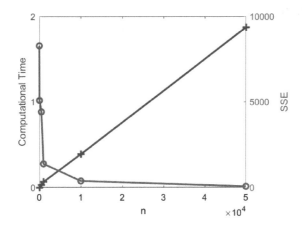

FIGURE 6.12 k-NN computational complexity.

σ (noise)	0.001	0.1	10	100	1000
SSE	8273.75	8274.82	8532.55	24,522.34	1,536,805.66

Note that as noise level increases in the training data, the predictions on the test data worsens. The best SSE is obtained for data having the least noise level.

Example 6.49. *Comment on the effect of number of training samples on computational complexity of k-NN by generating different number of training samples and testing on the data given in Example 6.47. Use 7 neighbors for predicting.*

Solution 6.49. *Different samples of "height" were generated from a uniform distribution with bounds [1, 2]. Similarly, various samples of "weight" were generated from uniform distribution with bounds [20, 110]. The true y values are calculated corresponding to each sample using the relation $y = \frac{w}{h^2}$. The results obtained for various numbers of training samples are provided below.*

n	12	50	500	1000	10,000	50,000
SSE	8273.74	5086.34	4413.31	1385.2	376.59	61.59
Computational time (s)	0	0	0.028	0.028	0.39	1.87

Computational time was calculated using an Intel core i7-8550U CPU with 8 GB RAM. Figure 6.12 shows the trend of both SSE and computational time with n. It is clear that the computational time increases linearly with n.

Example 6.50. *Consider the following data where Age group A indicates age ≥ 60, group B indicate $45 \leq$ age < 60, and group C indicates $18 \leq$ age < 45.*

State	A	A	A	B	B	C	C	D	D	B	D	C
Age-group	A	B	C	A	B	C	A	B	C	C	A	B
Vaccination ratio	0.91	0.74	0.67	0.85	0.63	0.35	0.69	0.77	0.69	0.56	0.93	0.47

Assume that the first 9 samples are used as training data and the last 3 are used for testing. Build a k-NN model to predict vaccination ratio using $k = 4$.

Solution 6.50. *In the given example, both the input features are categorical. Now, if we want to use Euclidean distance, these features have to be converted to numerical values. Else, we have to use other distance metrics, such as Hamming distance. We will show both methods here.*

a) Euclidean distance (sum squared difference between two vectors)
For converting to integer values, we can use different approaches. Here we will show the simplest one: "integer encoding". Different classes in the features are encoded using integers. For example, state A will be considered as 1, State B as 2, and so on. Now, we have the following dataset:

x_1	0	0	0	1	1	2	2	3	3	1	3	2
x_2	0	1	2	0	1	2	0	1	2	2	0	1
y	0.91	0.74	0.67	0.85	0.63	0.35	0.69	0.77	0.69	0.56	0.93	0.47

Now, we can implement k-NN similar to previous example. By implementing k-NN with $k = 4$, we get the following predictions.

y	0.56	0.93	0.47
$\hat{y}$	0.6	0.75	0.61

The corresponding SSE is 0.053.
Although the SSE is low, note that when we are using Euclidean distance, we are assuming that the order of classes matters. That is, state A is closer to state B than states C and D and so on. This is true for age groups as well. Since the given features do not have a sense of order between classes, Euclidean distance might not suit this data. Another choice is the "Hamming distance" as described next.

b) Hamming distance
This is a distance metric for strings. It is the number of positions in which 2 strings (of equal number of characters) differ. For example, the distance between "A23D" and "A13E" is 2 as the second and fourth characters are different.
The given data can be used as it is if Hamming distance were to be used. Implementing k-NN with 4 neighbors, we get the following predictions.

y	0.56	0.93	0.47
$\hat{y}$	0.625	0.805	0.6025

The corresponding SSE is 0.037.

Example 6.51. *Consider the following data where x_1 is the class (or section) in a school and x_2 is the teacher codes for 3 courses. To assess the performance of students and teachers a model relating these 2 features to the average marks gained by the student is to be built. For this purpose, build a k-NN model with 4 neighbors. Use the first 10 samples for training and the last 3 samples for testing.*

x_1	3a	3b	3c	4a	4c	5a	5c	6a	7a	8b	3a	4c	6b
x_2	ABC	AEF	DEC	DBF	GHI	AHC	GEI	AEF	GHI	GBC	DEC	ABC	GHI
y	75.4	71.2	63.7	68.6	67.5	74.3	63.4	73.4	70.2	62.4	65.5	72.5	61.2

Solution 6.51. *In the given data, both the input features are strings. Hamming distance may be used as discussed in Example 6.50. Hamming distances between the test sample of interest and each of the training samples are computed and 4 neighbors are chosen based on the 4 smallest Hamming distances. For example, for the test sample [3a, DEC] the Hamming distance with the training sample [3b, AEF] is the sum of Hamming distance between "3a" and "3b" and between "DEC" and "AEF".*

Distance between "3a" and "3b" is 1 as they differ only in 1 character. Similarly, the distance between "DEC" and "AEF" is 2 as they differ in 2 places. So, the distance between test sample [3a, DEC] and training sample [3b, AEF] is 3. Similarly, the distances with all training samples are identified. Samples 3 and 1 have distances 1 and 2 respectively. Then, there are 4 samples with distance 3 (samples 2,4,6, and 8).

For 4 neighbors, we have to either pick 2 out of 4 samples or have some weightage to all 4 samples. In this example, 2 samples in the order of samples are picked. Picking samples 2 and 4, the neighbors of the test sample ["3a", "DEC"] are samples 1 to 4 and the prediction is the average of the marks of samples 1 to 4.

$$\hat{y} = \frac{1}{4}\,(75.4 \,+\, 71.2 \,+\, 63.7 \,+\, 68.6) = 69.72$$

The same process is repeated for all test samples to get the following predictions.

y	65.5	72.5	61.2
$\hat{y}$	69.725	68.8	65.875

Note that the predictions are quite close to the true values and the corresponding SSE is 53.4. It should be noted that certain assumptions are implicit based on the manner in which the Hamming distance is defined. For example, if the teacher codes are random, then the calculation of Hamming distance based on the differences in character locations in the strings might not make much sense. Hence, some domain knowledge about the input features should inform the distance definition. One approach is to codify this information through weights in a weighted Hamming distance. In some cases, the weights can also be learnt as a part of the training process.

Example 6.52. *Consider the following data where x_1 is the star rating of a hotel and x_2 is the user rating of the corresponding hotel, x_3 is the best amenity/facility based on the user-rating for the hotel and y is the actual cost of room per day. Build a k-NN model using 3 neighbors to predict y given x_1, x_2 and x_3.*

Training Set:

x_1	1	4	2	1	4	4	3	4	2	4
x_2	3.2	2.2	4.5	4.8	1.9	4.0	2.6	2.8	4.6	0.4
x_3	Balcony	Elevator	Pool	Spa	Pool	Bonfire	Outdoor sports	City center	City center	Spa
y	675	1050	825	1300	1240	3460	2710	2500	1750	500

Test Set:

x_1	2	4	1
x_2	3.4	2.5	3.5
x_3	Balcony	Elevator	Pool
y	775	1150	725

Solution 6.52. *In the given data, x_1 is an integer, x_2 is continuous and x_3 is a string with unequal length. For x_1 and x_2, any distance metric for continuous variable would work. However, x_3 is a string of unequal lengths. For the sake of illustration, two common methods: integer encoding and "one-hot encoding" are described.*

a) *Method 1: Integer Encoding*
Here, each class in x_3 is given a separate integer code. For example, "balcony" is replaced with its category code, say 1. Similarly if "spa" is allotted code 4, the elements with x_3 as "spa" would be replaced with 4. Let us say the integer encoding for training and test data is as follows:

$$\text{Transformed } x_3 \text{ for training set} = [0, 3, 5, 6, 5, 1, 4, 2, 2, 6]$$

$$\text{Transformed } x_3 \text{ for test set} = [0, 3, 5]$$

Now that we have the numerical values for all features, distance metrics like Euclidean distance (L_2 norm), Manhattan distance (L_1 norm) etc can be used. Accordingly, we get the following results:

Test samples (input features)	True y	Predicted y using Euclidean distance	Predicted y using Manhattan distance
[2, 3.4, 0]	775	1961.67	1961.67
[4, 2.5, 3]	1150	2086.67	2086.67
[1, 3.5, 5]	725	1611.67	1611.67
		SSE=3071700	**SSE**=3071700

For the given data, both distance metrics provide the same predictions (as they find the same 3 neighbors). However, this need not happen in general. Further, both the Euclidean and Manhattan distances implicitly assume that there is an ordering between the classes which might not be valid. An alternate approach is one-hot encoding; this is discussed next.

b) *Method 2: One-hot Encoding*
Here, the third feature is divided into multiple binary features corresponding to each class. Since there are 7 different classes ("balcony", "spa", "city center", "bonfire", "pool", "elevator", "outdoor sports"), 7 different binary features are used to define x_3. This will result in 9 input features.

There are now 10 samples of 9 input features in the training data, and 3 samples of 9 input features in the test data. As before, any distance metric can be used. The results using Euclidean and Manhattan distances are provided below.

Test samples (input features)	True y	Predicted y (Euclidean distance)	Predicted y (Manhattan distance)
[2, 3.4, 1, 0, 0, 0, 0, 0, 0]	775	1083.33	1083.33
[4, 2.5, 0, 0, 0, 1, 0, 0, 0]	1150	1596.67	1596.67
[1, 3.5, 0, 0, 0, 0, 0, 1, 0]	725	933.33	933.33
		SSE=337983.33	**SSE**=337983.33

6.3.2 Decision Trees

Decision trees are designed to mimic the human decision making process. Most of us consider several factors before making decisions. Consider the problem of predicting how long it will take to go the airport from your house. One might first start by considering if it is a weekday or weekend and then if it is peak time or non-peak time and so on. Based on the answers to these questions, one will arrive at an estimate of the time it will take to reach the airport. Notice how the model and the result will depend on the answers to the various questions.

If one were to conceptualize this process in a tree like structure, one would start with the original problem, choose a variable, check its value and based on the value pose the next question and so on until some termination criterion

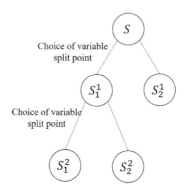

FIGURE 6.13 General scheme of decision tree generation.

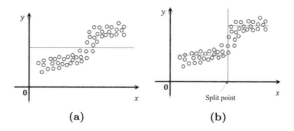

FIGURE 6.14 Mechanism by which prediction error decreases.

is met. This process is codified in decision tree algorithms as shown in Figure 6.13. Decisions are made based on the information stored in the terminal nodes.

Consider a multivariate function approximation problem with n inputs, one output and m samples. Since function approximation uses input variables to predict the output variable, the decisions at nodes of the tree are based on the input variables. At each node, one variable is chosen from all the input variables and a decision is made based on the value of the variable. As a result, the key aspects of the algorithm are related to how one chooses the variable to consider at a node and how one makes a decision based on the value of the variable. This process can be stopped at any level with any number of nodes. A stopping criteria is also needed to terminate the tree generation.

The choice of the node and the splitting decision at the node has to serve the underlying objective function that the model building is aiming to minimize. In function approximation, the objective is the error in prediction and the branching choices will attempt to minimize this error. This is illustrated in Figure 6.14. The figure depicts a simple one variable case. Assume that we start with data as shown in the left side of the figure. The simplest model is that the predicted output is the average of all output values. This is the first

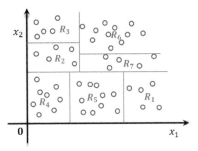

FIGURE 6.15 Rectangular partitions in the input space.

node or the top node of the decision tree. Now we split this node into two nodes. Since there is only one variable here, there is no choice; however, the split point is important. The figure on the right depicts the split point and two averages in these two regions model the data. Mathematically,

$$\hat{y} = \mu_1 \quad \forall\, x \le x_1; \qquad \hat{y} = \mu_2 \quad \forall\, x > x_1$$

Notice that the partitions are identified only in the input space and the outputs are used in computing the predicted value in the partitioned regions. Consider Figure 6.15 that visualizes the same approach for two input variables (x_1 and x_2). Now, at each node of the decision tree, there are two choices for variables (x_1 or x_2) and once a choice is made, then a split point needs to be chosen based on the decrease in the prediction error that is achieved. It can be seen that this process will carve the input space into multiple rectangles. When one traverses down the decision tree and comes to a terminal node, it is equivalent to being in one of the rectangles. The prediction at that point will be the average of the output of all the datapoints in that rectangle.

An alternate viewpoint that is worthwhile is to conceptualize the data as being split into multiple sets (S_j^i) as a decision tree is built with each set being associated with a node as shown in Figure 6.13. As a result, the information stored in each of the nodes is the subset of data. Any modeling exercise will then produce a local model with this subset of data. The most popular model for function approximation is to simply average the data; however, higher order models are also possible.

Another important idea in decision trees is that in many cases one can keep growing the tree to a very large number of nodes. This could be understood if one looks at Figure 6.15 and thinks about carving the input space into smaller rectangles. While this will keep improving the model performance on training data, there could be considerable over-fitting as minute irrelevant features might be captured and the size of the tree might become computationally unwieldy. As a result, in decision trees, one makes choices that limit the complexity of the tree growth such as: (i) tree depth (Figure 6.16), (ii)

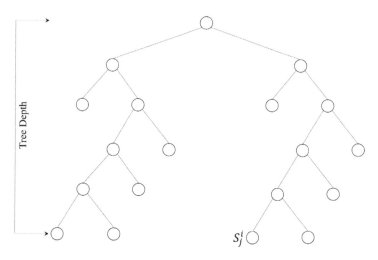

FIGURE 6.16 Understanding complexity of decision trees. S_j^i represents the jth node at level i.

number of nodes in the tree, (iii) minimum size of the node (as measured by the cardinality of the set or a hyper-volume of the set).

Example 6.53. *Consider the following data*

Sample No	1	2	13	4	5	6	7	8	9	10
Weight (w)	91.3	67.6	71.1	103.3	26.4	27.8	21.8	94.9	90	98.3
BMI (y)	40.58	23.39	27.77	45.91	13.47	10.86	11.12	26.29	22.5	50.15

Find the BMI corresponding to weights $35.2, 28,$ *and* 81.7 *using a decision tree with a single decision,* $w \leq 50$*. Find the prediction error if the true BMI's are* $10.86, 28,$ *and* 31.91*.*

Solution 6.53. *The decision tree is provided below.*

Tree:	S_1^1	S_2^1
Samples:	$5, 6, 7$	$1, 2, 3, 4, 8, 9, 10$

$$\text{Model:} \quad \bar{y}_1 = \frac{13.47 + 10.86 + 11.12}{2} \qquad \bar{y}_2 = \frac{1}{7} \sum_{y^i \in S_2^1} y^i$$

$$= 11.817 \qquad\qquad = 33.8$$

The results for test samples are complied below. Total prediction error (SSE) is 266.378.

Weight	Rectangle	$\hat{y}$ (Predicted BMI)	Absolute Error $=\|y - \hat{y}\|$	Squared Error$= (y - \hat{y})^2$
35.2	S_1^1	11.817	0.957	0.916
18	S_1^1	11.817	16.183	261.89
81.7	S_2^1	33.8	1.89	3.572
				SSE=266.378

Example 6.54. *For the training data given in Example 6.53, find the optimal split point (w_s) among the given samples based on minimum error.*

Solution 6.54. *For any split point (w_s), samples that satisfy $w^i < w_s$ will be in S_1^1 for which y will be approximated by $\bar{y}_{w^i \in S_1^1}$ (or simply $\bar{y}_1$) and the remaining samples $w^i \geq w_s$ will be in S_2^1 for which y will be approximated using $\bar{y}_{w^i \in S_2^1}$ (or $\bar{y}_2$). The total SSE is*

$$\sum_{i=1}^{n_1} \left(y_1^i - \bar{y}_1\right)^2 + \sum_{j=1}^{n_2} \left(y_2^i - \bar{y}_2\right)^2$$

where y_1^i and y_2^i are samples in S_1^1 and S_2^1, respectively. n_1 and n_2 are the total number of samples in S_1^1 and S_2^1.

To find the optimal point, there are various methods. Here, for easy demonstration, we describe a simple method where we find the SSE corresponding to each sample being the split point and find the point with the least SSE. Note that this is a crude method and there are much better approaches to finding the split points. Here, since we have only 10 samples, the method shown works well; however, with a large number of training datapoints, it will be computationally expensive. The results are provided in Table 6.16. Note that w=91.3 when used as the split point has the lowest SSE and hence the best split point. Corresponding rectangles are shown in Figure 6.17.

Example 6.55. *Consider the following data*

Samples	1	2	3	4	5	6	7	8	9	10
Weight (w)	91.3	67.6	71.1	103.3	26.4	27.8	21.8	94.9	90	98.3
Height (h)	1.5	1.7	1.6	1.5	1.4	1.6	1.4	1.9	2	1.4
BMI (y)	40.58	23.39	27.77	45.91	13.47	10.86	11.12	26.29	22.5	50.15

TABLE 6.16

Split Points and SSE for Example 6.54

Split Point (w)	Samples in S_1^1	Samples in S_2^1	$\bar{y}_{S_1^1}$	$\bar{y}_{S_2^1}$	SSE
91.3	2,3,5,6,7,9	1,4,8,10	18.185	40.7325	587.49
67.6	5,6,7	1,2,3,4,8,9,10	11.8167	33.7986	792.9
71.1	2,5,6,7	1,3,4,8,9,10	14.71	35.5333	766.96
103.3	1,2,3,5,6,7,8,9,10	4	25.1256	45.91	1418.83
26.4	7	1,2,3,4,5,6,8,9,10	11.12	28.9911	1520.18
27.8	5,7	1,2,3,4,6,8,9,10	12.295	30.9313	1251.9
21.8	-	1,2,3,4,5,6,7,8,9,10	-	27.204	1807.6
94.9	1,2,3,5,6,7,9	4,8,10	21.3843	40.7833	1017.3
90	2,3,5,6,7	1,4,8,9,10	17.322	37.086	831.0856
98.3	1,2,3,5,6,7,8,9	4,10	21.9975	48.03	723.3192

Find BMI for a person with weight 35.2 and height 1.8 using a decision tree with split points $w = 50$ ($w \leq 50$ and $w > 50$) and $h = 1.5$ ($h \leq 1.5$ and $h > 1.5$) respectively. Find the prediction error if true BMI is 10.97.

Solution 6.55. *The given decision tree is*

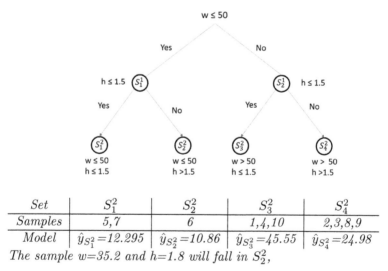

Set	S_1^2	S_2^2	S_3^2	S_4^2
Samples	5,7	6	1,4,10	2,3,8,9
Model	$\hat{y}_{S_1^2}=12.295$	$\hat{y}_{S_2^2}=10.86$	$\hat{y}_{S_3^2}=45.55$	$\hat{y}_{S_4^2}=24.98$

The sample $w=35.2$ and $h=1.8$ will fall in S_2^2,

$$\hat{y} = 10.86; \qquad SSE = (10.86 - 10.97)^2 = 0.0121$$

Example 6.56. *Consider the data provided in Example 6.55. Build a decision tree with maximum depth=2 and maximum leaf nodes=8. Predict BMI for the following test data and comment on the performance.*

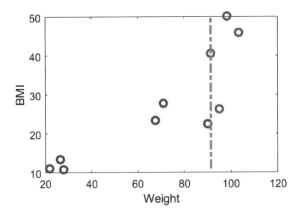

FIGURE 6.17 Sets corresponding to optimum split point for Example 6.54.

Sample No	1	2	3	4	5	6	7	8
Weight (w)	35.2	28	81.7	105.8	20.4	66.1	93.1	75.1
Height (h)	1.8	1	1.6	1.7	1.5	1.2	1.2	1.8
BMI (y)	10.86	28	31.91	36.61	9.07	45.9	64.65	23.18

Solution 6.56. *The decision tree corresponding to the given training data is built using inbuilt packages that are available. The decision tree obtained is shown in Figure 6.18a. Sum squared error (SSE) for the training data is 60.041 and SSE for the test data is 1253.486. Figure 6.18b shows the comparison of true BMI and predicted BMI values for both training and test data.*

Example 6.57. *Consider the data below:*

Class (x_1)	3	3	3	4	4	5	5	6	7	8
Section (x_2)	a	b	c	a	c	a	c	a	a	b
Average Mark (y)	75.4	71.2	63.7	68.6	67.5	74.3	63.4	73.4	70.2	62.4

a) *Build a decision tree of maximum depth=3 and maximum leaf nodes=4 to predict marks given class and section.*

b) *Find the optimal value of maximum depth and leaf nodes based on performance of decision tree on the following test data.*

Class (x_1)	3	4	6
Section (x_2)	a	c	b
Average Mark (y)	65.5	72.5	61.2

Solution 6.57. *a) The input features in the given data are both categorical. Among them, x_1 is numeric and ordinal whereas x_2 is nominal and non-numeric. Hence, to proceed, we first need to convert x_2 to numerical values. As discussed in k-NN examples, there are many ways of doing this. Two*

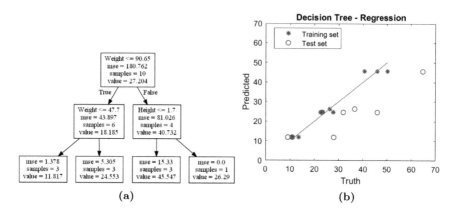

FIGURE 6.18 Decision tree results for Example 6.56.

most common methods are integer encoding and one-hot encoding. We will discuss results using both approaches.

i) *Integer encoding: Assigning section A as 0, section B as 1 and section C as 2, we have both x_1 and x_2 as numerical values. Now, decision tree of maximum depth=3 and maximum leaf nodes=4 is implemented in a simulation platform. Figure 6.19a shows the decision tree obtained. The sum squared error for the training data is 40.61. For the test data, the predicted values of marks are $\left[72.38, 65.6, 62.9\right]$. The corresponding sum squared error is 97.834.*

ii) *One-hot encoding: One-hot encoding was performed on x_2. Since there are three classes in x_2, we get 3 features instead of x_2. Including x_1, we have 4 input features: x_1, "Section A", "Section B", and "Section C". The last three input features are binary variables. We can implement decision tree just like before. The decision tree obtained using one-hot encoding is shown in Figure 6.19b. Although this tree looks different from that shown in Figure 6.19a, one can notice that all the leaf nodes are the same. Hence, while predicting, we get the same results. Thus, the sum squared error is 40.61 for the training data and 97.834 for the test data.*

b) *We use integer encoding in this section. Without setting a maximum for leaf nodes and varying maximum depth, the following results are obtained.*

Maximum depth	1	2	3	4	5
SSE (Training set)	86.54	50.114	16.51	0.405	0
SSE (Test set)	114.11	126.23	150.46	124.45	127.85

For each depth, the maximum number of leafs that are possible are generated. When the maximum depth reaches 5, the number of leaf nodes is 10 which is same as the total number of samples in the training set. Hence,

*further increase in depth will result in the same results as that of depth =
5.
Now, for depth = 5, what happens if we start limiting the number of leaf
nodes that are possible? The following table shows the results for various
restrictions on maximum leaf nodes while keeping the maximum depth as
5.*

Maximum leaf nodes	2	3	4	5	6	7	8	9
SSE (Training set)	86.54	61.51	40.61	29.21	17.01	8.13	0.9	0.4
SSE (Test set)	114.11	75.56	97.83	148.51	148.51	148.51	125.9	127.85

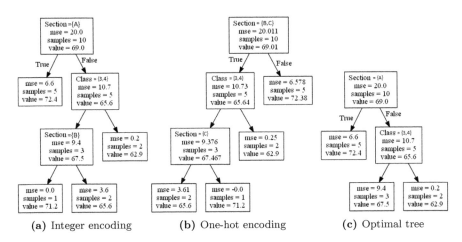

(a) Integer encoding   (b) One-hot encoding   (c) Optimal tree

FIGURE 6.19 Decision tree results for Example 6.57.

*Note that the case where maximum depth = 5 and maximum leaf nodes
= 3 gave the minimal SSE for the test data. In this example if we didn't
restrict the depth and only constrained the maximum leaf nodes as 3, we
would have gotten the same result. The decision tree obtained is shown in
Figure 6.19c and has depth = 2.*

Remark. *The inbuilt package used for building decision tree treats all the
variables as continuous. Since the method has only subset selection, the
results would be the same even if we treat the categorical variables as categorical during implementation. Figures 6.19a, 6.19b, and 6.19c show the
results in terms of the categorical variables.*

Example 6.58. *A travel company wants to study user-data to be able to
predict the rating that a user would give for a particular hotel. The following
data was collected.*

Star Rating	User Rating	Best Amenity	Price Per Room Per Day
1	3.2	balcony	675
4	2.2	elevator	1050
2	4.5	pool	825
1	4.8	spa	1300
4	1.9	pool	1240
4	4.0	bonfire	3460
3	2.6	outdoor sports	2710
4	2.8	city center	2500
2	4.6	city center	1750
4	0.4	spa	500
2	3.4	balcony	775
4	2.5	elevator	1150
1	3.5	pool	725

Build a decision tree using the first 10 samples to predict the user rating given the star rating, price and best amenity. Maximum depth of the tree should be 8. Predict the user ratings for the last three samples and comment on the performance.

Solution 6.58. *The input features for this example are star rating $(X[0])$, price $(X[1])$, and "best amenity". Since "best amenity" is a nominal categorical variable with 7 categories, we first used one-hot encoding to convert them to 7 binary features: "balcony" $(X[2])$, "elevator" $(X[3])$, "pool" $(X[4])$, "spa" $(X[5])$, "bonfire" $(X[6])$, "outdoor sports" $(X[7])$, and "city centre" $(X[8])$. The input data for training is as follows:*

$$
\begin{bmatrix}
1 & 675 & 1 & 0 & 0 & 0 & 0 & 0 & 0 \\
4 & 1050 & 0 & 0 & 0 & 1 & 0 & 0 & 0 \\
2 & 825 & 0 & 0 & 0 & 0 & 0 & 1 & 0 \\
1 & 1300 & 0 & 0 & 0 & 0 & 0 & 0 & 1 \\
4 & 1240 & 0 & 0 & 0 & 0 & 0 & 1 & 0 \\
4 & 3460 & 0 & 1 & 0 & 0 & 0 & 0 & 0 \\
3 & 2710 & 0 & 0 & 0 & 0 & 1 & 0 & 0 \\
4 & 2500 & 0 & 0 & 1 & 0 & 0 & 0 & 0 \\
2 & 1750 & 0 & 0 & 1 & 0 & 0 & 0 & 0 \\
4 & 500 & 0 & 0 & 0 & 0 & 0 & 0 & 1
\end{bmatrix}
$$

The output feature is user rating. Using inbuilt packages in a simulation tool, various decision trees were built with different maximum leaf nodes. The results for both the training and test data are given below:

Maximum leaf nodes	2	3	4	5	6	7	8	9	10
SSE (Training set)	8.595	4.19	2.075	0.534	0.112	0.067	0.025	0.005	0
SSE (Test set)	1.4	1.41	1.38	1.34	1.53	1.414	1.82	1.82	1.82

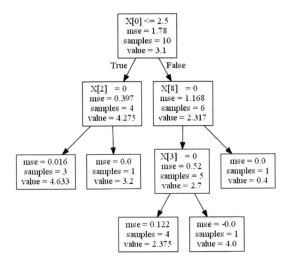

FIGURE 6.20 Decision tree for Example 6.58.

Note that by setting maximum leaf nodes to 5, we get the least SSE for the test data. Hence, we chose the decision tree with maximum depth = 5 and maximum leaf nodes = 5. The resulting tree is shown in Figure 6.20. The predicted user ratings using this decision tree for the test samples are 3.2, 2.2, and 4.8, respectively.

6.3.3 Random Forests

Random forests are a simple extension of decision trees. Forests comprise of many trees; equivalently, random forests consists of many decision trees. Random forests are motivated by the lack of robustness in decision trees. Minor changes in the tree construction procedure and/or minor changes in the data being modeled can lead to considerably different trees being built. This can lead to different results when these decision trees are used in a final application. To avoid this, multiple trees are built from the same datapoints. This is realized by choosing subsets of datapoints (D_i in Figure 6.21) or variables (V_i in Figure 6.21) and presenting the sub-selected (both variables and rows) data matrices to each of the decision trees to model. One can appreciate that using multiple such subsets, several trees can be generated. When a new datapoint is presented to the random forest, predictions from each of the decision trees are collected and in the function approximation case, an average of all of these predictions is the predicted output from the random forest model.

Example 6.59. *A food delivery company wants to predict the time taken for delivery. Table 6.17 provides the data collected. Develop a random forest regressor with variable selection (no data partition) using first 10 samples. Use*

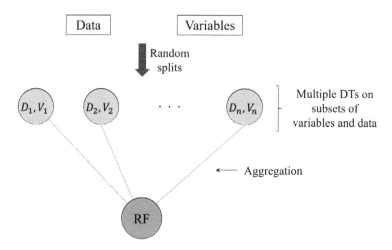

FIGURE 6.21 Random forest—a collection of decision trees.

9 decision trees in the random forest. For each decision tree, set maximum depth to 3 and maximum leaf nodes to 6. Test the developed random forest regressor on the last 4 samples.

Solution 6.59. *Nine different decision trees are built using various subsets of features. Inbuilt packages available are used for developing the decision trees. Different trees receive different subsets of feature information. Among the features used as input, all variables need not be used in the decision process. Figure 6.22 depicts the 9 decision trees built. Notice that the decision trees differ in variables but have the same number of samples.*

When a new sample is presented, predicted value from each decision tree is obtained. The average of these values is the prediction from the random forest. Figure 6.23a shows the predictions for both training and test data. A sum squared error of 4.03 is obtained for the training data. Predictions from each decision tree for test data is shown in Table 6.18 along with the averages.

The average values are the final predictions using random forest. The corresponding sum squared error is 283.12.

Example 6.60. *Develop a random forest regressor with data partition (no feature selection) for the data given in Example 6.59. Use 9 decision trees in the random forest. For each decision tree, set maximum depth to 3 and maximum leaf nodes to 6.*

Solution 6.60. *Since we are interested in sample selection instead of variable selection in this example, 9 different decision trees were built using a subset of samples (with at least 5 samples). Each decision tree is given all the feature information. Out of the given features, each decision tree may or may not use* all features. *The decision trees obtained are shown in Figure 6.24. Note that each decision tree has different number of samples.*

$$\hat{\mathbf{y}} = [45.57, 27.235, 35.95, 44.08]$$

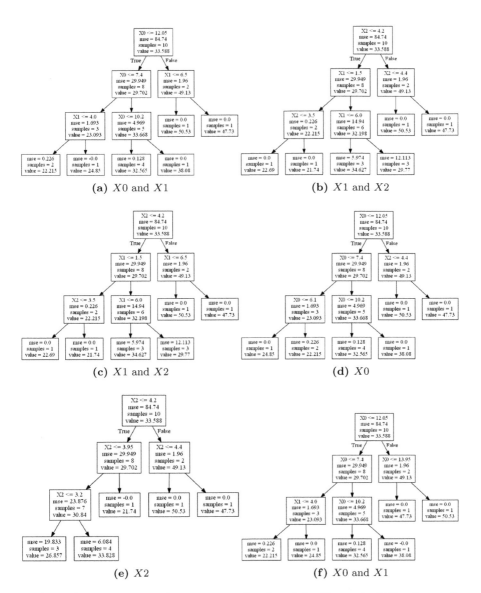

FIGURE 6.22 Various decision trees developed for Example 6.59 using variable selection.

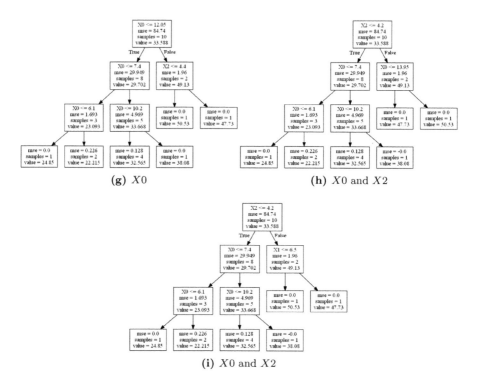

FIGURE 6.22 Various decision trees developed for Example 6.59 using variable selection (contd.).

The predictions for both training and test data are compared with their true values in Figure 6.23b. Sum squared error for the training data is 38.02 and for the test data is 143.94.

Example 6.61. *Develop a random forest regressor with both variable selection and data partition for the data given in Example 6.59. Use 9 decision trees in the random forest. For each decision tree, set maximum depth to 3 and maximum leaf nodes to 6.*

Solution 6.61. *As both feature and data partition is allowed, we use the inbuilt packages for implementing random forest that allows for bootstrapping and set the number of estimators to 9. Maximum depth and maximum leaf nodes have been set to 3 and 6 respectively. The decision trees that result are shown in Figure 6.25.*

Note that different trees have different number of features and samples as expected. The predictions for both the training and test data are shown in Figure 6.23c. The SSE for training set is 28.158 and SSE for the test set is 235.87.

TABLE 6.17

Dataset for Example 6.59

Distance from Hotel	Number of Items Ordered	Delivery Boy Rating	Time for Delivery
8.2	7	3.5	32.35
10.7	5	3.6	38.08
9.	5	2.7	33.03
8.2	7	3.8	32.11
6.4	1	4.1	21.74
9.7	2	3.3	32.77
6.6	1	2.9	22.69
13.4	7	4.5	47.73
14.5	6	4.3	50.53
5.8	7	3.1	24.85
12.7	1	2.8	44.1
5.2	2	2.5	37.32
11.3	9	4.1	40.7
12.5	1	4.4	39.9

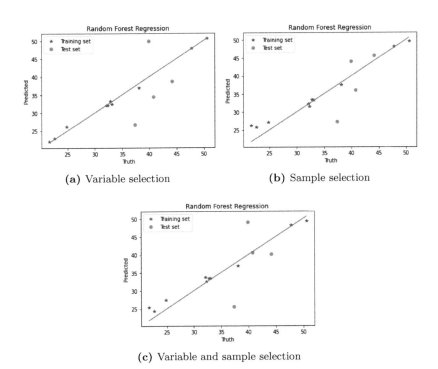

(a) Variable selection

(b) Sample selection

(c) Variable and sample selection

FIGURE 6.23 Predictions using random forests for 6.59.

TABLE 6.18

Predictions of Individual Decision Trees for Example 6.59

	Test sample 1	Test sample 2	Test sample 3	Test sample 4
	50.53	22.215	38.08	50.53
	22.69	34.63	29.77	50.53
	22.69	34.63	29.77	50.53
	50.53	24.85	38.08	50.53
	47.73	22.215	38.08	47.73
	26.86	26.86	21.74	50.53
	50.53	24.85	38.08	50.53
	38.08	24.85	38.08	47.73
	38.08	24.85	38.08	50.53
Average	38.635	26.66	34.42	49.91

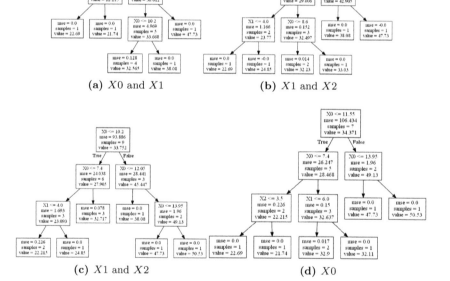

(a) $X0$ and $X1$

(b) $X1$ and $X2$

(c) $X1$ and $X2$

(d) $X0$

FIGURE 6.24 Various decision tress developed for Example 6.60.

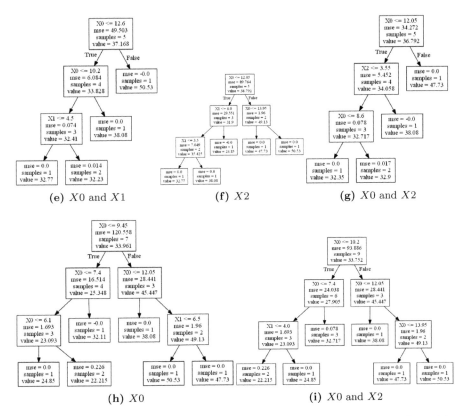

(e) $X0$ and $X1$ (f) $X2$ (g) $X0$ and $X2$

(h) $X0$ (i) $X0$ and $X2$

FIGURE 6.24 Various decision tress developed for Example 6.60 (contd.).

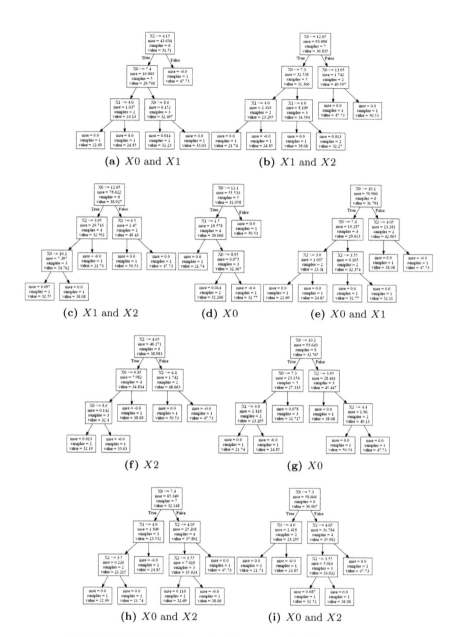

FIGURE 6.25 Various decision trees for Example 6.61.

7 Classification Methods

In this chapter, we will describe a selected set of classification methods that are widely used in data science. All the underlying theory required to understand these techniques have already been described in the three foundational chapters on linear algebra, statistics, and optimization.

7.1 TYPES OF CLASSIFICATION PROBLEMS

In Chapter 2, we had described binary, multi-class, and multi-dimensional classification problems. The most common of these are usually binary or multi-class problems and hence the techniques will be described from the viewpoint of these problems. In many cases, multi-class problems can solved using binary classification techniques and multi-dimensional problems can be solved using multi-class techniques.

7.2 PARAMETRIC METHODS

We will start with parametric methods. Parametric methods are ones where the classifier is a decision function characterized by parameters that are derived from data. These parameters are different from the tuning parameters that are chosen *a priori* as algorithm parameters by the user of the technique. In contrast, non-parametric methods described later will not derive any parameters from data for classification.

7.2.1 Naive Bayes Classifier

Naive Bayes classifier is a simple method to solve classification problems. Consider a simple classification problem, one of labeling a fruit as either an apple or an orange based on its dimensions, such as its radius. Naive Bayes classifier solves this problem mathematically by calculating the probability of the fruit being an apple given its radius, and similarly calculating the probability of the fruit being an orange given its radius. The fruit is then classified as either an apple or an orange depending on whichever probability is higher. More formally, we calculate $P(orange|radius)$ and $P(apple|radius)$, both conditional probabilities, and decide the class to which the fruit belongs. We will start with the simplest possible case, where the data has only one attribute and the domain of that attribute has either finite or countably infinite cardinality (this is referred to as the discrete case) and then build up to the general case.

Consider the following example problem. Given the temperature of an object (food), one would like to classify it either as soup (S) or ice cream (IC), assuming that, *a priori*, the object being either ice cream or soup are equally

DOI: 10.1201/b23276-7

likely. Notice that the domain of the attribute, temperature, has 3 elements, namely: hot, cold, and lukewarm. Hence this problem belongs to the discrete case. Also note that $P(IC) = P(S) = 1/2$ as outcomes are equally likely. Let us say the temperature of the new object is "Hot". Now, the task at hand is to calculate $P(S|Hot)$ and $P(IC|Hot)$. From Bayes theorem, we know that

$$P(A|B)P(B) = P(B|A)P(A)$$
$$\text{Therefore, } P(S|Hot)P(Hot) = P(Hot|S)P(S)$$
$$\text{or } P(S|Hot) = P(Hot|S)P(S)/P(Hot)$$
$$\text{Similarly, } P(IC|Hot) = P(Hot|IC)P(IC)/P(Hot)$$

Given that probabilities are all positive, one can ignore the denominator and compare only $P(Hot|IC)P(IC)$ and $P(Hot|S)P(S)$. If from data $P(Hot|IC) = 0$ and $P(Hot|S) = 0.6$, then $P(Hot|S)P(S) = 0.6 \times 1/2 = 0.3$ and $P(Hot|IC)P(IC) = 0$. Thus, the new object ("Hot") belongs in class Soup. What if we had another attribute as well? Let us say we add another attribute "Consistency" to the above data, with the domain of Consistency being thick, thin. Now, when we get a new object, it is characterised by a tuple (Temperature, Consistency) instead of a single value. The general method remains the same, we calculate $P(S|(T, C))$ and $P(IC|(T, C))$ and classify as before. What does $P(S|(T, C))$ actually mean? To make this clearer, let us consider an example. $P(S|(Hot, Thin))$ is the probability that the object is soup given that it is hot and thin.

$$P(S|Hot \cap Thin) = P(S \cap Hot \cap Thin)/P(Hot \cap Thin)$$
$$P(IC|Hot \cap Thin) = P(IC \cap Hot \cap Thin)/P(Hot \cap Thin)$$

While comparing the two terms, one can safely ignore the denominators as probabilities are positive. Therefore, we are looking for $\max(P(IC \cap Hot \cap Thin), P(S \cap Hot \cap Thin))$. Now, we can modify $P(S \cap Hot \cap Thin)$ to a simpler form to make it easier to handle.

$$P(S \cap Hot \cap Thin) = P(Hot|S \cap Thin)P(S \cap Thin)$$
$$\Rightarrow P(S \cap Hot \cap Thin) = P(Hot|S \cap Thin)P(Thin|S)P(S)$$

To simplify this expression, the naive Bayes classifier makes a very important assumption that all attributes are independent, that is, temperature and consistency are independent variables. Therefore, $P(Hot|S \cap Thin) = P(Hot|S)$ and hence the expression reduces to $P(S \cap Hot \cap Thin) = P(Hot|S)P(Thin|S)P(S)$. The terms $P(Hot|S)$ and $P(Thin|S)$ can easily be calculated from data. Similarly, $P(IC \cap Hot \cap Thin)$ can also be simplified and calculated from data.

This method can easily be generalised to datasets with n attributes and m class labels. Let the m classes be $C_1 \ldots C_m$ and the n attributes be $x_1 \ldots x_n$. Just like before, we have to compare $P(C_1|x_1 \ldots x_n)$, $P(C_2|x_1 \ldots x_n)$, $\ldots$

$P(C_m | x_1 \dots x_n)$. If we assume that all the attributes are mutually independent, the intersection can be split as before and their values calculated from
data. Is the assumption of independence appropriate? In the real world, in
most cases, probably not. For example, in the example problem we were considering, the consistency of soup will most likely depend on its temperature.
However, the independence assumption greatly simplifies problems and provides us with a reasonable first solution to start with.

Example 7.1. *Consider the following data*

Program of study (x_1)	B	M	M	P	B	P	M	B	P	B
State (x_2)	1	1	2	3	3	1	3	2	4	4
Fully vaccinated ? (y)	1	0	1	1	1	0	0	1	0	0

B = Bachelors, M = Masters, P = PhD
*Predict whether students with the following details are fully vaccinated or
not using a Naive Bayes Classifier (assuming x_1 and x_2 are independent).*

Program of study (x_1)	B	M	P
State (x_2)	2	4	1

Solution 7.1. *Assumption : x_1 and x_2 are independent.*

(a) *For the first test sample, we need to compare $P(y = 1 | (x_1 = B, x_2 = 2))$
and $P(y = 0 | (x_1 = B, x_2 = 2))$, which is equivalent to comparing
$P_1 = P((y = 1) \cap (x_1 = B) \cap (x_2 = 2))$ and $P_2 = P((y = 0 \cap
(x_1 = B) \cap (x_2 = 2))$. Since x_1 and x_2 are assumed to be independent,*

$$P_1 = P(x_1 = B | y = 1) \, P(x_2 = 2 | y = 1) P(y = 1)$$

*Since 5 samples out of 10 have $y = 1$, $P(y = 1) = 5/10 = 0.5$. Since 3
instances with $y = 1$ have $x_1 = B$ and 2 instances with $y = 1$ have $x_2 = 2$,*

$$P(x_1 = B | y = 1) = \frac{3}{5} = 0.6; \qquad P(x_2 = 2 | y = 1) = \frac{2}{5} = 0.4$$
$$\implies P_1 = 0.6 \times 0.4 \times 0.5 = 0.12$$

Similarly,

$$P_2 = P(x_1 = B | y = 0) P(x_2 = 2 | y = 0) P(y = 0) = \frac{1}{5} \times 0 \times \frac{1}{2} = 0$$

*Since $P_1 > P_2$, we can predict the sample to have $y = 1$, that is, fully
vaccinated.*

(b) *For the second sample,* $x_1 = M$ *and* $x_2 = 4$

$$P_1 = P((y = 1) \cap (x_1 = M) \cap (x_2 = 4))$$

$$= P(x_1 = M|y = 1)P(x_2 = 4|y = 1)P(y = 1) = \frac{1}{5} \times 0 \times \frac{1}{2} = 0$$

$$P_2 = P((y = 0) \cap (x_1 = M) \cap (x_2 = 4))$$

$$= P(x_1 = M|y = 0) P(x_2 = 4|y = 0)P(y = 0) = \frac{2}{5} \times \frac{2}{5} \times \frac{1}{2} = 0.08$$

Since $P_2 > P_1$, *we predict* $y = 0$ *for this sample, that is, not fully-vaccinated.*

(c) *For the third sample,*

$$P_1 = P(y = 1) \cap (x_1 = P) \cap (x_2 = 1)$$

$$= P(x_1 = P|y = 1) P(x_2 = 1|y = 1)P(y = 1) = \frac{1}{5} \times \frac{1}{5} \times \frac{1}{2} = 0.02$$

$$P_2 = P(y = 0) \cap (x_1 = P) \cap (x_2 = 1)$$

$$= P(x_1 = P|y = 0) P(x_2 = 1|y = 0)P(y = 0) = \frac{2}{5} \times \frac{2}{5} \times \frac{1}{2} = 0.08$$

Since $P_1 < P_2$, *we predict* $y = 0$ *for this sample i.e., not fully-vaccinated.*

Example 7.2. *Consider the following data*

x_1	B	M	M	P	B	P	M	B	P	B
x_2	1	2	2	3	1	3	2	1	3	1
y (1 - full vaccinated, 0 - not fully vaccinated)	1	0	1	1	1	0	0	1	0	0

Program: Bachelors(B), Masters(M), PhD(P)
Age-group: 17–21(1), 22–23(2), 23–28(3)

a) *Find* $P(y=0)$ *and* $P(y=1)$
b) *Find the conditional probabilities* $P(x_1 = B|y = 1))$, $P(x_1 = B|y = 0))$, $P(x_2 = 3|y = 1))$ *and* $P(x_2 = 3|y = 0))$.
c) *Assuming that* x_1 *and* x_2 *are independent, find whether a student with* $x_1 = B$ *and* $x_2 = 1$ *can be considered to be fully vaccinated or not.*
d) *Assuming that* x_1 *and* x_2 *are dependent, find whether a student with* $x_1 = B$ *and* $x_2 = 2$ *can be considered to be fully vaccinated or not. Compare with the results obtained in (c).*

Solution 7.2.

a) $P(y = 0) = \frac{1}{2}$; $\quad P(y = 1) = \frac{1}{2}$
b) $P(x_1 = B|y = 1) = \frac{3}{5}$; $\quad P(x_1 = B|y = 0) = \frac{1}{5}$;
$\quad P(x_2 = 3|y = 1) = \frac{1}{5}$; $\quad P(x_2 = 3|y = 0) = \frac{2}{5}$

c)

$$P_1 = P\left((y = 1) \cap (x_1 = B) \cap (x_2 = 3)\right)$$
$$= P\left(x_1 = B|y = 1\right) P\left(x_2 = 3|y = 1\right) P\left(y = 1\right) = \frac{3}{5} \times \frac{2}{5} \times \frac{1}{2} = 0.06$$

$$P_2 = P\left((y = 0) \cap (x_1 = B) \cap (x_2 = 3)\right)$$
$$= P\left(x_1 = B|y = 0\right) P\left(x_2 = 3|y = 0\right) P\left(y = 0\right) = \frac{1}{5} \times \frac{2}{5} \times \frac{1}{2} = 0.04$$

Since $P_1 > P_2$, we classify the sample with $x_1 = B$ and $x_2 = 3$ as fully vaccinated.

d) Since x_1 and x_2 are dependent,

$$P_3 = P\left((y = 1) \cap (x_1 = B) \cap (x_2 = 3)\right)$$
$$= P\left(x_1 = B|(y = 1) \cap (x_2 = 3)\right) P\left(x_2 = 3|y = 1\right) P\left(y = 1\right)$$

$$P\left(x_1 = B|(y = 1) \cap (x_2 = 3)\right) = 0 \Rightarrow P_3 = 0$$

(There is only one sample with y=1 and $x_2 = 3$, but that sample has $x_1 = P$)

$$P_4 = P\left((y = 0) \cap (x_1 = B) \cap (x_2 = 3)\right)$$
$$= P\left(x_1 = B|(y = 0) \cap (x_2 = 3)\right) P\left(x_2 = 3|y = 0\right) P\left(y = 0\right) = 0$$

(since $P\left(x_1 = B|(y = 0) \cap (x_2 = 3)\right) = 0$ as only 2 samples have both y=0 and $x_2 = 3$, and both of these two samples have $x_1 = P$.)

Since both P_3 and P_4 are zero, we cannot classify them as either fully vaccinated or not fully vaccinated. Both classes have equal probability (0.5). This is very different from the result obtained in part c where we had assumed that x_1 and x_2 are independent. But from the data it is very clear that x_1 and x_2 are dependent and hence, the assumption in part c is invalid. This shows that whenever the input features are dependent, assumption of independence might lead to poor results.

Example 7.3. Consider the following data.

Hours studied (x_1)	10.5	2.5	14	8.2	10.4	5	6.7	14.7
% Classes missed (x_2)	0.095	0.238	0.357	0.071	0	0	0.17	0.33
Pass/Fail (y)	1	0	0	1	1	1	0	1

Build a Naive Bayes classifier and predict whether a student with $x_1 = 3$ and $x_2 = 0.51$ will pass or fail in the course.

Solution 7.3. *Since the input features are continuous variables, one needs to assume a probability distribution to find conditional probabilities. For each x_i in each class, Gaussian pdf is assumed. Let us first look at $f(x_1|y = 1)$.*

$$f(x_1|y = 1) = \frac{1}{\sqrt{2\pi}\sigma_{x_1x_1,1}} \exp \frac{-(x_1 - \mu_{x_1,1})^2}{2\sigma_{x_1x_1,1}^2}$$

To define the pdfs, values of $\sigma_{x_1x_1,1}$ and $\mu_{x_1,1}$ are required. For this, we look at the samples having $y = 1$. The mean value of all x_1 samples having $y = 1$ is used as $\mu_{x_1,1}$ and the standard deviation of all such samples is used as $\sigma_{x_1x_1,1}$. For the given data,

$$\mu_{x_1,1} = \frac{10.5 + 8.2 + 10.4 + 5 + 14.7}{5} = 9.76$$

$$\sigma_{x_1x_1,1} = \sqrt{\frac{\Sigma(x_{1,i} - \mu_{x_1,1})^2}{n_1 - 1}} = 3.55 \qquad \forall \ i \ s.t. \ y_i = 1$$

Similarly, we can find the pdfs of x_1 given $y = 0$, x_2 given $y = 1$, and x_2 given $y = 0$.

$$f(x_1|y = 0) = \frac{1}{\sqrt{2\pi}\mu_{x_1,0}} \exp \frac{-(x_1 - \mu_{x_1,0})^2}{2\sigma_{x_1x_1,0}^2}$$

$$f(x_2|y = 1) = \frac{1}{\sqrt{2\pi}\mu_{x_2,1}} \exp \frac{-(x_2 - \mu_{x_2,1})^2}{2\sigma_{x_2x_2,1}^2}$$

$$f(x_2|y = 0) = \frac{1}{\sqrt{2\pi}\mu_{x_2,0}} \exp \frac{-(x_2 - \mu_{x_2,0})^2}{2\sigma_{x_2x_2,0}^2}$$

From data, we find,

$$\mu_{x_1,0} = 7.73; \quad \mu_{x_2,0} = 0.255; \quad \mu_{x_2,1} = 0.099;$$
$$\sigma_{x_1x_1,0} = 5.82; \quad \sigma_{x_2x_2,0} = 0.095; \quad \sigma_{x_2x_2,1} = 0.136;$$

Now, to find the class corresponding to $x_1 = 3$ and $x_2 = 0.51$, we find $f_0 = f((y = 0) \cap (x_1 = 3) \cap (x_2 = 0.51))$ and $f_1 = f((y = 1) \cap (x_1 = 3) \cap (x_2 = 0.51))$. Note that these are not final probabilities but pdf values. We assume that x_1 and x_2 are independent. Hence,

$$f_0 = f((x_1 = 3|y = 0))f((x_2 = 0.51|y = 0))f_y(y = 0)$$
$$f_1 = f((x_1 = 3|y = 1))f((x_2 = 0.51|y = 1))f_y(y = 1)$$
$$f_y(y = 0) = \frac{3}{8}; \qquad f_y(y = 1) = \frac{5}{8}$$
$$f((x_1 = 3|y = 0)) = 0.0492; \qquad f((x_1 = 3|y = 1)) = 0.0184$$
$$f((x_2 = 0.51|y = 0)) = 0.1119; \qquad f((x_2 = 0.51|y = 1)) = 0.0303$$
$$\implies \quad f_0 = f((y = 0), (x_1 = 3), (x_2 = 0.51)) = 0.0021$$
$$f_1 = f((y = 1), (x_1 = 3), (x_2 = 0.51)) = 3.47 \times 10^{-4}$$

Since $f_0 > f_1$, $y = 0$ is the predicted class for $x_1 = 3$ and $x_2 = 0.51$.

7.2.2   Linear Discriminant Analysis (LDA)

In linear discriminant analysis, the data for the classes are assumed to come from multivariable distributions, and conditional probabilities are used to perform classification. Let us illustrate this in a two variable case for a binary classification problem. The means and covariance matrices for the two classes are needed for the definition of multivariate Gaussian pdfs. Let the means and covariance matrices for the two classes be $\{\mu_1, \mu_2\}$ and $\{V_1, V_2\}^1$, respectively. These quantities are derived from the data using following relations.

$$\mu_1 = \frac{1}{n_1} \sum_{i \ s.t. \ \mathbf{x^i} \in \ C_1} \mathbf{x^i}; \qquad \mu_2 = \frac{1}{n_2} \sum_{i \ s.t. \ \mathbf{x^i} \in \ C_2} \mathbf{x^i} \qquad (7.1)$$

$$V_1(j, k) = \frac{1}{n_1 - 1} \sum_{i \ s.t. \ \mathbf{x^i} \in \ C_1} \left(x_j^i - \mu_{1,j}\right)\left(x_k^i - \mu_{1,k}\right) \qquad (7.2)$$

$$V_2(j, k) = \frac{1}{n_2 - 1} \sum_{i \ s.t. \ \mathbf{x^i} \in \ C_2} \left(x_j^i - \mu_{1,j}\right)\left(x_k^i - \mu_{1,k}\right) \qquad (7.3)$$

Now, when a new datapoint $\mathbf{x^{new}}$ is presented to the classifier, a decision on $\mathbf{x^{new}}$ belonging to class 1 or 2 has to be made. This decision depends on the conditional probabilities $P(C_1|\mathbf{x^{new}})$ and $P(C_2|\mathbf{x^{new}})$. $\mathbf{x^{new}}$ is assigned to the class with the larger conditional probability. Mathematically, the decision function becomes $h(\mathbf{x}) = P(C_2|\mathbf{x^{new}}) - P(C_1|\mathbf{x^{new}})$. If $h(\mathbf{x^{new}}) > 0$, then $\mathbf{x^{new}}$ is assigned to Class-2 and to Class-1 otherwise. Since probabilities are always greater than zero and the log function is an increasing function, the previous $h(\mathbf{x})$ can be replaced by $h(\mathbf{x}) = log(P(C_2|\mathbf{x})) - log(P(C_1|\mathbf{x}))$. Now,

$$P(C_2|\mathbf{x}) = P(\mathbf{x}|C_2)P(C_2)/P(\mathbf{x}) \qquad (7.4)$$
$$P(C_1|\mathbf{x}) = P(\mathbf{x}|C_1)P(C_1)/P(\mathbf{x}) \qquad (7.5)$$

Equations 7.4 and 7.5 are simply statements of Bayes theorem, also used in the naive Bayes classifier approach. $P(C_1)$ and $P(C_2)$ represent the *a priori* probabilities for classes 1 and 2 respectively. We make the assumption that the data belonging to classes 1 and 2 (represented by $P(\mathbf{x}|C_1)$ and $P(\mathbf{x}|C_2)$) can be represented by Gaussian distributions with the corresponding means and covariance matrices. Now,

$$h(\mathbf{x}) = log(P(C_2|\mathbf{x})) - log(P(C_1|\mathbf{x})) = log(P(\mathbf{x}|C_2)P(C_2)) - log(P(\mathbf{x}|C_1)P(C_1))$$

$$= log(P(C_2)/P(C_1)) + log\left(\frac{1}{2\pi^{n/2}|V_2|^{1/2}}\right) - 1/2(\mathbf{x} - \mu_2)^T V_2^{-1}(\mathbf{x} - \mu_2)$$

$$- log\left(\frac{1}{2\pi^{n/2}|V_1|^{1/2}}\right) + 1/2(\mathbf{x} - \mu_1)^T V_1^{-1}(\mathbf{x} - \mu_1) \qquad (7.6)$$

---

[1] One could divide by n_1 instead of $n_1 - 1$ for calculating V_1, and n_2 instead of $n_2 - 1$ for V_2. These are biased estimators for sample variance.

In the widely used LDA formula, a further assumption that the covariance matrices for both the classes are the same, that is, $V_1 = V_2$ is made, see Figure 7.1. With this assumption, equation 7.6 becomes

$$\mathbf{x}^T V^{-1}(\boldsymbol{\mu}_2 - \boldsymbol{\mu}_1) + \log(P(C_2)/P(C_1)) + \frac{1}{2}\boldsymbol{\mu}_1^T V^{-1}\boldsymbol{\mu}_1 - \frac{1}{2}\boldsymbol{\mu}_2^T V^{-1}\boldsymbol{\mu}_2 \qquad (7.7)$$

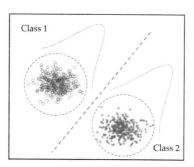

FIGURE 7.1 Equal covariance assumption in LDA.

It can be seen from equation 7.7 that the decision function is linear in $\mathbf{x}$ and hence this approach is called the linear discriminant analysis. If there is no reason to believe that Class-1 is more likely than Class-2, then $P(C_1) = P(C_2)$, which will further simplify equation 7.7. This approach can be readily extended to multi-class problems. In this case, $\mathbf{x}^{\mathbf{new}}$ is assigned to the Class-k with the maximum *a posteriori* probability $P(C_k|\mathbf{x}^{\mathbf{new}})$.

Example 7.4. *Consider the data given below containing the details of employment growth rate (x_1) and economic growth rate (x_2) of hypothetical countries categorized as Category A $(y = 0)$ and Category B $(y = 1)$. The data for both the classes were generated from the joint normal distribution with variance $\left(\Sigma = \begin{bmatrix} 3.2 & -0.5 \\ -0.5 & 5.5 \end{bmatrix}\right)$. Mean of the joint distribution for Class-0 is $\begin{bmatrix} 1 & 2 \end{bmatrix}$ and that for Class-1 is $\begin{bmatrix} 2.9 & 8 \end{bmatrix}$. Fit a linear discriminant model based on the training data (assuming unknown means and variances) and determine the model's performance in classifying the test data for new countries.*

Training data:

Country	Employment growth rate, x_1	Economy growth rate, x_2	Category, y
1	-2.138	11.994	0
2	1.713	8.254	1
3	3.722	4.759	1
4	4.579	2.484	1
5	2.618	9.995	0
6	-1.633	8.152	1
7	3.915	6.978	1
8	4.624	-2.558	0
9	0.754	4.528	0
10	-2.246	2.676	0

Test data:

Country	Employment growth rate, x_1	Economy growth rate, x_2	Category, y
11	1.377	14.326	1
12	-5.487	-2.407	0
13	1.751	11.197	1
14	-1.848	8.191	0

Solution 7.4. *The given data consists of 5 instances of Class-0 and 5 instances of Class-1. LDA assumes equal covariance matrix for both the classes. The class-wise estimated mean and covariance matrix is*

$$Class\text{-}0\ Mean,\ \boldsymbol{\mu_0} = \begin{bmatrix} \mu_{x_{10}} & \mu_{x_{20}} \end{bmatrix}^T = \begin{bmatrix} 0.722 & 5.327 \end{bmatrix}^T$$

$$Class\text{-}1\ Mean,\ \boldsymbol{\mu_1} = \begin{bmatrix} \mu_{x_{11}} & \mu_{x_{21}} \end{bmatrix}^T = \begin{bmatrix} 2.46 & 6.126 \end{bmatrix}^T$$

$$Class\text{-}0\ Covariance\ Matrix,\ V_0 = \begin{bmatrix} 8.95 & -8.285 \\ -8.285 & 34.02 \end{bmatrix}$$

$$Class\text{-}1\ Covariance\ Matrix,\ V_1 = \begin{bmatrix} 6.38 & -4.52 \\ -4.52 & 6.123 \end{bmatrix}$$

$$Shared\ Covariance\ Matrix,\ V = \frac{(n_0 - 1)V_0 + (n_1 - 1)V_1}{n_1 + n_2 - 2} = \begin{bmatrix} 7.66 & -6.4 \\ -6.4 & 20.072 \end{bmatrix}$$

where $n_0 = 5$ and $n_1 = 5$ are the number of samples in Class-0 and Class-1, respectively. The decision function of LDA for Class-k is

$$\delta_k(\mathbf{x}) = \log(P(\mathbf{x}|C_k)P(C_k))$$

$$= \log(P(C_k|\mathbf{x})P(\mathbf{x})) = \mathbf{x}^T V^{-1} \boldsymbol{\mu_k} \quad - \frac{1}{2}\boldsymbol{\mu_k}^T V^{-1} \boldsymbol{\mu_k} \quad + \log(P(C_k))$$

For binary classification, $h(x) = \delta_1(\mathbf{x}) - \delta_0(\mathbf{x})$, where $\delta_1(\mathbf{x})$ is the decision function for Class-1 and $\delta_0(\mathbf{x})$ is the decision function for Class-0. Consider the test data sample $x_1 = 1.377$ and $x_2 = 14.326$.

For Class-0,

$$\delta_0(\mathbf{x}) = \log(P(\mathbf{x}|C_0)P(C_0)) = \mathbf{x}^T V^{-1} \boldsymbol{\mu_0} \quad - \frac{1}{2}\boldsymbol{\mu_0}^T V^{-1}\boldsymbol{\mu_0} \quad + \log(P(C_0))$$

$$= [1.377 \ 14.326] \times \begin{bmatrix} 7.66 & -6.4 \\ -6.4 & 20.072 \end{bmatrix}^{-1} \begin{bmatrix} 0.722 \\ 5.327 \end{bmatrix}$$

$$-\frac{1}{2}\begin{bmatrix} 0.722 \\ 5.327 \end{bmatrix}^T \times \begin{bmatrix} 7.66 & -6.4 \\ -6.4 & 20.072 \end{bmatrix}^{-1} \times \begin{bmatrix} 0.722 \\ 5.327 \end{bmatrix} + \log\left(\frac{5}{10}\right) = 4.44$$

For Class-1,

$$\delta_1(\mathbf{x}) = \log(P(\mathbf{x}|C_1)P(C_1)) = \mathbf{x}^T V^{-1}\boldsymbol{\mu_1} \quad - \frac{1}{2}\boldsymbol{\mu_1}^T V^{-1}\boldsymbol{\mu_1} \quad + \log(P(C_1))$$

$$= [1.377 \ 14.326] \times \begin{bmatrix} 7.66 & -6.4 \\ -6.4 & 20.072 \end{bmatrix}^{-1} \begin{bmatrix} 2.46 \\ 6.126 \end{bmatrix}$$

$$-\frac{1}{2}\begin{bmatrix} 2.46 \\ 6.126 \end{bmatrix}^T \times \begin{bmatrix} 7.66 & -6.4 \\ -6.4 & 20.072 \end{bmatrix}^{-1} \times \begin{bmatrix} 2.46 \\ 6.126 \end{bmatrix} + \log\left(\frac{5}{10}\right) = 5.68$$

$$h(\mathbf{x}) = \delta_1(\mathbf{x}) - \delta_0(\mathbf{x}) = 5.68 - 4.44 = 1.24$$

Since $h(\mathbf{x})$ is greater than 0, the given data sample belongs to Class-1, i.e., country 11 belongs to Category A. The parameters of the LDA decision boundary $\mathbf{w}^T\mathbf{x} + b = 0$ where $\mathbf{x} = [x_1, \ x_2]^T$ can be found as shown below:

$$\mathbf{w} = \begin{bmatrix} w_1 \\ w_2 \end{bmatrix} = V^{-1}(\boldsymbol{\mu_1} - \boldsymbol{\mu_0}) = \begin{bmatrix} 0.354 \\ 0.152 \end{bmatrix}$$

$$b = \log(P(C_1)/P(C_0)) + \frac{1}{2}\boldsymbol{\mu_0}^T V^{-1}\boldsymbol{\mu_0} - \frac{1}{2}\boldsymbol{\mu_1}^T V^{-1}\boldsymbol{\mu_1} = -1.44$$

The predicted posterior probabilities for the test data using the LDA model are tabulated below:

Country	Employment growth rate, x_1	Economy growth rate, x_2	y_{true}	$P(C_0|\mathbf{x})$	$P(C_1|\mathbf{x})$	y_{pred}
11	1.377	14.326	1	0.225	0.775	1
12	−5.487	−2.407	0	0.977	0.023	0
13	1.751	11.197	1	0.291	0.709	1
14	−1.848	8.191	0	0.699	0.301	0

The accuracy score for the classification of test data is 1 since all test datapoints are correctly classified using the model built. Figure 7.2 shows the predicted joint normal distributions and the linear decision boundary developed based on LDA. The points marked using star symbol correspond to Class-0 and those marked by triangle symbol correspond to Class-1.

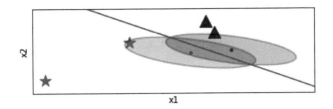

FIGURE 7.2 LDA for classes with equal variance.

Example 7.5. *Consider the data given below where both the categories have unequal variance. The samples for Class-0 are generated from a joint Gaussian distribution with mean* $\begin{bmatrix} 1 & 2 \end{bmatrix}$ *and covariance matrix* $\begin{bmatrix} 3.2 & -0.5 \\ -0.5 & 5.5 \end{bmatrix}$. *Similarly, samples for Class-1 are generated from a joint Gaussian distribution with mean* $\begin{bmatrix} 1.3 & 2.2 \end{bmatrix}$ *and covariance matrix* $\begin{bmatrix} 1.08 & -0.03 \\ -0.03 & 4.2 \end{bmatrix}$. *Fit a linear discriminant model based on the training data (assuming unknown means and variances) and determine the model's performance in classifying the test data for new countries.*

Training data:

Country	Employment growth rate, x_1	Economy growth rate, x_2	Category, y
1	1.750	2.542	1
2	−3.813	11.209	0
3	1.712	7.296	1
4	0.390	9.196	0
5	1.212	−5.459	1
6	3.053	5.934	1
7	6.319	7.931	0
8	2.430	−5.609	0
9	0.032	9.893	1
10	0.395	0.861	0

Test data:

Country	Employment growth rate, x_1	Economy growth rate, x_2	Category, y
11	0.663	8.216	1
12	0.511	6.459	1
13	3.970	−5.651	0
14	2.302	8.443	1

Solution 7.5. *LDA assumes that both classes have the same covariance matrix. The estimated means and variance matrix using simulation tools are*

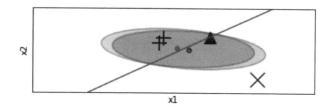

FIGURE 7.3 LDA for classes with unequal variance.

provided below:

$$\text{Class-0 Mean, } \boldsymbol{\mu_0} = \begin{bmatrix} \mu_{x_{10}} & \mu_{x_{20}} \end{bmatrix}^T = \begin{bmatrix} 1.144 & 4.72 \end{bmatrix}^T$$

$$\text{Class-1 Mean, } \boldsymbol{\mu_0} = \begin{bmatrix} \mu_{x_{11}} & \mu_{x_{21}} \end{bmatrix}^T = \begin{bmatrix} 1.55 & 4.04 \end{bmatrix}^T$$

$$V_0 = \begin{bmatrix} 13.53 & -7.33 \\ -7.33 & 48.51 \end{bmatrix}; \quad V_1 = \begin{bmatrix} 1.186 & -0.65 \\ -0.65 & 35.233 \end{bmatrix}$$

$$\text{Shared Covariance Matrix, } V = \frac{(n_0 - 1)V_0 + (n_1 - 1)V_1}{n_1 + n_2 - 2} = \begin{bmatrix} 7.36 & -3.99 \\ -3.99 & 41.87 \end{bmatrix}$$

The parameters of the LDA model $w^T x + b$ where $x = [x_1, \; x_2]^T$ are $w_1 = 0.049$, $w_2 = -0.0115$ and $b = -0.0161$. The predicted posterior probabilities for the test data using the LDA model are tabulated below:

Country	Employment growth rate, x_1	Economy growth rate, x_2	y_{true}	$P(C_0\|\mathbf{x})$	$P(C_1\|\mathbf{x})$	y_{pred}
11	0.663	8.216	1	0.52	0.48	0
12	0.511	6.459	1	0.516	0.484	0
13	3.970	-5.651	0	0.44	0.56	1
14	2.302	8.443	1	0.4999	0.5001	1

The accuracy score for the classification of test data is 0.25 since three out of the 4 test datapoints are wrongly classified using the model built. Note that all the probabilities are close to 0.5, implying less confidence in predictions. Figure 7.3 shows the predicted distributions and decision boundary. The misclassified Class-1 points are indicated by the symbol "+" while the symbol "×" represent the misclassified Class-0 sample. The classification performance is summarized below:

True Label	Predicted Label	
	Class-0	Class-1
Class-0	0	1
Class-1	2	1

	Precision	Recall	F1-score
Class-0	0	0	0
Class-1	0.5	0.33	0.4

7.2.3 Quadratic Discriminant Analysis (QDA)

Quadratic Discriminant Analysis (QDA) follows the same approach as LDA in terms of assuming that class data come from multivariate Gaussian distributions. However, QDA does not make the assumption of equal covariance

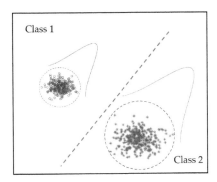

FIGURE 7.4 Unequal covariance assumption in QDA.

matrices that is made in LDA (Figure 7.4). As a result, simplification of equation 7.6 to equation 7.7 is not possible. Hence the decision function for QDA
is given by equation 7.6. Notice that the equation has quadratic terms in X
and hence the name QDA.

Example 7.6. *Consider the data given in Example 7.4 where both the categories have equal variance. Fit a quadratic discriminant model based on the
training data and determine the model's performance in classifying the test
data for new countries.*

Solution 7.6. *For the given data, QDA assumes that the classes have different covariance matrix. The estimated class-wise means and covariance matrices are:*

$$\mu_0 = \begin{bmatrix} 0.722 & 5.33 \end{bmatrix} ; \quad \mu_1 = \begin{bmatrix} 2.46 & 6.12 \end{bmatrix}$$

$$V_0 = \begin{bmatrix} 8.95 & -8.28 \\ -8.28 & 34.02 \end{bmatrix} ; \quad V_1 = \begin{bmatrix} 6.38 & -4.52 \\ -4.52 & 6.12 \end{bmatrix}$$

*The estimated posterior probabilities and the corresponding predicted class
for the test dataset are tabulated below.*

Country	Employment growth rate, x_1	Economy growth rate, x_2	y_{true}	$P(C_0|\mathbf{x})$	$P(C_1|\mathbf{x})$	y_{pred}
11	1.377	14.326	1	0.999	0.001	0
12	−5.487	−2.407	0	1.000	0.000	0
13	1.751	11.197	1	0.803	0.197	0
14	−1.848	8.191	0	0.494	0.506	1

*The quadratic decision boundary developed using QDA is shown in Figure
7.5. Only one of the samples is classified correctly.*

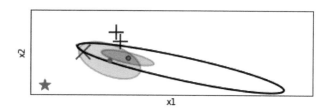

FIGURE 7.5 QDA for classes with equal variance.

Example 7.7. *Consider the data given in Example 7.5 where both the categories have unequal variance. Fit a quadratic discriminant model based on the training data and determine the model's performance in classifying the test data for new countries.*

Solution 7.7. *For the given data, QDA estimate of the mean and covariance matrix of the two classes are given below*

$$\mu_0 = \begin{bmatrix} 1.144 & 4.72 \end{bmatrix} \qquad \mu_1 = \begin{bmatrix} 1.55 & 4.04 \end{bmatrix}$$

$$V_0 = \begin{bmatrix} 13.53 & -7.33 \\ -7.33 & 48.51 \end{bmatrix} \qquad V_1 = \begin{bmatrix} 1.19 & -0.65 \\ -0.65 & 35.23 \end{bmatrix}$$

The predicted posterior probabilities for the test data using the LDA model are tabulated below:

Country	Employment growth rate, x_1	Economy growth rate, x_2	y_{true}	$P(C_1\|x)$	$P(C_2\|x)$	y_{pred}
11	0.663	8.216	1	0.281	0.719	1
12	0.511	6.459	1	0.295	0.705	1
13	3.970	-5.651	0	0.722	0.278	0
14	2.302	8.443	1	0.263	0.737	1

The classification performance is summarized below:

True Label	Predicted Label	
	Class-0	Class-1
Class-0	1	0
Class-1	0	3

	Precision	Recall	F1-score
Class-0	1	1	1
Class-1	1	1	1

The accuracy score of classification for the data with unequal class covariance is found to be 1 which is significantly better than the accuracy obtained using LDA for the same data. The predicted probability for the correct class is also higher when compared to LDA predictions. In LDA, all the probabilities were close to 0.5 implying that the predictions are poor. Figure 7.6 shows the decision boundary based on QDA and the the density functions of each class. All samples have been correctly identified.

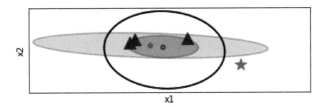

FIGURE 7.6 QDA for classes with unequal variance.

7.2.4 Logistic Regression

Logistic regression is another classification approach, which performs quite well for many binary classification problems. This technique can also be extended to multi-class problems. The most common logistic regression approaches use a linear model, which can be directly extended to polynomial models to handle nonlinear cases. Consider two classes C_1 and C_2. In traditional classification approaches, a new datapoint $\mathbf{x}^{\mathbf{new}}$ is categorically assigned to either one of the classes. However, assigning a probability that a datapoint belongs to a particular class might be more useful. In essence, the aim is to convert the discrete class output to a continuous probability output.

Logistic regression achieves this by modeling the probability that a datapoint belongs to a certain class. To do this, let us start with a notion of a hyperplane that separates the two classes $h(\mathbf{x}) = \mathbf{n}^T\mathbf{x} + b$. If $h(\mathbf{x}) > 0$, the datapoint belongs to class C_2 and vice versa. If this decision function has to be converted to a probability function, then a function $p(\mathbf{x})$ that transforms decision function values to be in the range of 0 and 1 is required. Using such a transformation, whenever the probability is greater than 0.5, that datapoint belongs to C_2 and when less than 0.5, to class C_1. That is, $p(\mathbf{x})$ is interpreted as the probability that the datapoint belongs to C_2. Logistic regression uses the following transformation:[2]

$$p(\mathbf{x}) = \frac{e^{h(\mathbf{x})}}{1 + e^{h(\mathbf{x})}} \tag{7.8}$$

Certain observations can directly be made based on $p(\mathbf{x})$. When $h(\mathbf{x}) = 0$, that is when the point is right on the hyperplane, $p(\mathbf{x}) = 0.5$ implying that it is not possible to decide to which class the datapoint belongs. When $h(\mathbf{x}) \gg 0$, then $p(\mathbf{x}) \approx 1$ (Class-2) and if $h(\mathbf{x}) \ll 0$, then $p(\mathbf{x}) \approx 0$ (Class-1). This is shown in Figure 7.7. Equation 7.8 can be rewritten as

$$h(\mathbf{x}) = \log\left(\frac{p(\mathbf{x})}{1 - p(\mathbf{x})}\right) \tag{7.9}$$

---

[2] Although P was used until now for probability consistent with Chapter 5, here we use p for probability as this is a scalar function.

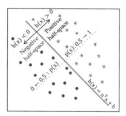

FIGURE 7.7 Schematic to explain LR.

As a result logistic regression can be thought of as modeling the log of odds ratio (probability of success divided by the probability of failure) as a linear function.

We now have a problem in training a logistic regression classifier. Each datapoint is simply labeled as belonging to Class-1 or Class-2 and we do not have the probabilities that these points belong to the two classes. One approach to address this problem is to assume that the output is one for all datapoints for Class-2 and 0 for all datapoints belonging to Class-1 and run a standard regression type algorithm. However, this is does not take cognizance of the notion of probabilities and hence is not used as an approach for training logistic regression classifiers. Logistic regression training follows a maximum likelihood approach that was discussed in the statistics section. Let $i \in S_1$ and $i \in S_2$ represent datapoints labeled as Class-1 and Class-2, respectively. An output $y^i = 0$ if $i \in S_1$ and $y^i = 1$ if $i \in S_2$. We would like $p(\mathbf{x}^i)$ to be large if $i \in S_1$ and $1 - p(\mathbf{x}^i)$ to be large if $i \in S_2$.

A likelihood function that captures this requirement is the following

$$\max_{\mathbf{n}, b} L = \prod_{i=1}^{n} p(\mathbf{x}^i)^{y^i} (1 - p(\mathbf{x}^i))^{(1-y^i)} \tag{7.10}$$

Maximizing L is equivalent to maximizing $\log(L)$. This maximization problem is an unconstrained nonlinear programming problem and techniques discussed in Chapter 4 are deployed to solve this optimization problem to identify optimal $\mathbf{n}$ and b.

This approach can be reasonably easily extended to nonlinear problems by changing $h(\mathbf{x})$. A simple extension would be use a quadratic function $h(\mathbf{x}) = \mathbf{x}^T Q \mathbf{x} + \mathbf{n}^T \mathbf{x} + b$ to handle certain types of nonlinear decision surfaces. Logistic regression approach can also be extended to multi-class problems by solving appropriate binary classification problems and then using a decision logic to process the results of the binary logistic regression classifiers. This can sometimes become cumbersome and other approaches that directly handle multi-class problems might be preferred.

Over-fitting in logistic regression can be avoided through regularization, where a penalty is attached to increasing the number of parameters. This is achieved by modifying the objective function to the following,

$$\max_{\mathbf{p}_L} J = \log(L) - C \, \|\mathbf{p}_L\| \qquad (7.11)$$

where C is a constant that can be used to increase or decrease the penalty, and $\mathbf{p}_L = [\mathbf{n}, b]$ in the linear case and additional parameters in the nonlinear case. The norm can be chosen to be any p-norm, however 1-norm and 2-norm are used the most.

Example 7.8. *Consider the following data*

No. of hours studied, x_1	10.5	2.5	14	8.2	10.4	5	6.7	14.7
No. of classes missed, x_2	4	10	15	3	0	0	7	14
y (Fail =0, Pass =1)	1	0	0	1	1	1	0	1

Given that the decision boundary is

$$h(x) = \beta_0 + \beta_1 x_1 + \beta_2 x_2$$

Calculate the predicted probabilities of passing the exam and predicted y with $p = 0.5$ as threshold using logistic function if

a) $\beta_0 = 1; \beta_1 = 5; \beta_2 = -10;$
b) $\beta_0 = 6.25; \beta_1 = 3.5; \beta_2 = -3;$

Compare the predicted probabilities with the truth and comment on which of the above parameter sets is a better choice for determining whether a student will pass or fail in the course given the number of hours studied and number of classes missed.

Solution 7.8.

$$p(x) = \frac{e^{(h(x))}}{1 + e^{(h(x))}}$$

a)

No. of hours studied, x_1	10.5	2.5	14	8.2	10.4	5	6.7	14.7
No. of classes missed, x_2	4	10	15	3	0	0	7	14
y (Fail =0, Pass =1)	1	0	0	1	1	1	0	1
h_x^l	13.5	-86.5	-79	12	53	26	-35.5	-65.5
p_x^l	0.99	0	0	0.99	1	1	0	3.6×10^{-29}
$\hat{y}$	1	0	0	1	1	1	0	0

Note that the p(x) evaluated is quite close to the true class except for the last sample. The corresponding likelihood can be evaluated as:

$$\Sigma(p_i)^{y_i}(1 - p_i)^{1-y_i} = 6.999$$

The predictions can be compared with the truth using a confusion matrix as given below.

	Predicted Label	
	0	1
True label 0	3	0
1	1	4

b)

No. of hours studied x_1	10.5	2.5	14	8.2	10.4	5	6.7	14.7
No. of classes missed	4	10	15	3	0	0	7	14
y (Fail $=0$, Pass $=1$)	1	0	0	1	1	1	0	1
h_x^{II}	31	-15	10.25	25.95	42.65	23.75	8.7	15.67
p_x^{II}	1	3×10^{-7}	0.99	1	1	1	0.99	0.99
$\hat{y}$	1	0	1	1	1	1	1	1

$$\Sigma(p_i)^{y_i}(1 - p_i)^{1-y_i} = 6.001$$

The predictions can be compared with the truth using a confusion matrix as given below.

	Predicted Label	
	0	1
True label 0	1	2
1	0	5

Here, although most of the classes have been correctly identified, 3rd and 7th samples have been wrongly identified. So it is clear that the model used in part a) is better as 7 out of 8 samples are predicted correctly.

One could also check the value of $\Sigma(p_i)^{y_i}(1 - p_i)^{1-y_i}$ in both cases. Since $6.999 >> 6.001$, parameters in part a) are better.

Example 7.9. Consider the data containing the soil organic carbon content x_1, soil pH x_2 and base saturation x_3 of two types of soils Type A $(y = 0)$ and Type B $(y = 1)$. Build a logistic regression model for binary classification and comment on the influence of cut-off probability on the model's performance. Also perform classification on the given test data based on the optimum value of cut-off probability chosen.

Training data:

Organic carbon x_1	0.448	0.449	0.439	0.439	0.448	0.499	0.448
pH x_2	5.682	5.85	6.511	5.963	6.03	5.841	6.211
Base saturation x_3	33.8	41.5	21.1	45.7	85.5	61.4	6.5
Soil type y	0	0	1	0	0	0	1
Organic carbon x_1	0.439	0.499	0.448	0.448	0.448	0.448	428
pH x_2	5.998	5.966	5.786	6.77	6.069	5.399	7.024
Base saturation x_3	21.4	30.2	33.3	2.9	40	95.3	15.8
Soil type y	1	1	0	1	0	0	1

Test data:

Organic carbon x_1	0.538	0.439	0.448	0.499	0.448	0.428
pH x_2	6.096	6.115	5.602	5.933	6.169	6.595
Base saturation x_3	96.9	63	62	68.2	6.6	21.8
Soil type y	0	0	0	0	1	1

Solution 7.9. *Logistic regression model provides the probability that a given datapoint belongs to Class-2 using the transformation*

$$p(\mathbf{x}) = \frac{e^{h(\mathbf{x})}}{1 + e^{h(\mathbf{x})}} = \frac{e^{(b + w_1 x_1 + w_2 x_2 + w_3 x_3)}}{1 + e^{(b + w_1 x_1 + w_2 x_2 + w_3 x_3)}}$$

A decision is made regarding the category of the datapoint using the following logic:

$$Datapoint \in \begin{cases} Class\,1, & if\ p(\mathbf{x}) < p_{threshold}(\mathbf{x}) \\ Class\,2, & if\ p(\mathbf{x}) \geq p_{threshold}(\mathbf{x}) \end{cases}$$

Using the given training data, the model coefficients are determined to be

$$b = -0.4589;\ w_1 = 0.0958;\ w_2 = 0.9948;\ w_3 = -1.249$$

The estimated probabilities using the logistic regression model are tabulated below.

x_1	0.448	0.449	0.439	0.439	0.448	0.499	0.448
x_2	5.682	5.85	6.511	5.963	6.03	5.841	6.211
x_3	33.8	41.5	21.1	45.7	85.5	61.4	6.5
Estimated probability $p(x)$	0.226	0.271	0.789	0.2383	0.053	0.123	0.792
x_1	0.439	0.499	0.448	0.448	0.448	0.448	428
x_2	5.998	5.966	5.786	6.77	6.069	5.399	7.024
x_3	21.4	30.2	33.3	2.9	40	95.3	15.8
Estimated probability $p(x)$	0.52	0.458	0.277	0.945	0.354	0.008	0.94

The best threshold value for $p(x)$ is chosen such that true predictions are high and false predictions are low. Here, we test the same using balanced accuracy.

$$Balanced\ accuracy = \frac{Sensitivity + Specificity}{2}$$

$$Sensitivity = \frac{True\ positives}{True\ positives + False\ negatives}$$

$$Specificity = \frac{True\ negatives}{False\ positives + True\ negatives}$$

For this example, if we set $p_{threshold} = 0.52$, all samples except sample 9 (false negative) are correctly classified. Five out of six positives are correctly identified as positive and all 8 negatives are correctly identified as negative. Hence, sensitivity $= 5/6$, specificity $= 8/8 = 1$, and balanced accuracy $= 0.916$. Similarly, balanced accuracy for various threshold values are estimated and

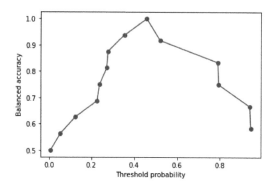

FIGURE 7.8 Sensitivity vs Threshold probability.

presented in Figure 7.8. From Figure 7.8, it is evident that the sensitivity and selectivity are maximum when threshold probability is chosen as 0.458. Applying the logistic regression model and using this threshold for the classification of test data

x_1	x_2	x_3	y_{true}	Predicted probability $p(x)$	y_{pred}
0.538	6.096	96.9	0	0.052	0
0.439	6.115	63	0	0.164	0
0.448	5.602	62	0	0.0589	0
0.499	5.933	68.2	0	0.112	0
0.448	6.169	6.6	1	0.775	1
0.428	6.595	21.8	1	0.809	1

Example 7.10. *Consider the data given below. Build logistic regression models with polynomial features (degrees 1 to 3) and comment on the effect of polynomial degree on classification performance.*

Training data

No. of hours studied x_1	10.5	2.5	14	8.2	10.4	5	6.7	14.7
No. of classes missed	4	10	15	3	0	0	7	14
y (Fail =0, Pass =1)	1	0	0	1	1	1	0	1

Test data

	Sample 1	Sample 2	Sample 3	Sample 4
No. of hours studied x_1	4.9	5.5	15.1	12.1
No. of classes missed	8	5	7	10
y (Fail =0, Pass =1)	0	0	1	1

Solution 7.10. *The data contains two input features x_1 and x_2.*

$$p(\mathbf{x}) = \frac{e^{h(\mathbf{x})}}{1 + e^{(h(\mathbf{x}))}}; \qquad \mathbf{x} = \begin{bmatrix} x_1 & x_2 \end{bmatrix}$$

a) Polynomial Degree 1: The identified model is

$$h(\mathbf{x}) = b + w_1 x_1 + w_2 x_2$$

The optimum parameters of the linear model are:

$$b = -11.773; \quad w_1 = 16.60; \quad w_2 = -15.778$$

b) *Polynomial Degree 2: The logistic regression model is*

$$h(\mathbf{x}) = b + w_1\,x_1 + w_2\,x_2 + w_3x_1^2 + w_4x_1x_2 + w_5x_2^2$$

The optimum model parameters are

$$b = -0.1767; \quad w_1 = -0.7264; \quad w_2 = -0.9208;$$
$$w_3 = 1.373 \; w_4 = -0.1673; \quad w_5 = -1.1319$$

c) *Polynomial Degree 3: The logistic regression model is*

$$h(\mathbf{x}) = b + w_1\,x_1 + w_2\,x_2 + w_3x_1^2 + w_4x_1x_2 + w_5x_2^2 + w_6x_1^3$$
$$+ w_7x_1^2x_2 + w_7x_1x_2^2 + w_8x_2^3$$

The optimum model parameters are

$$b = -0.0016; \quad w_1 = -0.008; \quad w_2 = -0.0125; \quad w_3 = -0.02 \; w_4 = -0.054;$$
$$w_5 = -0.072; \quad w_6 = 0.343; \quad w_7 = 0.0384; \quad w_8 = -0.1; \quad w_9 = -0.264$$

The following table summarizes the results.

Sample	y_{true}	Simple logistic regression			Poly. features (degree 2)			Poly. features (degree 3)		
		Fail	Pass	y_{pred}	Fail	Pass	y_{pred}	Fail	Pass	y_{pred}
1	Fail	1	0	Fail	1	0	Fail	1	0	Fail
2	Fail	0.345	0.654	Pass	0.533	0.466	Fail	0.001	0.999	Pass
3	Pass	0	1	Pass	0	1	Pass	0	1	Pass
4	Pass	0	1	Pass	0	1	Pass	0	1	Pass

The second sample is incorrectly classified as Class-1 using a linear model but with the inclusion of quadratic features, the sample is correctly classified. But as higher powers are included, the model may become too complex and overfit the training data resulting in incorrect classification (like sample 2 in this example).

Example 7.11. *Consider the data given below consisting of three input features GDP (x_1), economic growth rate (x_2), and population growth rate (x_3) of cities across the world categorized into two classes: Category A ($y = 1$) and Category B ($y = 0$). Examine the impact of L_1 and L_2 regularization in the logistic regression model based on the training data. Also comment on its influence on the predictions performed for the test data.*

Training data:

GDP, x_1	Economic growth rate, x_2	Population growth rate, x_3	Category, y
1137	−1.587	−6.607	1
1275	−3.098	-10.460	1
830	−1.442	3.292	1
885	0.030	−1.051	1
78	0.248	7.643	0
776	−5.903	6.568	1
62	4.258	11.296	0
83	1.866	7.776	0
1238	−1.094	2.307	1
1331	0.224	−0.521	1
919	0.895	4.774	1
646	1.162	10.292	0
1046	0.889	2.264	1

Test data:

GDP, x_1	Economic growth rate, x_2	Population growth rate, x_3	Category, y
684	−2.076	10.822	0
1287	−3.993	2.622	1
801	−0.913	−2.011	1
37	3.628	0.813	0
36	−1.131	1.903	0
1306	−1.224	1.748	1

Solution 7.11. *In order to build a logistic regression model based on the training data, the input features are normalized ($z = \frac{x-\mu}{\sigma}$ where μ = estimated mean of training sample, σ = estimated standard deviation of the training samples). Models are built by considering different regularization strengths and analysed for their generalization on test data. The results are tabulated in Table 7.1.*

From Table 7.1 it can be observed that regularization has an impact on model bias, prediction errors and the number of features retained.

Example 7.12. *Consider the data below consisting of sepal and petal lengths of three types of Iris plant species: Iris Virginica (0), Iris Setosa (1) and Iris Versicolor (2). Build a logistic regression model to classify the plant species based on sepal and petal lengths.*

Training data:

Sepal Length (x_1)	5.1	4.9	4.7	4.6	4.8	6.8	6.7	6.7	6.3	5.9	5.3	5.5
Petal Length (x_2)	2.4	1.4	1.3	1.5	1.3	5.9	5.7	5.2	5	3.1	2.9	3.4
Species (y)	2	1	1	1	1	0	0	0	0	2	2	2

Test data:

Sample	1	2	3	4
Sepal Length (x_1)	4.5	6.2	5	5.2
Petal Length (x_2)	1.2	5.4	3.4	2.2
Species (y)	1	0	2	2

TABLE 7.1

Logistic Regression with Regularization: Example 7.11

Regularization Parameter, C	L_1 Regularization			L_2 Regularization		
	Training Data Accuracy	Model Parameters	Test Data Accuracy	Training Data Accuracy	Model Parameters	Test Data Accuracy
20	0.307	$b = 0$ $w_1 = 0$ $w_2 = 0$ $w_3 = 0$	0.5	0.692	$b = 0.845$ $w_1 = 0.207$ $w_2 = -0.137$ $w_3 = -0.157$	0.5
10	0.307	$b = 0$ $w_1 = 0$ $w_2 = 0$ $w_3 = 0$	0.5	0.846	$b = 0.897$ $w_1 = 0.347$ $w_2 = -0.222$ $w_3 = -0.258$	0.5
5	0.923	$b = 0$ $w_1 = 0.077$ $w_2 = 0$ $w_3 = 0$	1	0.923	$b = 0.994$ $w_1 = 0.5367$ $w_2 = -0.3292$ $w_3 = -0.3919$	0.667
2	1	$b = 0.1227$ $w_1 = 1.218$ $w_2 = -0.007$ $w_3 = 0$	1	1	$b = 1.1994$ $w_1 = 0.855$ $w_2 = -0.497$ $w_3 = -0.628$	0.667
1	1	$b = 0.52$ $w_1 = 1.843$ $w_2 = -0.289$ $w_3 = 0$	0.833	1	$b = 1.413$ $w_1 = 1.137$ $w_2 = -0.636$ $w_3 = -0.868$	0.833
0.1	1	$b = 2.29$ $w_1 = 3.494$ $w_2 = -1.105$ $w_3 = -2.08$	1	1	$b = 2.543$ $w_1 = 2.262$ $w_2 = -1.194$ $w_3 = -2.186$	0.833
0.05	1	$b = 2.943$ $w_1 = 4.24$ $w_2 = -1.310$ $w_3 = -2.839$	1	1	$b = 3.025$ $w_1 = 2.659$ $w_2 = -1.370$ $w_3 = -2.744$	1

Solution 7.12. *The data contains 3 classes of plant species and hence this is a multi-class classification problem. Logistic regression algorithm being an inherently binary classifier, this multi-class classification can be solved as multiple binary classification problems. This is known as the one versus the rest approach where a binary classifier for each class versus the rest of the classes is built.*

Classifier 1 - Iris Virginica vs (Iris Setosa & Iris Versicolor)
Classifier 2 - Iris Setosa vs (Iris Virginica & Iris Versicolor)
Classifier 3 - Iris Versicolor vs (Iris Virginica & Iris Setosa)

Using the training data, each binary classifier is trained such that data from a specific class is treated as a member of the class while data from all the other classes are considered as non-members. Each model is then used to predict the probability for a particular class and a class is assigned to the data sample based on the maximum value of the predicted probability values for different classes. The results are summarized below:

| Sample | y_{true} | Predicted Probability | | | y_{pred} |
		Class-0	Class-1	Class-2	
1	1	0.0061	0.667	0.3268	1
2	0	0.7493	0.005	0.2449	0
3	2	0.2351	0.240	0.524	2
4	2	0.047	0.529	0.422	1

7.2.5 Clustering Techniques

Clustering techniques are unsupervised techniques in the sense that the technique is presented with unlabeled data and the intent is to group data so that some meaningful insights can be derived. Since there is no labeling to classify data, other ideas to group data are needed. One approach is to look at the similarity between datapoints and group datapoints that are similar together (similarity-based learning).

7.2.5.1 k-Means Clustering

The goal of any clustering algorithm is to partition datapoints into different clusters in a meaningful manner. The key idea is to group datapoints with high similarity in the same cluster. Philosophically, this can be contrasted with supervised learning, where the aim is to separate datapoints based on how dissimilar they are. This is why supervised algorithms are sometimes called dissimilarity-based learning algorithms. Intuitively, a similarity measure can be based on distances between datapoints. Datapoints are similar if they are close to each other and dissimilar if they are not. Maximizing similarity is equivalent to minimizing distances in clustering algorithms. We describe how this is done collectively for a group of datapoints.

Given k clusters, the aim of the algorithm is to partition data into these k clusters (see Figure 7.9) [27]. The figure shows five clusters. These clusters are characterized by the mean of the data in each individual cluster and hence the name k-means clustering. The algorithm comprises of the steps outlined below:

Step 1: Randomly initialize the means of k clusters
Step 2: Find the distances between all the datapoints and cluster centers
Step 3: (Re)Assign datapoints to their closest cluster centers
Step 4: Recompute cluster centers as the mean of all datapoints assigned to that cluster
Step 5: Go back to Step 2 and repeat till the cluster centers do not change and/or there are no reassignments

Example 7.13. *Consider the data containing the sepal and petal lengths of three types of Iris species—Iris virginica, Iris setosa, Iris versicolor. Perform k-means clustering.*

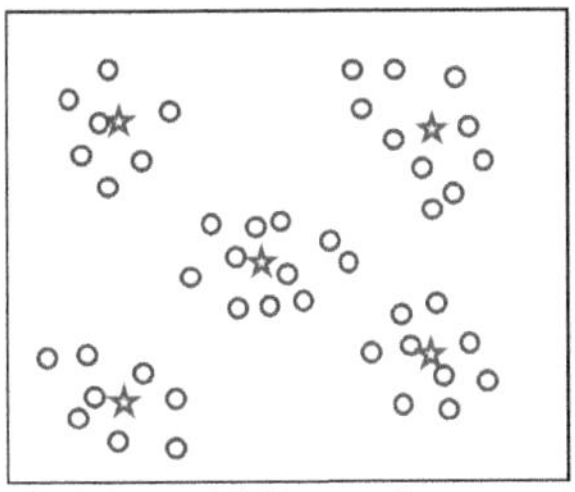

☆ Mean of different clusters
○ Samples

FIGURE 7.9 Datpoints and cluster centers.

Sepal length, x_1	5.1	4.9	4.7	4.6	4.8	6.8	6.7	6.7	6.3	5.9	5.3	5.5
Petal length, x_2	2.4	1.4	1.3	1.5	1.3	5.9	5.7	5.2	5	3.1	2.9	3.4

Solution 7.13. *Given that the data contains features of 3 different species, we consider $k = 3$ for clustering. The clustering process is initialized by randomly choosing 3 cluster centres $C_1^{init} = (4.7, 1.3)$, $C_2^{init} = (5.9, 3.1)$, and $C_3^{init} = (4.9, 1.4)$.*

Iteration 1: Euclidean distance is computed between three the cluster centers and all datapoints and each sample is assigned to the closest cluster center. The distances are computed using the formula:

$$d = \sqrt{(x - x_c)^2 + (y - y_c^2)}$$

For example, for the sample $[5.1, 2.4]$,

$$\text{Distance from } C_1 = \sqrt{(5.1 - 4.7)^2 + (2.4 - 1.3)^2} = 1.17$$
$$\text{Distance from } C_2 = \sqrt{(5.1 - 5.9)^2 + (2.4 - 3.1)^2} = 1.06$$
$$\text{Distance from } C_3 = \sqrt{(5.1 - 4.9)^2 + (2.4 - 1.4)^2} = 1.02$$

Note that the sample $[5.1, 2.4]$ is closest to the third cluster center, C_3. Hence sample $[5.1, 2.4]$ is assigned to the third cluster. This process is repeated for all the samples and cluster centers are assigned:

Cluster 1	Cluster 2	Cluster 3
$[4.7, 1.3]$, $[4.6, 1.5]$, $[4.8, 1.3]$	$[6.8, 5.9]$, $[6.7, 5.7]$, $[6.7, 5.2]$, $[6.3, 5]$, $[5.9, 3.1]$, $[5.3, 2.9]$, $[5.5, 3.4]$	$[5.1, 2.4]$ $[4.9, 1.4]$

The cluster centres are recomputed as the centroid of all samples assigned to a cluster. The new cluster center for the first cluster is $[(4.7+4.6+4.8)/3, (1.3+1.5+1.3)/3] = [4.7, 1.37]$. Similary, the new cluster centers for the second and third clusters are $[6.17, 4.45]$ and $[5, 1.9]$.

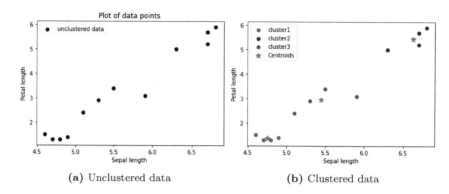

(a) Unclustered data   (b) Clustered data

FIGURE 7.10 k-means clustering: Example 7.13.

This process is repeated until the cluster centers do not change. Although the maximum number of iterations was set to be 100, the algorithm converged in 4 iterations. The final cluster centers are $C_1 = (4.75, 1.375)$, $C_2 = (6.625, 5.45)$, and $C_3 = (5.45, 2.95)$.

The final clusters are presented in Figure 7.10 and Table 7.2. From Figure 7.10, it is evident that the samples that are close to each other are assigned to the same cluster.

TABLE 7.2

Cluster assignment: Example 7.13

x_1	x_2	Distance from, C_1	Distance from, C_2	Distance from, C_3	Cluster Assigned
5.1	2.4	1.0831	3.41	0.651	3
4.9	1.4	0.152	4.402	1.644	1
4.7	1.3	0.090	4.574	1.812	1
4.6	1.5	0.195	4.438	1.680	1
4.8	1.3	0.090	4.533	1.773	1
6.8	5.9	4.967	0.482	3.244	2
6.7	5.7	4.744	0.261	3.020	2
6.7	5.2	4.293	0.261	2.573	2
6.3	5	3.942	0.555	2.219	2
5.9	3.1	2.073	2.459	0.474	3
5.3	2.9	1.621	2.873	0.158	3
5.5	3.4	2.159	2.338	0.452	3

From an optimization perspective, k-means algorithm attempts to minimize the total sum of squared distances (TSSD) or within cluster sum of squared distances (WCSSD). The objective function is (where S_i is the set of sample

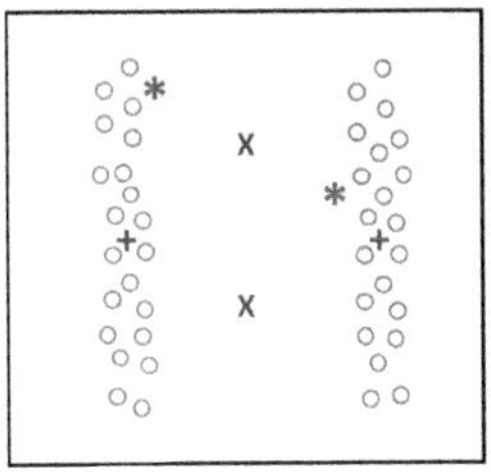

FIGURE 7.11 Importance of initialization and stochastic gradient improvements.

indices belonging to cluster i)

$$\min_{\mathbf{c}_1...\mathbf{c}_K} \sum_{j=1}^{K} \sum_{i \in S_i} ||\mathbf{x}^i - \mathbf{c}_j||^2 \tag{7.12}$$

k-means clustering algorithm is a local algorithm and hence, the final result will depend on initialization. This is illustrated in Figure 7.11. If the clustering procedure is initialized with two cluster centers marked by "x", then the cluster centers are likely to move very little and converge to about the same location because one of the cluster centers is the centroid for the data in the upper part of the diagram and the other for the lower part. There are two ways in which this problem may be avoided. The first approach is to attempt different cluster initializations. If the two cluster centers initialized are either the ones marked by "+" or "$*$", then the algorithm, on termination, is likely to describe the data better.

Another approach is the stochastic optimization approach described in Chapter 4. Assume that the algorithm starts with the two cluster locations marked by "x". If instead of presenting all the data, the upper half of the data on the right side and the lower half of the data on the left side were presented first, the two centers will move to the centroids of the respective data. Subsequently, when the remaining data is presented, the centers will move to the centroids of the left and right side data respectively.

Example 7.14. *Consider the following data.*

x_1	5	5	5	5	5	5	6	6	6	6	6	6
x_2	2	2.5	3	3.5	4	4.5	2	2.5	3	3.5	4	4.5

Perform k-means clustering starting from the cluster centers
a) $\begin{bmatrix} 5.5 & 4.5 \end{bmatrix}^T$ *and* $\begin{bmatrix} 5.5 & 2 \end{bmatrix}^T$ *b)* $\begin{bmatrix} 6.1 & 3.25 \end{bmatrix}^T$ *and* $\begin{bmatrix} 5.0 & 3.5 \end{bmatrix}^T$

Solution 7.14. *a) The k-means algorithm converges to cluster centers* $\begin{bmatrix} 5.5 & 4 \end{bmatrix}^T$ *and* $\begin{bmatrix} 5.5 & 2.5 \end{bmatrix}^T$. *The top 6 samples are clustered together while the bottom 6 samples form another cluster as shown in Figure 7.12a.*

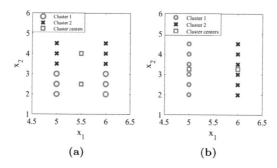

FIGURE 7.12 k-means clustering: Example 7.14.

b) The k-means algorithm converges to cluster centers $\begin{bmatrix} 5 & 3.25 \end{bmatrix}^T$ and $\begin{bmatrix} 6 & 3.25 \end{bmatrix}^T$. The 6 samples on the left are clustered together while the remaining 6 samples on the right form another cluster as shown in Figure 7.12b.

It is important to choose a correct number of clusters while using k-means algorithm. This number is identified through what is called a "knee-plot". Figure 7.13a shows an example of data with a single cluster. The cluster center will converge to be in the middle of the two distinguishable clusters. Figure 7.13b shows the same data with two cluster centers. One can see that the objective function value will come down considerably because the distances will all be smaller than the previous single cluster case. If one were to increase the number of clusters to three, the result is shown in Figure 7.13c; the TSSD will definitely be less than the two cluster case but the decrease will be marginal. This is evident in the "knee-plot" shown in Figure 7.13d. There is a definite slope change at $k = 2$. This slope change ("knee-point") can be used to identify the optimal number of cluster centers.

Example 7.15. *Consider the data given in Example 7.13. Determine the optimum number of clusters using the "knee-plot".*

Solution 7.15. *For the given data, k-means clustering is performed for different number of clusters (k varying from 1 to 8). We initialize the clustering process by choosing the first k cluster centres from the following list:*

C_i	1	2	3	4	5	6	7	8
x_1	4.7	5.9	4.9	4.8	5.3	4.6	6.7	6.8
x_2	1.3	3.1	1.4	1.3	2.9	1.5	5.7	5.9

The results obtained are tabulated below:

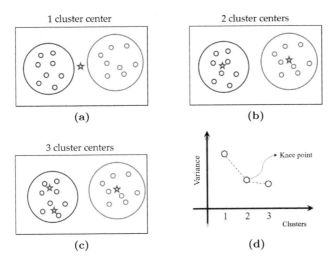

FIGURE 7.13 Identifying optimal number of clusters.

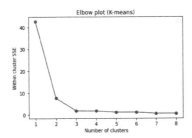

FIGURE 7.14 Elbow-Plot.

k	1	2	3	4	5	6	7	8
WCSSD	*42.598*	*7.576*	*1.635*	*1.593*	*0.983*	*0.958*	*0.405*	*0.38*

WCSSD or TSSD is maximum when all the datapoints are predicted to belong to one single cluster indicating the need to consider more clusters. A sudden change in slope of the WCSSD curve is observed at $k = 3$, and for all $k > 3$, there is marginal decrease in WCSSD. Hence, the optimal number of clusters based on the "knee-plot" is 3.

7.2.6 Neural Networks

In classification problems, neural networks classify the input as one of q classes. Consider a neural network that has q output nodes or the output is a q-dimensional vector. The choice of loss function depends on the activation function used at the output layer. There are a number of loss function-output activation pairs that can be used in neural networks for classification.

7.2.6.1 Softmax Layer

Assume that an input $\mathbf{x}$ has to be classified into one of the two classes, Class-1 and Class-2. One way to do this is to estimate the probability that the input will belong to either of the classes. For example, if we estimate that $P(\mathbf{x} \in$ Class-1$) = 0.8$ and $P(\mathbf{x} \in$ Class-2$) = 0.2$, then the input $\mathbf{x}$ can be classified as belonging to Class-1. For the neural network to predict the probability that the input belongs to one of the classes, the output should follow the rules of probability. The rules that the output of the neural network should obey are:

1. All outputs must be a positive number
2. Sum of all outputs must be 1

In general, output of a neural network is unbounded, which means that it can take any value because the weights and biases are unconstrained. To ensure that the above rules are satisfied, an additional layer is introduced at the output of the neural network. One option is to use a "hardmax" function. This function returns a vector such that the value is set to 1 at the index where the output is maximum and all other values are set to 0. For example, for a binary classification, if the output vector is [100, 1000], then hardmax function would return the vector [0, 1], where 1 is set at the index which holds the maximum value in the output vector. As another example, if output vector is [0, 100, 200], hardmax will return the vector [0, 0, 1]. This clearly obeys the rule of probability but has two problems.

1. This function is not a differentiable function. In neural networks, smooth functions are preferred because of the need to compute gradients for training.
2. Consider the case when the output is [100, 100.0001]. This function will then output a vector of [0, 1] which means that the probability that the input belongs to Class-1 is 0 and Class-2 is 1. But from the output values, we would intuitively prefer to have almost equal probabilities for both classes since output values are almost the same.

To address these issues, define a function called softmax in which each element of the output vector is defined as,

$$\text{softmax}(f_j) = \frac{\exp(f_j)}{\sum_{r=1}^{q} \exp(f_r)} \qquad (7.13)$$

Here, f_j represents the jth output of the neural network before softmax layer and q is the number of classes. It can be clearly seen that each $\text{softmax}(f_j) \in [0, 1]$ and sum of all softmax outputs ($j \in [1, q]$) is 1. Further, the output can be considered to represent the relative probability that an input belongs to a particular class. For example, in the case of $[100,100.001]$, softmax outputs $[0.4998, 5.002]$ and in the case of $[0,100,200]$, softmax outputs approximately $[0, 0, 1]$.

7.2.6.2 Cross-Entropy Loss

One would like to define a loss function such that when the true probability and estimated probability match, the loss function is minimized. Cross-entropy defined by equation 7.14 is one candidate loss function.

$$H(y_r^i, \hat{y}_r^i) = -\sum_{i=1}^{m}\sum_{r=1}^{q} y_r^i \log(\hat{y}_r^i) \qquad (7.14)$$

where $\hat{y}_r^i$ is the predicted probability that the input i belongs to a Class-r and y_r^i is the true probability that the input i belong to Class-r. For example, in a binary classification, if input sample 1 is labelled as Class-2, then the true probabilities are $y_1^1 = 0$ and $y_2^1 = 1$.

One way to arrive at this loss function is using maximum likelihood. The likelihood can be written as,

$$\text{Likelihood } (y_r^i, \hat{y}_r^i) = \prod_{i=1}^{m}\prod_{r=1}^{q} (\hat{y}_r^i)^{I(y_r^i==1)} \qquad (7.15)$$

where I is the indicator function which evaluates to 1 if the condition is true or to 0 otherwise. As observed from equation 7.15, only terms in which the indicator function evaluates to 1 have an effect on likelihood calculation. Hence we have to find the probability function $\hat{y}_r^i$ that maximizes the function given by equation 7.15. One can take log to simplify equation 7.15 since log is a monotonous function and will not affect the optima. Further, we can negate the function and minimize instead of maximizing the original function. Hence, equation 7.15 becomes Negative-Log-Likelihood (NLL),

$$NLL(y_r^i, \hat{y}_r^i) = -\sum_{i=1}^{m}\sum_{r=1}^{q} I(y_r^i == 1) \log(\hat{y}_r^i) \qquad (7.16)$$

This is easily reduced to equation 7.17 by applying the Indicator function.

$$H(y_r^i, \hat{y}_r^i) = \sum_{i=1}^{m}\sum_{r=1}^{q} (y_r^i) \log(\hat{y}_r^i) \qquad (7.17)$$

This function is at a minimum value of 0 if y_r^i and $\hat{y}_r^i$ are the same for all the training points.

Backpropagation – Classification

$n-$ input features $k-$ layer no $q-$ output features
$i-$ sample index $p-$ layers $r-$ node in output layer
$m-$ samples $l-$ node in k^{th} layer $j-$ node in $(k+1)^{th}$ layer
r_k- number of nodes in k^{th} layer

$$\omega_{k+1,lj}^{new} = \omega_{k+1,lj} + \eta \sum_{i=1}^{m}\left(-\frac{\partial H^i}{\partial \omega_{k+1,lj}}\right) = \omega_{k+1,lj} + \eta \sum_{i=1}^{m}\left(-\frac{\partial H^i}{\partial o_{k+1,j}}\frac{\partial o_{k+1,j}}{\partial \omega_{k+1,j}}\right)$$

$$= \omega_{k+1,lj} - \eta \sum_{i=1}^{m}\delta_{k+1,j}^i h_{k,l}^i$$

$$\delta_{p,j}^i = \frac{\partial H^i}{\partial o_{p,j}} = \sum_{r=1}^{q}\frac{\partial H^i}{\partial \hat{y}_r^i}\frac{\partial \hat{y}_r^i}{\partial o_{p,j}}\quad\left(\because H^i = -\sum_{r=1}^{q}y_r^i\log\left(\hat{y}_r^i\right)\right)$$

$$= -\sum_{r=1}^{q}\frac{y_r^i}{\hat{y}_r^i}\frac{\partial A\left(o_{p,j}^i\right)}{\partial o_{p,j}^i}\quad\left(\because \hat{y}_r^i = h_{p,r}^i \text{ and } A\left(o_{p,j}\right) \text{ is soft max}\right)$$

$$\delta_{k,l}^i = \frac{\partial H^i}{\partial o_{k,l}^i} = \sum_{j=1}^{r_{k+1}}\frac{\partial H^i}{\partial o_{k+1,j}^i}\frac{\partial o_{k+1,j}^i}{\partial o_{k,l}^i} = \sum_{j=1}^{r_{k+1}}\delta_{k+1,j}^i\frac{\partial A'\left(o_{k,l}^i\right)\omega_{k+1,lj}}{\partial o_{k,l}^i}$$

$$= \frac{\partial A\left(o_{k,l}^i\right)}{\partial o_{k,l}^i}\sum_{j=1}^{r_{k+1}}\delta_{k+1,j}^i\omega_{k+1,lj}\quad\left(A' \text{ is the first derivative of A}\right)$$

7.2.6.3 Summary

We discussed the motivation and theory behind cross-entropy loss function for classification. To summarize, the output of the neural network is first converted into probabilities by the use of a softmax layer. These probabilities are then used in the cross-entropy loss to train the network by minimizing the loss function with respect to learnable parameters of the neural network.

Example 7.16. *Given 3 input features $x_1 = -3.2$, $x_2 = 4.1$, and $x_3 = 0.163$, find the binary class y if a single layer NN with softmax activation function is used. Use the following weights for summation. $P(y = 0) = h_{1,1}$ and $P(y = 1) = h_{1,2}$.*

$$\omega_{11} = 1;\quad \omega_{12} = 2;\quad \omega_{21} = 2;\quad \omega_{22} = 1;\quad \omega_{31} = 1;\quad \omega_{32} = 1$$

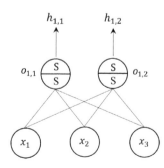

Solution 7.16.

$$o_{1,1} = x_1 + 2x_2 + x_3 = 5.163; \quad o_{1,2} = 2x_1 + x_2 + x_3 = -2.137$$

$$h_{1,1} = \frac{e^{5.163}}{e^{5.163} + e^{-2.137}} = 0.9993; \quad h_{1,2} = \frac{e^{-12.137}}{e^{5.163} + e^{-2.137}} = 6.75 \times 10^{-4}$$

The predicted class is 0 since $h_{1,1}$ is higher.

Example 7.17. *Given 3 input features $x_1 = -3.2$, $x_2 = 4.1$, and $x_3 = 0.163$, calculate output values if a neutral network with 1 hidden and 1 output layer is setup for binary classification. Use softmax activation for output layer. Use the following weights:*

$\omega_{1,11} = 1$ $\omega_{1,12} = 2$ $\omega_{1,21} = 2$ $\omega_{1,22} = 1$ $\omega_{1,31} = 1$
$\omega_{1,32} = 1$ $\omega_{2,11} = 3$ $\omega_{2,12} = 1$ $\omega_{2,21} = 1$ $\omega_{2,22} = 2$
Use 2 nodes in hidden layer and

a) Signoid activation function for the hidden layer
b) tanh activation function for the hidden layer
c) ReLU activation function for the hidden layer

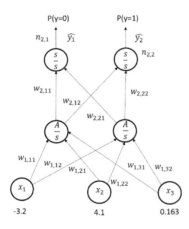

FIGURE 7.15 NN structure: Example 7.17.

Solution 7.17. *a) A= Signmoid activation function*

$$o_{1,1} = 5.163 \Rightarrow h_{1,1} = \frac{1}{1 + e^{-5.163}} = 0.9943$$

$$o_{1,2} = -2.137 \Rightarrow h_{1,2} = \frac{1}{1 + e^{+2.137}} = 0.1056$$

$$o_{2,1} = 0.9943 \times 3 + 0.1056 = 3.0885; \quad o_{2,2} = 0.9943 + 0.1056 \times 2 = 1.2055$$

$$h_{2,1} = \frac{e^{3.0885}}{e^{3.0885} + e^{1.2055}} = 0.8680; \quad h_{2,2} = \frac{e^{1.2055}}{e^{3.0885} + e^{1.2055}} = 0.132$$

b) A = tanh activation function

$$`o_{1,1} = 5.163; \quad o_{1,2} = -2.137$$

$$h_{1,1} = \frac{e^{5.163} - e^{-5.163}}{e^{5.163} + e^{-5.163}} = 0.9999; \quad h_{1,2} = \frac{e^{-2.137} - e^{+2.137}}{e^{-2.137} + e^{2.137}} = -0.9725$$

$$o_{2,1} = 0.9999 \times 3 - 0.9725 = 2.0273; \quad o_{2,2} = 0.9999 - 2 \times 0.9725 = -0.9451$$

$$h_{2,1} = \frac{e^{2.0272}}{e^{2.0272} + e^{-0.9451}} = 0.9513; \quad h_{2,2} = \frac{e^{-0.9451}}{e^{2.0272} + e^{-0.9451}} = 0.0487$$

c) A = ReLU activation function

$$o_{1,1} = 5.163; \quad o_{1,2} = -2.137; \quad h_{1,1} = 5.163; \quad h_{1,2} = 0$$

$$o_{2,1} = 3 \times 5.163 = 15.489; \quad o_{2,2} = 5.163$$

$$h_{2,1} = \frac{e^{15.489}}{e^{15.489} + e^{5.163}} = 1; \quad h_{2,2} = \frac{e^{5.163}}{e^{15.489} + e^{5.163}} = 3.277 \times 10^{-5} \approx 0$$

Example 7.18. *Given 3 input features x_1, x_2, and x_3 to predict class y (binary class) and a neural network having 1 hidden layer with 2 neurons, show the 1st iteration of parameter optimization using steepest descent. Use step length as 0.5. Use sigmoid activation function in the hidden layer. Use weights given in Example 7.17 as initial guess. Given sample,*

$$x_1 = -3.2; \quad x_2 = 4.1; \quad x_3 = 0.163; \quad y = 1$$

Use $h_{2,1}$ as probability of y=0 and $h_{2,2}$ as probability of y=1.

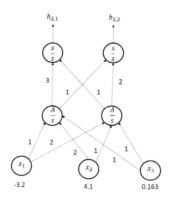

Solution 7.18. *Objective function*

$$\min_{\omega} \ H = -\sum_{i=1}^{m} \sum_{r=1}^{2} \left(y_r^i\right) \log\left(\hat{y}_r^i\right)$$

Since there is only 1 sample and y=1 for that sample

$$H = -\log \hat{y}_2 = -\log h_{2,2} = -\log \left(\frac{e^{o_{2,2}}}{e^{o_{2,2}} + e^{o_{2,1}}} \right)$$

$$\frac{\partial H}{\partial o_{2,2}} = \frac{-1}{h_{2,2}} \frac{\partial h_{2,2}}{\partial o_{2,2}} = \frac{-1}{h_{2,2}} h_{2,2} (1 - h_{2,2}) = h_{2,2} - 1$$

$$\frac{\partial H}{\partial o_{2,1}} = \frac{-1}{h_{2,2}} \frac{\partial h_{2,2}}{\partial o_{2,1}} = \frac{-1}{h_{2,2}} \frac{-e^{o_{2,2}} e^{o_{2,1}}}{\left(-e^{o_{2,2}} + e^{o_{2,1}}\right)^2} = \frac{-1}{h_{2,2}} (-h_{2,1} h_{2,2}) = h_{2,1}$$

$$\frac{\partial H}{\partial \omega_{2,11}} = \frac{\partial H}{\partial o_{2,1}} \frac{\partial o_{2,1}}{\partial \omega_{2,11}} + \frac{\partial H}{\partial o_{2,2}} \frac{\partial o_{2,2}}{\partial \omega_{2,11}}$$

Since $o_{2,1} = \omega_{2,11} h_{1,1} + \omega_{2,21} h_{1,2}$ and $o_{2,2} = \omega_{2,12} h_{1,1} + \omega_{2,22} h_{1,2}$,

$$\frac{\partial H}{\partial \omega_{2,11}} = h_{2,1} h_{1,1} + (h_{2,2} - 1) \times 0 = h_{2,1} h_{1,1}$$

$$\frac{\partial H}{\partial \omega_{2,12}} = \frac{\partial H}{\partial o_{2,1}} \frac{\partial o_{2,1}}{\partial \omega_{2,12}} + \frac{\partial H}{\partial o_{2,2}} \frac{\partial o_{2,2}}{\partial \omega_{2,12}} = (h_{2,2} - 1) h_{1,1}$$

$$\frac{\partial H}{\partial \omega_{2,21}} = \frac{\partial H}{\partial o_{2,1}} \frac{\partial o_{2,1}}{\partial \omega_{2,21}} + \frac{\partial H}{\partial o_{2,2}} \frac{\partial o_{2,2}}{\partial \omega_{2,21}} = h_{2,1} h_{1,2} + 0$$

$$\frac{\partial H}{\partial \omega_{2,22}} = \frac{\partial H}{\partial o_{2,1}} \frac{\partial o_{2,1}}{\partial \omega_{2,22}} + \frac{\partial H}{\partial o_{2,2}} \frac{\partial o_{2,2}}{\partial \omega_{2,22}} = 0 + (h_{2,2} - 1) h_{1,2}$$

$$\frac{\partial H}{\partial \omega_{1,11}} = \frac{\partial H}{\partial o_{2,1}} \frac{\partial o_{2,1}}{\partial \omega_{1,11}} + \frac{\partial H}{\partial o_{2,2}} = h_{2,1} \omega_{2,11} \frac{\partial h_{1,1}}{\partial \omega_{1,11}} + (h_{2,2} - 1) \omega_{2,12} \frac{\partial h_{1,1}}{\partial \omega_{1,11}}$$

$$= ((h_{2,2} - 1) \omega_{2,12} + h_{2,1} \omega_{2,11}) \frac{\partial h_{1,1}}{\partial o_{1,1}} \frac{\partial o_{1,1}}{\partial \omega_{1,11}}$$

$$= ((h_{2,2} - 1) \omega_{2,12} + h_{2,1} \omega_{2,11}) \frac{e^{-o_{1,1}}}{\left(1 + e^{-o_{1,1}}\right)^2} x_1$$

$$\frac{\partial H}{\partial \omega_{1,21}} = \frac{\partial H}{\partial o_{2,1}} \frac{\partial o_{2,1}}{\partial \omega_{1,21}} + \frac{\partial H}{\partial o_{2,2}} \frac{\partial o_{2,2}}{\partial \omega_{1,21}}$$

$$= h_{2,1} \omega_{2,11} \frac{\partial h_{1,1}}{\partial \omega_{1,21}} + (h_{2,2} - 1) \omega_{2,12} \frac{\partial h_{1,1}}{\partial \omega_{1,21}}$$

$$= (h_{2,1} \omega_{2,11} + (h_{2,2} - 1) \omega_{2,12}) \frac{e^{-o_{1,1}}}{\left(1 + e^{-o_{1,1}}\right)^2} x_2$$

$$\frac{\partial H}{\omega_{1,31}} = (h_{2,1} \omega_{2,11} + (h_{2,2} - 1) \omega_{2,12}) \frac{e^{-o_{1,1}}}{\left(1 + e^{-o_{1,1}}\right)^2} x_3$$

$$\frac{\partial H}{\partial \omega_{1,12}} = \frac{\partial H}{\partial o_{2,1}} \frac{\partial o_{2,1}}{\partial \omega_{1,12}} + \frac{\partial H}{\partial o_{2,2}} \frac{\partial o_{2,2}}{\partial \omega_{1,12}}$$

$$= (h_{21} \omega_{2,21} + (h_{2,2} - 1) \omega_{2,22}) \frac{e^{-o_{1,2}}}{\left(1 + e^{-o_{1,2}}\right)^2} x_1$$

$$\frac{\partial H}{\partial w_{1,22}} = (h_{2,1}w_{2,21} + (h_{2,2} - 1)w_{2,22})\frac{e^{-o_{1,2}}}{(1 + e^{-o_{1,2}})^2}x_2$$

$$\frac{\partial H}{\partial w_{1,32}} = (h_{2,1}w_{2,21} + (h_{2,2} - 1)w_{2,22})\frac{e^{-o_{1,2}}}{(1 + e^{-o_{1,2}})^2}x_3$$

For steepest descent,

$$w_{k,lj}^{q+1} = w_{k,lj}^q - \alpha\left.\frac{\partial H}{\partial w_{k,lj}}\right|_{w^q}; \qquad w_{1,11}^1 = w_{1,11}^0 - 0.5\left.\frac{\partial H}{\partial w_{1,11}}\right|_{w^0}$$

$$w_{1,11}^0 = 1, \; w_{1,12}^0 = 2, \; w_{1,21}^0 = 2, \; w_{1,22}^0 = 1, w_{1,31}^0 = 1$$

$$w_{1,32}^0 = 1, \; w_{2,11}^0 = 3, \; w_{2,12}^0 = 1, \; w_{2,21}^0 = 1, \; w_{2,22}^0 = 2$$

$$o_{1,1} = 5.163, \; o_{1,2} = -2.137, \; o_{2,1} = 3.0885, \; o_{2,2} = 1.2055$$

$$h_{1,1} = 0.9943, \; h_{1,2} = 0.1056, \; h_{2,1} = 0.868, \; h_{2,2} = 0.132$$

$$x_1 = -3.2, \; x_2 = 4.1, \; x_3 = 0.163$$

$$w_{1,11}^1 = 1 - 0.5\left[0.868 \times 3 + (0.132 - 1)\right]\frac{e^{-5.163}}{(1 + e^{-5.163})^2} \times (-3.2) = 1.0157$$

$$w_{1,21}^1 = 2 - 0.5\left[0.868 \times 3 + (0.132 - 1)\right]\frac{e^{-5.163}}{(1 + e^{-5.163})^2} \times 4.1 = 1.9799$$

$$w_{1,31}^1 = 1 - 0.5\left[0.868 \times 3 + (0.132 - 1) \times 2\right]\frac{e^{-5.163}}{(1 + e^{-5.163})^2} \times 0.163 = 0.9992$$

$$w_{1,12}^1 = 2 - 0.5\left[0.868 \times 1 + (0.132 - 1) \times 2\right]\frac{e^{2.137}}{(1 + e^{2.137})^2} (-3.2) = 1.8689$$

$$w_{1,22}^1 = 1 - 0.5\left[0.868 + 2(0.132 - 1)\right]\frac{e^{2.317}}{(1 + e^{2.137})^2} \times 4.1 = 1.168$$

$$w_{1,32}^1 = 1 - 0.5\left[0.868 + 2(0.132 - 1)\right]\frac{e^{2.137}}{(1 + e^{2.137})^2} \times 0.163 = 1.0067$$

$$w_{2,11}^1 = 3 - 0.5(0.868 \times 0.9943) = 2.5685$$

$$w_{2,12}^1 = 1 - 0.5(0.132 - 1)0.9943 = 1.4315$$

$$w_{2,21}^1 = 1 - 0.5(0.868 \times 0.1056) = 0.9542$$

$$w_{2,22}^1 = 2 - 0.5 \times 0.1056(0.132 - 1) = 2.0458$$

Now that all the weights have been calculated, we can evaluate the new output value.

$$o_{1,1}^1 = 1.0157(-3.2) + 1.9799(4.1) + 0.9992(0.163) = 5.0302$$

$$o_{1,2}^1 = 1.8689(-3.2) + (1.168 \times 4.1) + (1.0067 \times 0.163) = -1.0276$$

$$h^1_{1,1} = \frac{1}{1 + e^{-5.0303}} = 0.9935; \quad h^1_{1,2} = \frac{1}{1 + e^{1.0276}} = 0.2635$$

$$o^1_{2,1} = 2.8033; \quad o^1_{2,2} = 1.9614; \quad h^1_{2,1} = 0.6989; \quad h^1_{2,2} = 0.3011$$

Note that the probability of y=1 has increased from 0.132 to 0.3011 indicating that we have improved the prediction from the previous iteration.

Example 7.19. For the problem described in Example 7.18, show the 1st iteration of parameter optimization using backpropagation. Use $\eta = 0.5$.

Solution 7.19. Objective function: $\min_w H = -\sum_{i=1}^{m} \sum_{r=1}^{2} y_r^i \log (\hat{y}_r^i)$
Since there is only 1 sample and $y = 1$ for that sample,

$$H = -\log \hat{y}_2 = -\log h_{2,2}; \quad w^{new}_{2,11} = w^0_{2,11} + \eta \left(\frac{-\partial H}{\partial w_{2,11}}\right) \text{ and so on}$$

$$\delta_{2,j} = \frac{\partial H}{\partial o_{2,j}} = -\sum_{r=1}^{2} \frac{y_r}{\hat{y}_r} \frac{\partial A(o_{p,r})}{\partial o_{p,j}}$$

Since $y = 1$, $y_1 = 0$ and $y_2 = 1$. For softmax function,

$$h_{p,r} = A(o_{p,r}) = \frac{e^{o_{p,r}}}{\sum_{r=1}^{q} e^{o_{p,r}}}$$

$$A'(o_{p,r}) = \frac{\partial A(o_{p,r})}{\partial o_{p,j}} = \begin{cases} h_{p,r}(1 - h_{p,r}) \text{ if } r = j \\ -h_{p,r} \times h_{p,j} \text{ if } r \neq j \end{cases}$$

$$\delta_{2,1} = -\sum_{r=1}^{2} \frac{y_r}{\hat{y}_r} \frac{\partial A(o_{p,r})}{\partial o_{p,1}} = -\left[\frac{y_2}{\hat{y}_2} \frac{\partial A(o_{p,2})}{\partial o_{p,1}}\right]$$

$$= \frac{-y_2}{\hat{y}_2}(-h_{p,2}h_{p,1}) = \frac{1}{0.132} \times 0.868 \times 0.132 = 0.868$$

$$\delta_{2,2} = \frac{-y_2}{\hat{y}_2}h_{p,2}(1 - h_{p2}) = -y_2(1 - h_{p2}) = h_{p2} - 1 = 0.132 - 1 = -0.868$$

$$\delta_{1,l} = \sum_{j=1}^{2} \delta_{2,j} \frac{\partial A(o_{1,l})}{\partial o_{1,l}} w_{22,lj}$$

For a sigmoid, $\frac{\partial A(o_{1,l})}{\partial o_{1,l}} = \frac{e^{-o_{1,l}}}{(1+e^{-o_{1,l}})^2}$.

$$\delta_{1,1} = \frac{e^{-o_{1,1}}}{(1 + e^{-o_{1,1}})^2}(\delta_{2,1}w_{2,11} + \delta_{2,2}w_{2,12})$$

$$= \frac{e^{-5.163}}{(1 + e^{-5.163})^2}(0.868 \times 3 + (-0.868)) = 0.0098$$

$$\delta_{1,2} = \frac{e^{-o_{1,2}}}{\left(1 + e^{-o_{1,2}}\right)^2} \left(\delta_{2,1}\omega_{2,21} + \delta_{2,2}\omega_{2,22}\right)$$

$$= \frac{e^{2.137}}{\left(1 + e^{2.137}\right)^2} \left(0.868 \times 1 - 0.868 \times 2\right) = -0.0819$$

$\omega_{2,11}^{new} = \omega_{2,11} - 0.5\,\delta_{2,1}h_{1,1} = 2.5685;$ $\omega_{2,12}^{new} = \omega_{2,12} - 0.5\,\delta_{2,2}h_{1,1} = 1.4315$

$\omega_{2,21}^{new} = \omega_{2,21} - 0.5\,\delta_{2,1}h_{1,2} = 0.9542;$ $\omega_{2,22}^{new} = \omega_{2,22} - 0.5\,\delta_{2,2}h_{1,2} = 2.0458$

$\omega_{1,11}^{new} = \omega_{1,11} - 0.5\,\delta_{1,1}h_{0,1} = 1.0157;$ $\omega_{1,12}^{new} = \omega_{1,12} - 0.5\,\delta_{1,2}h_{0,1} = 1.8689$

$\omega_{1,21}^{new} = \omega_{1,21} - 0.5\,\delta_{1,1}h_{0,2} = 1.9799;$ $\omega_{1,22}^{new} = \omega_{1,22} - 0.5\,\delta_{1,2}h_{0,2} = 1.168$

$\omega_{1,31}^{new} = \omega_{1,31} - 0.5\,\delta_{1,1}h_{0,3} = 0.9992;$ $\omega_{1,32}^{new} = \omega_{1,32} - 0.5\,\delta_{1,2}h_{0,3} = 1.0067$

Using the updated weights

$$h_{1,1}^{new} = 0.9935; \quad o_{1,1}^{new} = 5.03; \quad h_{1,2}^{new} = 0.2635; \quad o_{1,2}^{new} = 1.0276$$

$$o_{2,1}^{new} = 2.8033; \quad \hat{y}_1^{new} = h_{2,1}^{new} = 0.6989; \quad o_{2,2}^{new} = 1.9614; \quad \hat{y}_2^{new} = h_{2,2}^{new} = 0.3011$$

Note that $\hat{y}_2$ has increased from 0.132 to 0.3011. Also, $H^{new} = -\log 0.3011 = 1.2003$ and $H^{old} = 2.025$.

Example 7.20. *Consider the following data.*

x_1	x_2	x_3	Class, y
-3.2	4.1	0.163	1
-1.7	3.9	1.35	0
-4.16	4.33	0.78	1
-2.73	3.14	1.07	0
-3.3	4.0	0.15	1
-1.8	3.8	1.3	0

Assume that a neural network with 1 hidden layer having 2 neurons is used to predict y. Show 1 epoch (with whole data as one batch) of parameter optimization using backpropagation. Use $\eta = 0.5$. Hidden layer uses a sigmoid activation function. Use the following weights as initial guess.

$$\omega_{1,11} = 1; \quad \omega_{1,12} = 2; \quad \omega_{1,21} = 2; \quad \omega_{1,22} = 1; \omega_{1,31} = 1;$$

$$\omega_{1,32} = 1; \quad \omega_{2,12} = 1; \quad \omega_{2,21} = 1; \omega_{2,22} = 2; \quad \omega_{2,11} = 3$$

Solution 7.20. *Probability of $y = 0$ is represented as y_1 and probability of $y = 1$ as y_2.*

Using the given weights $\eta_{1,lj}$, we get the values of $o_{1,1}$ and $o_{1,2}$. Now, using sigmoid activation function, values of $h_{1,1}$ and $h_{1,2}$ for each sample are evaluated. Using the given $\eta_{2,lj}$, we evaluated $o_{1,1}$ and $o_{1,2}$ for each sample. Then, using softmax activation function we can evaluate $h_{2,1}$ ($\hat{y}_1$) and $h_{2,2}$ ($\hat{y}_2$). The values obtained for $\hat{y}_i$ for each sample are given in Table 7.3 (iteration 0).

Now, we will show the 1st iteration of parameter optimization using back-propagation. First δ_2 for each sample was calculated.

$$\delta_{2,1} = \left[\delta_{2,1}^i\right] = [0.868, -0.243, 0.876, -0.147, 0.871, -0.235]$$

$$\delta_{2,2} = \left[\delta_{2,2}^i\right] = [-0.868, 0.243, -0.876, 0.147, -0.871, 0.235]$$

Now, using δ_2 values, δ_1 values for each sample were calculated.

$$\delta_{1,1} = \left[\delta_{1,1}^i\right] = [0.0098, 0.001, 0.0088, 0.0168, 0.0134, 0.0014]$$

$$\delta_{1,2} = \left[\delta_{1,2}^i\right] = [0.023, 0.0286, 0.0091, 0.0421, 0.0178, 0.0363]$$

Updating the weights using $\omega_{k+1,lj}^{new} = \omega_{k+1,lj} - \eta \sum_{i=1}^{m} \delta_{k+1,j}^i \, h_{k,l}^i$, we get

$$\omega_{1,11}^{new} = 1.0810; \quad \omega_{1,21}^{new} = 1.9031; \quad \omega_{1,31}^{new} = 0.9842; \quad \omega_{1,12}^{new} = 2.1994; \quad \omega_{1,22}^{new} = 0.7069$$

$$\omega_{1,32}^{new} = 0.9279; \quad \omega_{2,11}^{new} = 2.0128; \quad \omega_{2,21}^{new} = 1.12; \quad \omega_{2,12}^{new} = 1.9872; \quad \omega_{2,22}^{new} = 1.88$$

Using these weights, new output values are calculated just like iteration 0. The values are reported in Table 7.3.

$$\text{Cross-entropy value at iteration } 0 = -\sum_{i=1}^{6}\sum_{r=1}^{2} y_r^i . \log\left(\hat{y}_r^i\right) = 6.858$$

$$\text{Cross-entropy value at iteration } 1 = 4.601$$

Note that cross-entropy has decreased with 1 iteration.

Example 7.21. *For the problem given in Example 7.20, perform 1 epoch of parameter optimization if data is split into 2 batches.*

Solution 7.21. *The initial $\hat{y}$ corresponding to the weights given will be same as that given in Table 7.3. Initial cross-entropy is 6.858. Here 1 epoch will have 2 iterations, 1 iteration for each batch.*
Batch 1: (-3.2, 4.1, 0.163),   (-1.7, 3.9, 1.35),   (-4.16,4.33,0.78)
Batch 2: (-2.73, 3.14, 1.07),   (-3.3, 4, 0.15),   (-0.18, 3.8, 1.3)

Iteration 1
For the 1st batch, using only these 3 samples, we proceed in the same manner as described in Example 7.20 and the new weights are obtained and noted below:

$$\omega_{1,11}^1 = 1.0348; \quad \omega_{1,21}^1 = 1.9589; \quad \omega_{1,31}^1 = 0.9951; \quad \omega_{1,12}^1 = 2.0799; \quad \omega_{1,22}^1 = 0.8776$$

$$\omega_{1,32}^1 = 0.9753; \quad \omega_{2,11}^1 = 2.2545; \quad \omega_{2,21}^1 = 1.0423 \quad \omega_{2,12}^1 = 1.7455; \quad \omega_{2,22}^1 = 1.9577$$

Using these weights class probabilities are predicted for each sample (both batch 1 and batch 2) and are reported in Table 7.3 in column iteration 1. Here cross-entropy value is 4.494. Now, we will use batch 2 data for optimizing parameters further.

TABLE 7.3

Backpropagation Using 1 Batch

x_1	x_2	x_3	True y_1	True y_2	Iteration 1 $\hat{y}_1$	Iteration 1 $\hat{y}_2$	Iteration 2/Epoch 1 $\hat{y}_1$	Iteration 2/Epoch 1 $\hat{y}_2$
−3.2	4.1	0.163	0	1	0.868	0.132	0.503	0.497
−1.7	3.9	1.35	1	0	0.757	0.243	0.400	0.600
−4.16	4.33	0.78	0	1	0.876	0.124	0.505	0.495
−2.73	3.14	1.07	1	0	0.853	0.147	0.495	0.505
−3.3	4	0.15	0	1	0.870	0.130	0.504	0.496
−1.8	3.8	1.3	1	0	0.765	0.235	0.415	0.585

TABLE 7.4

Backpropagation Using 2 Batches

x_1	x_2	x_3	True y_1	True y_2	Iteration 0 $\hat{y}_1$	Iteration 0 $\hat{y}_2$	Epoch 1 Iteration 1 $\hat{y}_1$	Epoch 1 Iteration 1 $\hat{y}_2$	Epoch 1 Iteration 2 $\hat{y}_1$	Epoch 1 Iteration 2 $\hat{y}_2$
−3.2	4.1	0.163	0	1	0.868	0.132	0.612	0.388	0.687	0.313
−1.7	3.9	1.35	1	0	0.757	0.243	0.451	0.549	0.600	0.400
−4.16	4.33	0.78	0	1	0.876	0.124	0.620	0.380	0.693	0.307
−2.73	3.14	1.07	1	0	0.853	0.147	0.594	0.406	0.674	0.326
−3.3	4	0.15	0	1	0.870	0.130	0.615	0.385	0.689	0.311
−1.8	3.8	1.3	1	0	0.765	0.235	0.466	0.534	0.606	0.394

Iteration 2

Using the samples in batch 2 alone and proceeding the same way as described in Example 7.20, we get the following weights:

$$w_{1,11}^2 = 1.018; \quad w_{1,21}^{[2]} = 1.9792; \quad w_{1,31}^2 = 0.999; \quad w_{1,12}^2 = 2.0268; \quad w_{1,22}^2 = 0.9627$$

$$w_{1,32}^1 = 1.0014; \quad w_{2,11}^2 = 2.4171; \quad w_{2,21}^2 = 1.2446; \quad w_{2,12}^2 = 1.5829; \quad w_{2,22}^2 = 1.7554$$

Using the updated weights, probabilities for each class are computed for each sample. The predicted classes are provided in Table 7.3. For the updated prediction, cross-entropy is 4.915.

7.3 NON-PARAMETRIC METHODS

In this section, several non-parametric methods for classification will be described.

7.3.1 *k*-NN

We have already described k-NN in detail in the previous chapter. When k-NN is implemented for classification problems, the procedure remains the same;

however, instead of averaging the outputs of the k neighbors to predict an output for $\mathbf{x}^{\text{new}}$, a majority voting of the k neighbors is used to assign a class to $\mathbf{x}^{\text{new}}$. The k-NN approach can be used quite easily for binary and multi-class classification problems. Just like in function approximation, in classification problems also, when the number of datapoints increase, identifying the k nearest neighbors can become computationally cumbersome.

Example 7.22. *Consider the Iris species data consisting of sepal length, sepal width, petal length and petal width of three species of flowers : Iris Setosa (0), Iris Versicolor (1) and Iris Viriginica (2). Perform classification of the given test data using k-NN. Comment on the effect of the choice of number of neighbours on the performance of k-NN algorithm.*

Training data:

Sepal Length (x_1)	Sepal Width (x_2)	Petal Length (x_3)	Petal Width (x_4)	Species Class (y)
7.4	2.8	6.1	3	2
5.5	2.4	3.8	1.9	1
5.6	3	4.5	1.5	1
4.8	3	1.4	0.1	0
5	3.4	1.5	0.2	0
6.2	3.4	5.4	2.3	2
7.9	3.8	6.4	2	2
6.4	2.8	5.6	2.2	2
5.4	3.9	1.7	0.4	0
4.9	3.6	1.4	0.1	0
5.7	2.8	4.5	1.3	1
5.1	3.5	1.4	0.3	0
6.5	3	5.8	2.2	2
6.8	2.8	4.8	1.4	1
5.5	4.2	1.4	0.2	0
5.8	2.7	5.1	1.9	2
4.8	3	1.4	0.3	0

Test data:

Sepal Length (x_1)	Sepal Width (x_2)	Petal Length (x_3)	Petal Width (x_4)	Species Class (y)
6.3	2.5	4.9	1.5	1
4.9	3.1	1.5	0.1	1
6.3	2.3	4.4	1.3	1
6.5	3	5.2	2	2
6.4	2.7	5.3	1.9	2
6.9	3.1	4.9	1.5	1
5.7	3.8	1.7	0.3	0
7.2	3	5.8	1.6	2

Solution 7.22. *For the given test data, let us consider the case where 3 nearest neighbours from the training data are selected for performing majority voting. Consider the first data sample in the test data $x_{test}^1 = (6.3, 2.5, 4.9, 1.5)$. We then compute the Euclidean distance between the data sample and all the datapoints in the training set. For example, the distance between the given test sample x_{test}^1 and the first data sample of training data $x_{training}^1 = (7.4, 2.8, 6.1, 3)$ is*

$$d = \sqrt{(7.4 - 6.3)^2 + (2.8 - 2.5)^2 + (6.1 - 4.9)^2 + (3 - 1.5)^2} = 2.234$$

Similarly, distances of the test data sample x_{test}^1 to all the training samples are computed and the 3 closest neighbours are tabulated below:

Sl. No.	Nearest training data samples	Distance	Class Label
1	(6.8,2.8,4.8,1.4)	0.6	1
2	(5.7,2.8,4.5,1.3)	0.806	1
3	(5.6,3,4.5,1.5)	0.95	1

Since all 3 neighbors belong to Class-1, x_{test}^1 is assigned to Class-1 based on majority voting. This process is repeated for all the test samples.

$$\hat{y}_{test} = \begin{bmatrix} 1 & 0 & 1 & 2 & 2 & 2 & 0 & 2 \end{bmatrix}$$

The classification performance is summarized below:

True Label	Predicted Label		
	Iris Setosa	Iris Versicolor	Iris Virginica
Iris Setosa	1	0	0
Iris Versicolor	1	2	1
Iris Virginica	0	0	3

	Precision	Recall	F1-score
Iris Setosa	0.5	1	0.67
Iris Versicolor	1	0.5	0.67
Iris Virginica	0.75	'1	0.86

To understand the impact of choice of k on classification, predictions for k = 3, 5, 7, 9, 11 were performed and the results are reported in Table 7.5.

TABLE 7.5

k-NN Performance Using Euclidean and Manhattan Distances (Ref. Examples 7.22 and 7.23)

k	Euclidean Distance					Manhattan Distance				
	Confusion Matrix			Accuracy		Confusion Matrix			Accuracy	
3	1 0 0			0.75		1 0 0			0.875	
	1 2 1					1 3 0				
	0 0 3					0 0 3				
5	1 0 0			0.75		1 0 0			0.875	
	1 2 1					1 3 0				
	0 0 3					0 0 3				
7	1 0 0			0.875		1 0 0			0.625	
	1 3 0					1 2 1				
	0 0 3					0 1 2				
9	1 0 0			0.5		1 0 0			0.5	
	1 0 3					1 0 3				
	0 0 3					0 0 3				
11	1 0 0			0.5		1 0 0			0.5	
	1 0 3					1 0 3				
	0 0 3					0 0 3				

From the results, as the value of k increases, the accuracy score increases initially and then decreases. Initial increase in k favors performance by negating the influence of noise in data. As k increases further, the region around the sample considered for majority voting expands. This compromises the majority voting procedure by including neighbours that are located far away from the test sample.

Example 7.23. *Consider the data given in Example 7.22. Perform k-NN classification using Manhattan distance metric for computing the nearest neighbours.*

Solution 7.23. *Manhattan distance between two samples is the sum of absolute differences between them. For example, for the test sample* $x^1_{test} = (6.3, 2.5, 4.9, 1.5)$, *the Manhattan distance from the first training sample* $x^1_{train} = (7.4, 2.8, 6.1, 3)$ *is*

$$d = |7.4 - 6.3| + |2.8 - 2.5| + |6.1 - 4.9| + |3 - 1.5| = 1.1 + 0.3 + 1.2 + 1.5 = 4.1$$

For $k = 3$, *three nearest samples need to be found. For* x^1_{test}, *the nearest neighbors are*

Sl. No.	Nearest training data samples	Distance	Class Label
1	(6.8, 2.8, 4.8, 1.4)	1.0	1
2	(5.7, 2.8, 4.5, 1.3)	1.5	1
3	(5.6, 3, 4.5, 1.5)	1.6	1

Since all three neighbors are of Class-1, the predicted class for x_{test}^1 is Class-1. Similarly, the predicted classes for all the test samples based on Manhattan distance are given below.

$$\hat{y}_{test} = \begin{bmatrix} 1 & 0 & 1 & 2 & 2 & 1 & 0 & 2 \end{bmatrix}$$

k-NN classification using Manhattan distance was performed for various k values. The results are reported in Table 7.5. Here, the maximum accuracy of 0.875 is obtained for $k = 3$ and $k = 5$.

Example 7.24. *Consider the hotel star rating data below. In place of predicting the hotel room price, determine the hotel star rating using k-NN, given the information about user ratings, hotel amenities and price of the room. Use the first 10 samples as the training data and the remaining 3 samples as the test data. Use $k = 4$.*

x_1	675	1050	825	1300	1240	3460	2710	2500	1750	500	775	1150	725
x_2	3.2	2.2	4.5	4.8	1.9	4.0	2.6	2.8	4.6	0.4	3.4	2.5	3.5
x_3	B	E	P	S	P	Bo	O	C	C	S	B	E	P
y	1	4	2	1	4	4	3	4	2	4	4	1	2

B - Balcony; E - Elevator; P - Pool; S - Spa; Bo - Bonfire O - Outdoor sports
C - City center; x_1 - Room price; x_2 - User rating; x_3 - Amenities

Solution 7.24. *We can follow the one-hot encoding approach as previously done in Example 6.52 to handle the categorical input feature "Amenities" where we create 7 new features (each new feature represents a categorical element of x_3): x_{31} (Balcony), x_{32} (Elevator), x_{33} (Pool), x_{34} (Spa), x_{35} (Bonfire), x_{36} (Outdoor sports), x_{37} (City center). Now, each sample has 9 features: x_1, x_2 and x_{3i} for all $i \in [1, 7]$.*

Consider the test datapoint (775,3.4,1,0,0,0,0,0,0). The 4 nearest neighbors obtained using both Euclidean and Manhattan distance metric on the test sample are shown below:

k-NN (Euclidean distance)			k-NN (Manhattan distance)		
Neighbours	Distance	Label	Neighbours	Distance	Label
(825,4.5,0,0,0,0,0,1,0)	50.032	2	(825,4.5,0,0,0,0,0,1,0)	53.1	2
(675,3.2,1,0,0,0,0,0,0)	100	1	(675,3.2,0,0,0,0,0,0,1)	100.2	1
(1050,2.2,0,0,0,1,0,0,0)	275.01	4	(1050,2.2,0,0,0,1,0,0,0)	278.2	4
(500,0.4,0,0,0,0,0,0,1)	275.02	4	(500,0.4,0,0,0,0,0,0,1)	280.	4
(1240,1.9,0,0,0,0,0,1,0)	465.01	4	(1240,1.9,0,0,0,0,0,1,0)	468.5	4
Predicted label = 4			Predicted label = 4		

The table shows that the k-NN using both the distance metric produced similar results. The predicted class for the last 2 samples using both metrics are 1 and 4. The last sample has been misclassified and thus, the overall accuracy = 0.67.

7.3.2 Hierarchical Clustering

Hierarchical clustering (HC) is a simple unsupervised clustering algorithm that parallels some of the ideas of decision trees [44]. In both the cases, the basic concept is to split the data into multiple sets and then the sets are used to derive insights, predict outputs and so on. In fact, both the approaches present results in a tree like structure; however, the interpretation of the tree at different levels and how the tree is generated are quite different.

HC can be performed in two modes, either a bottom-up agglomerative approach (Figure 7.16) or a top-down splitting/divisive approach (Figure 7.23). In both the cases, the two ends of the tree are already decided. The bottom of the tree has all the datapoints separated out as individual clusters and the top of the tree has a single cluster which agglomerates all the datapoints. In between these two layers there can be up to a maximum of $m-2$ layers where m is the number of samples. The kth layer (from top to bottom) will represent the k clusters (represented by sets $S_1^k \ldots S_k^k$). Any algorithm that comes up with this tree like structure can be thought of as a hierarchical clustering algorithm.

While an optimization viewpoint can be taken to understand HC algorithms, it is sometimes worthwhile to take the approach that the clustering procedure may reflect any objective of a modeler, even if not precisely written down in a mathematical form. This includes definition of any valid distance measure and any agglomeration and/or splitting strategy. However, analyzing the quality of the solution and computational complexity would require a precise mathematical formulation. In summary, the solution to the clustering problem at each layer minimizes an optimization problem for that layer. For the given objective of the modeler, in a HC algorithm, heuristics need to be defined for merging (bottom-up) or splitting (top-down) clusters. For the definition of heuristics, typically, a distance measure between two datapoints and a distance measure between two sets of datapoints are defined. Once these are defined, then various approaches to splitting and merging can be proposed and these are illustrated using the examples that follow.

The distance metric between two points d_{ij} that is chosen can be Euclidean, Manhattan, Mahalanobis and so on. For data that is not numeric, metrics such as Hamming, Levenshtein and so on are used. The second distance that needs to be defined is between sets of data $d_{S_i S_j}$. There are several possibilities, two of which are shown below.

$$\text{Maximum Linkage: } d_{S_i S_j} = \max(d_{ij} \; \forall \; i \in S_i, \; j \in S_j)$$
$$\text{Minimum Linkage: } d_{S_i S_j} = \min(d_{ij} \; \forall \; i \in S_i, \; j \in S_j)$$

One specific example is the Ward's method where the distance between two sets is evaluated using equation 7.18. The distance metric for ward's linkage is the increase in within cluster sum squared error (WCSSE) after merging

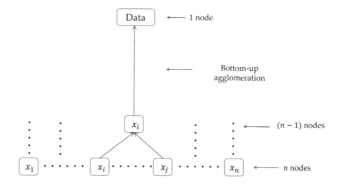

FIGURE 7.16 Bottom up data agglomeration approach.

two sets into one.

$$\text{Ward's Linkage: } d_{S_i,S_j} = SSE(S_i, S_j) - SSE(S_i) - SSE(S_j) \qquad (7.18)$$

$$SSE(S) = \sum_{i \in S} ||\mathbf{x^i} - \mathbf{x^c}||^2 \qquad (7.19)$$

where $\mathbf{x^c} = (\sum_{(i \in S)} \mathbf{x^i})/n_S$ is the cluster center for the set S and n_S is the total number of samples in the set. Using ward's linkage, two clusters to be merged are chosen such that WCSSE after merging is minimized.

Example 7.25. *Consider the following data:*

Samples	1	2	3	4
Sepal length (x)	6.6	6.7	6.41	5.9
Petal length (x)	5.3	5.2	5.3	3.1
Species (y)	0	0	0	1

Implement bottom-up hierarchical clustering using Euclidean distance as the distance metric between two points and "minimum" linkage (for distance between two sets). If number of clusters is 2, identify the cluster that the sample (5.7,3.2) would belong to.

Solution 7.25. *Since the bottom most level has each sample as a separate set, we have 4 sets. Since there is only one sample in each of these sets, distance between sets is equivalent to distance between samples. Euclidean distance between all samples are calculated and are reported below.*

$$d_{ij} = \begin{bmatrix} 0 & 0.14 & 0.19 & 2.31 \\ 0.14 & 0 & 0.31 & 2.24 \\ 0.19 & 0.31 & 0 & 2.26 \\ 2.31 & 2.24 & 2.26 & 0 \end{bmatrix}$$

Note that the lowest distance is 0.14 (between samples 1 and 2). Thus, for the next level, we merge samples 1 and 2 as one set as shown below.

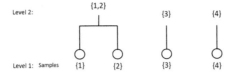

Now, to get level 3, we need to find the distances between 3 sets at level 2. Since we are using "minimum linkage", we need to find the euclidean distance between all samples and look at minimum distance between samples of 2 sets. Let S_i^j denote j^{th} set in level i. Hence, S_2^1 includes samples 1 and 2, and S_2^2 includes sample 3.

$$\text{Distance between } S_2^1 \text{ and } S_2^2 = \min\{d_{13} \text{ and } d_{23}\}$$
$$= \min\{0.19 \text{ and } 0.31\} = 0.19$$

Similarly,

$$\text{Distance between } S_2^1 \text{ and } S_2^3 = 2.24$$
$$\text{Distance between } S_2^2 \text{ and } S_2^3 = 2.26$$

The corresponding distance matrix is $\begin{bmatrix} 0 & 0.19 & 2.24 \\ 0.19 & 0 & 2.26 \\ 2.24 & 2.26 & 0 \end{bmatrix}$.

The smallest distance between sets in level 2 is 0.19 (distance between S_2^1 and S_2^2). Thus, for level 3, we merge S_2^1 and S_2^2.

Thus the dendogram is as shown in Figure 7.17.

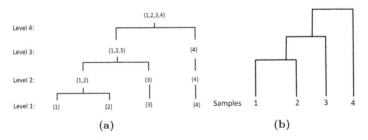

FIGURE 7.17 Dendograms for Example 7.25.

If we are interested in 2 clusters, level 3 needs to be considered. Distance of sample (5.7,3.2) from the training samples are as follows:

	Sample 1	Sample 2	Sample 3	Sample 4
Distance from (5.7,3.2)	2.28	2.24	2.216	0.224

Distance of sample (5.7,3.2) from S_3^1 is the minimum of the first three distances $\Rightarrow 2.216$. Distance from S_3^2 is 0.224. Since distance from S_3^2 is smaller than that from S_3^1, the given test sample is from cluster S_3^2.

Example 7.26. *Consider the data given in Example 7.25. Implement hierarchical clustering using "maximum" linkage for distance between two sets. Use "Euclidean" distance for distance between samples.*

Solution 7.26. *Since Euclidean distance is being used, d_{ij} matrix calculated in Example 7.25 can be used. Level 2 is the same as Example 7.25 as all sets in level 1 had just 1 sample.*
Level 2 has $S_2^1 = \{1,2\}, S_2^2 = \{3\}$ and $S_2^3 \{4\}$
Distance between S_2^1 and S_2^2 $= \max\{0.19 \text{ and } 0.31\} = 0.31$
Distance between S_2^1 and S_2^3 $= \max\{2.31 \text{ and } 2.24\} = 2.31$
Distance between S_2^2 and S_2^3 $= 2.26$
The smallest distance among these three is 0.31. We merge S_2^1 and S_2^2. Final dendogram is as shown in Figure 7.18.

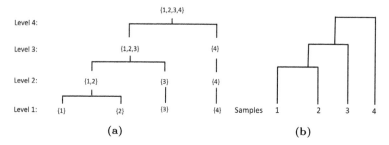

FIGURE 7.18 Dendograms for Example 7.26.

Example 7.27. *Consider the data given in Example 7.25. Implement a dendogram using "minimum" linkage and "Manhattan" distance as distance metrics between sets and samples, respectively.*

Solution 7.27. *Manhattan distance (L_1-norm) between given samples are*

$$
d_{ij} = \begin{bmatrix} 0 & 0.2 & 0.19 & 2.9 \\ 0.2 & 0 & 0.39 & 2.9 \\ 0.19 & 0.39 & 0 & 2.71 \\ 2.9 & 2.9 & 2.71 & 0 \end{bmatrix}
$$

Minimum distance is between samples 1 and 3 and hence are merged for level 2.
Distance between S_2^1 and S_2^2 $= \min\{d_{12}, d_{32}\} = \min\{0.2, 0.39\} = 0.2$
Distance between S_2^1 and S_2^3 $= \min\{d_{14}, d_{34}\} = \min\{2.9, 2.71\} = 2.71$
Distance between S_2^2 and S_2^3 $= d_{14} = 2.9$

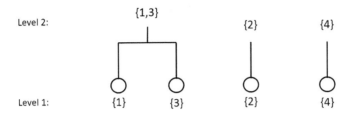

Since the minimum distance among the above three is 0.2, we merge S_2^1 and S_2^2. Thus, we get the dendogram as shown in Figure 7.19.

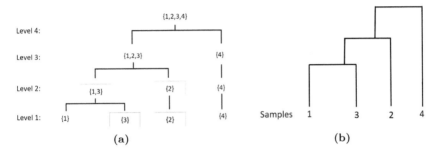

FIGURE 7.19 Dendograms for Example 7.27.

Example 7.28. *Consider the data in Example 7.25. Implement hierarchical clustering using "ward's" linkage to build 3 clusters.*

Solution 7.28. *Ward's linkage works differently from other linkages. Here, distance between two clusters S_1 and S_2 is found by*

$$d_{S_1,S_2} = SSE\left(\{S_1,S_2\}\right) - SSE\left(S_1\right) - SSE\left(S_2\right)$$

SSE is the sum squared error of samples in a set with respect to their centroid given by $SSE\left(S\right) = \sum_{i=1}^{n}||\mathbf{x^i} - \mathbf{x^c}||^2$ where set $S = \{x_i \;\forall\; i = 1:n\}$ and $\mathbf{x^c}$ is the centroid of all samples x_i in S $\left(C = \frac{1}{n}\sum_{i=1}^{n}x_i\right)$.
$SSE\left(\{S_1,S_2\}\right)$ is the sum squared error of the combined set $\{S_1,S_2\}$ from its centroid. For the given data, level 1 will have 4 sets (each having 1 sample). Since all sets S_1^j $(\forall\; j = 1:4)$ have only 1 sample, centroid is the sample itself and hence $SSE\left(S_1^j\right) = 0$.

To find which sets/samples have to be be merged to create level 2, we need to find distance between sets $\left(d_{S_1^j,S_1^k}\right)$.

$$d_{S_1^1,S_1^2} = SSE\left(\{S_1^1,S_1^2\}\right) - SSE\left(S_1^1\right) - SSE\left(S_1^2\right) = SSE\left(\{1,2\}\right)$$

Sample 1 is (6.6,5.3) and sample 2 is (6.7,5.2). Then, the centroid of $\left\{S_1^1, S_1^2\right\}$ is

$$\mathbf{x}^c = \left(\frac{1}{2}\left(6.6 + 6.7\right), \frac{1}{2}\left(5.3 + 5.2\right)\right) = (6.65, 5.25)$$

$$d_{S_1^1, S_1^2} = \left\| \begin{bmatrix} 6.6 \\ 5.3 \end{bmatrix} - \begin{bmatrix} 6.65 \\ 5.25 \end{bmatrix} \right\|^2 + \left\| \begin{bmatrix} 6.7 \\ 5.2 \end{bmatrix} - \begin{bmatrix} 6.65 \\ 5.25 \end{bmatrix} \right\|^2$$

$$= (6.6 - 6.65)^2 + (5.3 - 5.25)^2 + (6.7 - 6.65)^2 + (5.2 - 5.25)^2 = 0.01$$

Centroid of set $\left\{S_1^1, S_1^3\right\} = \left(\frac{1}{2}\left(6.6 + 6.41\right), \frac{1}{2}\left(5.3 + 5.3\right)\right) = (6.505, 5.3)$

Centroid of set $\left\{S_1^1, S_1^4\right\} = (6.25, 4.2)$

Centroid of set $\left\{S_1^2, S_1^3\right\} = (6.555, 5.25)$

Centroid of set $\left\{S_1^2, S_1^4\right\} = (6.3, 4.15)$

Centroid of set $\left\{S_1^3, S_1^4\right\} = (6.155, 4.2)$

Proceeding the same way as $d_{S_1^1, S_1^2}$, we can find the distances between all sets.

$$D = \left\{d_{S_1^j, S_1^k}\right\} = \begin{bmatrix} 0 & 0.01 & 0.018 & 2.665 \\ 0.01 & 0 & 0.47 & 2.525 \\ 0.018 & 0.047 & 0 & 2.55 \\ 2.665 & 2.525 & 2.55 & 0 \end{bmatrix}$$

Note that the smallest distance is between S_1^1 and S_1^2. Hence, we merge those two for the next level as shown in Figure 7.20.

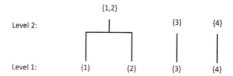

FIGURE 7.20 Dendograms for Example 7.28.

Thus, we get 3 clusters from the given data. We can proceed in the same manner to obtain other levels.

Example 7.29. *Consider the data provided in Table 7.6. Implement hierarchical clustering with "maximum" linkage and Euclidean distance on features x_1 and x_2.*

TABLE 7.6
Hierarchical Clustering Using Maximum Linkage – Example 7.29

Sample No.	x_1	x_2	y	Sample No.	x_1	x_2	y
0	0.67	0.74	0	15	−0.3	0.43	1
1	0.33	−0.37	0	16	0.91	0.41	0
2	0.67	−0.74	0	17	0.46	−0.2	0
3	−0.55	−0.87	1	18	−0.86	0.59	1
4	−0.1	−0.5	1	19	−0.15	0.99	1
5	−0.45	−0.29	1	20	0.15	0.48	0
6	0.15	−0.48	0	21	−0.1	0.5	1
7	−0.15	−0.99	1	22	−0.86	−0.59	1
8	0.33	0.37	0	23	−1.03	0.21	1
9	−0.54	0.1	1	24	−0.3	-0.43	1
10	−0.55	0.87	1	25	0.31	0.95	0
11	0.46	0.2	0	26	1	0	0
12	−1.03	−0.21	1	27	−0.45	0.29	1
13	0.91	−0.41	0	28	0.5	0	0
14	−0.54	−0.1	1	29	0.31	−0.95	0

Solution 7.29. *The given data has two classes as shown in Figure 7.21a. Note that the samples on the left and right are identified as two classes. Hierarchical clustering with "maximum" linkage and Euclidean distance was implemented in a programming platform to build two clusters.*

The clusters obtained can be seen in Figure 7.21b. The dendogram for this clustering is shown in Figure 7.21c. Since we are interested in two clusters, the top most split in the dendogram is considered. Samples 0, 16, 25, 8, 20, 26, 11, 17, 28, 2, 13, 29, 1, and 6 are identified as one cluster and the remaining samples as the next cluster. This is exactly the same as the true class labels.

Example 7.30. *Consider the data provided in Table 7.7. Implement hierarchical clustering with "minimum" linkage and Euclidean distance on features x_1 and x_2.*

Solution 7.30. *The given data has two classes as shown in Figure 7.22a. Note that the samples on the inner and outer circles are identified as two classes.*

Hierarchical clustering with "minimum" linkage and Euclidean distance was implemented in a programming platform to build two clusters. The clusters obtained can be seen in Figure 7.22b. The dendogram for this clustering is shown in Figure 7.22c. Since we are interested in two clusters, the top most split in the dendogram is considered. Samples 21, 4, 15, 27, 9, 14, 5, 24, 8, 20, 11, 17, 28, 1, and 6 are identified as one cluster and the remaining samples as the next cluster. This is exactly same as the true class labels. Note that maximum linkage for the same data resulted in completely different clusters.

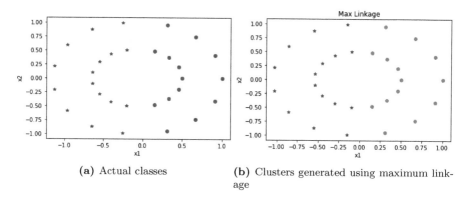

(a) Actual classes

(b) Clusters generated using maximum linkage

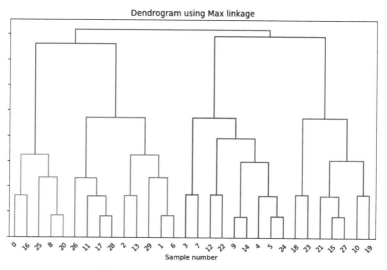

(c) Dendogram using maximum linkage

FIGURE 7.21 Hierarchical clustering using maximum linkage—Example 7.29.

7.3.3 Support Vector Machines

Support vector machine (SVM) is another popular method for both linear and nonlinear binary classification problems. As the name suggests, the algorithm identifies vectors that support the hyperplane decision function. This algorithm can also be extended to multi-class problems.

Let us start with the description of this algorithm for a binary classification problem that is linearly separable. A pictorial representation of such a problem can be seen in Figure 7.24a. Here datapoints belonging to the two classes

TABLE 7.7

Hierarchical Clustering Using Minimum Linkage – Example 7.30

Sample No.	x_1	x_2	y	Sample No.	x_1	x_2	y
0	0.67	0.74	0	15	−0.3	0.43	1
1	0.33	−0.37	1	16	0.91	0.41	0
2	0.67	−0.74	0	17	0.46	−0.2	1
3	−0.55	−0.87	0	18	−0.86	0.59	0
4	−0.1	−0.5	1	19	−0.15	0.99	0
5	−0.45	−0.29	1	20	0.15	0.48	1
6	0.15	−0.48	1	21	−0.1	0.5	1
7	−0.15	−0.99	0	22	−0.86	−0.59	0
8	0.33	0.37	1	23	−1.03	0.21	0
9	−0.54	0.1	1	24	−0.3	−0.43	1
10	−0.55	0.87	0	25	0.31	0.95	0
11	0.46	0.2	1	26	1	0	0
12	−1.03	−0.21	0	27	−0.45	0.29	1
13	0.91	−0.41	0	28	0.5	0	1
14	−0.54	−0.1	1	29	0.31	−0.95	0

represented by "o" (Class-0) and "x" (Class-1) are shown. This is a linearly separable problem and the dotted line is a hyperplane that can separate the classes. SVM develops this hyperplane in a particular manner. Consider the two hyperplanes on either side of the dotted hyperplane (h_1 and h_2). The key characteristic of these hyperplanes is that each of the hyperplanes have at least one of the datapoints located on the hyperplane (of course there could be many points also). The points that are located on the hyperplanes are called the support vectors (as they support the hyperplane). Notice that these hyperplanes are parallel to each other. Visually, it is clear that good separation of the classes will be attained if these two hyperplanes are separated as much as possible without any misclassification. This is the objective function that is used in the SVM formulation.

From Chapter 3, we know that $\mathbf{w}$ is a vector perpendicular to h_1. A unit vector in this direction is $\frac{\mathbf{w}}{||\mathbf{w}||}$. Consider a vector $\mathbf{x}_d = \mathbf{x}_o + \alpha \frac{\mathbf{w}}{||\mathbf{w}||}$ starting from any point $\mathbf{x}_o$ on h_1; this will be in a direction perpendicular to the hyperplane. When we find α such that $\mathbf{x}_o + \alpha \frac{\mathbf{w}}{||\mathbf{w}||}$ is in h_2, then α will give us the distance between the hyperplanes. Since $\mathbf{x}_d$ is on h_2

$$\mathbf{w}^T \mathbf{x}_d + b = 1; \quad \mathbf{w}^T \mathbf{x}_o + \alpha \frac{\mathbf{w}^T \mathbf{w}}{||\mathbf{w}||} + b = 1$$

Since $\mathbf{x}_o$ is on h_1, $\mathbf{w}^T \mathbf{x}_o + b = -1$, as a result,

$$\alpha = \frac{2||\mathbf{w}||}{\mathbf{w}^T \mathbf{w}} = \frac{2}{\sqrt{\mathbf{w}^T \mathbf{w}}}$$

The SVM algorithms look to maximize the separation between the hyperplanes ($\frac{2}{\sqrt{\mathbf{w}^T \mathbf{w}}}$), which is equivalent to minimizing $\frac{1}{2}\mathbf{w}^T \mathbf{w}$. Let there be a total

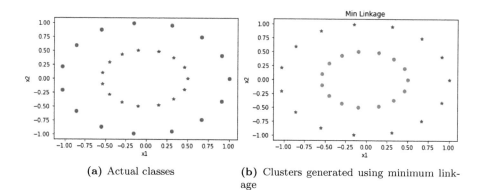

(a) Actual classes   (b) Clusters generated using minimum link-
                                      age

(c) Dendrogram using minimum linkage

FIGURE 7.22 Hierarchical clustering using minimum linkage—Example
7.30.

of m sample points in the training data, $S = [1 \dots m]$. Of those, let us collect
indices of all datapoints belonging to Class-1 as S_1 and indices belonging to
Class-2 as S_2. Now we can formulate an optimization problem

$$\min_{\mathbf{w},b} \quad \tfrac{1}{2}\mathbf{w}^T\mathbf{w}$$
$$\mathbf{w}^T\mathbf{x}^i + b \leq -1 \; \forall \, i \in S_1$$
$$\mathbf{w}^T\mathbf{x}^i + b \geq 1 \; \forall \, i \in S_2$$

The first set of constraints reflect the requirement that datapoints in Class-1
have to be in the negative half-plane of h_1 and the second set of constraints

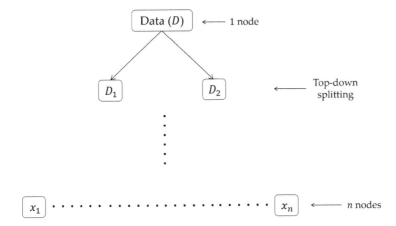

FIGURE 7.23 Top down data splitting approach.

reflect the requirement that all datapoints belonging to Class-2 have to be in the positive half of h_2. From Chapter 4, we know that this is a quadratic programming problem and there are several approaches to solving this problem. For example, the KKT conditions for the optimum solution will lead to a linear programming feasibility problem.

Example 7.31. *Consider the data:*

Sample No.	1	2	3	4	5	6
x_1	1	2	3	5	6	7
x_2	2	1	3	1	3	2
y	0	0	0	1	1	1

Perform SVM and find the best hyperplane that separates the data. Predict the classes based on the hyperplane.

Solution 7.31. *For SVM,*

$$\min_{w,b} \quad 0.5\mathbf{w}^T\mathbf{w} = 0.5(w_1^2 + w_2^2)$$

$$s.t. \quad \mathbf{w}^T\mathbf{x}^1 + b \leq -1$$

$$\mathbf{w}^T\mathbf{x}^2 + b \leq -1$$

$$\mathbf{w}^T\mathbf{x}^3 + b \leq -1$$

$$\mathbf{w}^T\mathbf{x}^4 + b \geq 1 \implies -\mathbf{w}^T\mathbf{x}^4 - b \leq -1$$

$$\mathbf{w}^T\mathbf{x}^5 + b \geq 1 \implies -\mathbf{w}^T\mathbf{x}^5 - b \leq -1$$

$$\mathbf{w}^T\mathbf{x}^6 + b \geq 1 \implies -\mathbf{w}^T\mathbf{x}^6 - b \leq -1$$

where $\mathbf{x}^i$ is the ith sample. Since the objective function is a convex quadratic function and constraints are linear, this is quadratic (convex) programming

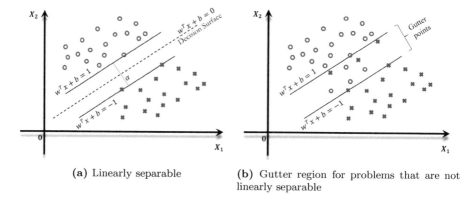

(a) Linearly separable

(b) Gutter region for problems that are not linearly separable

FIGURE 7.24 The concept of support vectors and hyperplane separation.

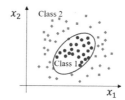

FIGURE 7.25 Lifting to higher dimensions for nonlinear classification problems.

problem. Being a convex programming problem, KKT conditions are both necessary and sufficient.

The Lagrangian of the above optimization problem is

$$\min_{w,b,\mu} \quad 0.5(w_1^2 + w_2^2) + \sum_{i=1}^{3} \mu_i(\mathbf{w}^T\mathbf{x}^i + b + 1) + \sum_{i=4}^{6} \mu_i(-\mathbf{w}^T\mathbf{x}^i - b + 1)$$

where $\mu = \begin{bmatrix} \mu_1 & \mu_2 & \mu_3 & \mu_4 & \mu_5 & \mu_6 \end{bmatrix}^T$. *KKT conditions are as follows:*

$$w_1 + \sum_{i=1}^{3} \mu_i x_1^i + \sum_{i=4}^{6} -\mu_i x_1^i = 0; \quad w_2 + \sum_{i=1}^{3} \mu_i x_2^i + \sum_{i=4}^{6} -\mu_i x_2^i = 0$$

$$\mu_i(\mathbf{w}^T\mathbf{x}^i + b + 1) = 0 \quad \forall \ i \in [1,3]; \quad \mu_i(-\mathbf{w}^T\mathbf{x}^i - b + 1) = 0 \quad \forall \ i \in [4,6]$$

$$\sum_{i=1}^{3} \mu_i + \sum_{i=4}^{6} -\mu_i = 0; \quad \mu_i \geq 0 \quad \forall \ i \in [1,6]$$

Expanding,

$$w_1 + \mu_1 + 2\mu_2 + 3\mu_3 - 5\mu_4 - 6\mu_5 - 7\mu_6 = 0 \qquad (7.20)$$
$$w_2 + 2\mu_1 + \mu_2 + 3\mu_3 - \mu_4 - 3\mu_5 - 2\mu_6 = 0 \qquad (7.21)$$
$$\mu_1(w_1 + 2w_2 + b + 1) = 0 \qquad (7.22)$$
$$\mu_2(2w_1 + w_2 + b + 1) = 0 \qquad (7.23)$$
$$\mu_3(3w_1 + 3w_2 + b + 1) = 0 \qquad (7.24)$$
$$\mu_4(-5w_1 - w_2 - b + 1) = 0 \qquad (7.25)$$
$$\mu_5(-6w_1 - 3w_2 - b + 1) = 0 \qquad (7.26)$$
$$\mu_6(-7w_1 - 2w_2 - b + 1) = 0 \qquad (7.27)$$
$$\mu_1 + \mu_2 + \mu_3 - \mu_4 - \mu_5 - \mu_6 = 0 \qquad (7.28)$$

a) **Case a:** $\mu_i = 0 \quad \forall \, i = 1 : 6$
From 7.20 and 7.21, we have $w_1 = w_2 = 0$. *Substituting in original inequality conditions, we get $b \leq -1$ from first 3 inequalities and $b \geq 1$ from the last 3 inequalities. These two equations are inconsistent and thus, this is not a feasible case.*

b) **Case b:** $\mu_i = 0 \quad \forall \, i = 4 : 6$ and $\mu_i \neq 0 \quad \forall \, i = 1 : 3$
From 7.22 to 7.24, we have

$$w_1 + 2w_2 + b + 1 = 0; \quad 2w_1 + w_2 + b + 1 = 0; \quad 3w_1 + 3w_2 + b + 1 = 0$$

Solving, we get $w_1 = w_2 = 0$ and $b = -1$. The last three inequality conditions become inconsistent and thus, this is not a feasible case.

c) **Case c:** $\mu_i = 0 \quad \forall \, i = 1 : 3$ and $\mu_i \neq 0 \quad \forall \, i = 4 : 6$
From 7.25 to 7.27, we have

$$-5w_1 - 1w_2 - b + 1 = 0; \quad -6w_1 - 3w_2 - b + 1 = 0;$$
$$-7w_1 - 2w_2 - b + 1 = 0$$

Solving, we get $w_1 = w_2 = 0$ and $b = 1$. The first three inequality conditions become inconsistent and thus, this is not a feasible case.

d) **Case d:** $\mu_i = 0 \quad \forall \, i = 2 : 5$ and $\mu_i \neq 0 \quad \forall \, i = 1, 6$
From 7.20 to 7.22 and 7.27, we have

$$w_1 + \mu_1 - 7\mu_6 = 0; \quad w_2 + 2\mu_1 - 2\mu_6 = 0; \quad \mu_1 - \mu_6 = 0$$
$$w_1 + 2w_2 + b + 1 = 0; \quad -7w_1 - 2w_2 - b + 1 = 0$$

Solving, we get $w_1 = 0.33$, $w_2 = 0$, $b = -1.33$, and $\mu_1 = \mu_6 = 0.0556$. The second to fifth inequality conditions become inconsistent and thus, this is not a feasible case.

e) **Case e:** $\mu_i = 0 \quad \forall \, i = 1, 6$ and $\mu_i \neq 0 \quad \forall \, i = 2 : 5$
From equations 7.23 to 7.26,

$$2w_1 + w_2 + b + 1 = 0; \quad 3w_1 + 3w_2 + b + 1 = 0$$
$$-5w_1 - 1w_2 - b + 1 = 0; \quad -6w_1 - 3w_2 - b + 1 = 0$$

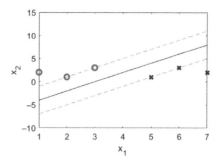

FIGURE 7.26 Example 7.31.

Solving, we get $w_1 = 0.6667$, $w_2 = -0.3333$, and $b = -2$. For these values, all the constraints are satisfied. Solving equations 7.20 to 7.28, we get $\mu_2 = 0$, $\mu_3 = 0.2778$, $\mu_4 = 0.1667$, and $\mu_5 = 0.1111$. All μ values are non-negative. Hence, this is a feasible solution. The points (2,1) and (3,3) are the support vectors for Class-0 and (5,1) and (6,3) are the support vectors for Class-1.

As the optimization problem is convex, KKT conditions are sufficient and hence, the best linear decision boundary is $0.6667x_1 - 0.3333x_2 - 2 = 0$. The decision boundary is shown in Figure 7.26. True Class-0 samples are shown as circles while the true Class-1 samples are shown as "×". Note that the predicted decision boundary is able to correctly distinguish between the classes.

Example 7.32. *Consider the example discussed in Example 7.31. Reformulate SVM using slack variables and solve for the best decision boundary.*

Solution 7.32. *The optimization problem discussed in Example 7.31 can be rewritten as*

$$\min_{w_1,w_2,b,\epsilon_i} \quad 0.5(w_1^2 + w_2^2)$$
$$s.t. \quad \mathbf{w}^T\mathbf{x}^1 + b = -1 + \epsilon_1$$
$$\mathbf{w}^T\mathbf{x}^2 + b = -1 + \epsilon_2$$
$$\mathbf{w}^T\mathbf{x}^3 + b = -1 + \epsilon_3$$
$$-\mathbf{w}^T\mathbf{x}^4 - b = -1 + \epsilon_4$$
$$-\mathbf{w}^T\mathbf{x}^5 - b = -1 + \epsilon_5$$
$$-\mathbf{w}^T\mathbf{x}^6 - b = -1 + \epsilon_6$$

where $\epsilon_i \leq 0 \quad \forall\, i = 1:6$ are slack variables.

In this reformulated optimization problem, all constraints except for the bounds are all equality constraints. Solving the above optimization problem, we get

$$w_1 = 0.6667; \quad w_2 = -0.3333; \quad b = -2$$

$$\epsilon_1 = -1; \quad \epsilon_6 = -1; \quad \epsilon_2 = \epsilon_3 = \epsilon_4 = \epsilon_5 = 0$$

Note that the solution obtained is the same as the one obtained in Example 7.31. Further, the values of slack variables (ϵ_i) are all non-positive.

Example 7.33. *Consider the data:*

Sample No.	1	2	3	4	5	6	7
x_1	1	2	3	5	6	7	2
x_2	2	1	3	1	3	2	2
y	0	0	0	1	1	1	1

Perform SVM and find the best hyperplane that separates the data. Predict the classes based on the hyperplane.

Solution 7.33. *For the given data, the optimization problem for SVM in terms of slack variables is as follows:*

$$\min_{w_1,w_2,b,\epsilon_i} \quad 0.5(w_1^2 + w_2^2)$$

$$s.t. \quad \mathbf{w}^T\mathbf{x}^1 + b = -1 + \epsilon_1$$

$$\mathbf{w}^T\mathbf{x}^2 + b = -1 + \epsilon_2$$

$$\mathbf{w}^T\mathbf{x}^3 + b = -1 + \epsilon_3$$

$$-\mathbf{w}^T\mathbf{x}^4 - b = -1 + \epsilon_4$$

$$-\mathbf{w}^T\mathbf{x}^5 - b = -1 + \epsilon_5$$

$$-\mathbf{w}^T\mathbf{x}^6 - b = -1 + \epsilon_6$$

$$-\mathbf{w}^T\mathbf{x}^7 - b = -1 + \epsilon_7$$

$$\epsilon_i \leq 0 \quad \forall \, i = 1 : 7$$

where x^i is the ith sample and ϵ_i is the ith slack variable. When we try to solve the above optimization problem, it is evident that there exists no solution. It is also evident from Figure 7.27a that the data is not linearly separable and hence, we cannot find a line that separates the classes exactly.

In data that are not linearly separable as in Figure 7.24b, there will always be some points in between h_1 and h_2, the so called gutter points. If one tries to solve the previously posed problem for such data, the optimization subroutine will return the result that the problem is infeasible. One way to address this is to move the points in class 1 in the gutter region to set S_{g1} and points in class 2 in the gutter region to S_{g2}. Now we can excuse these points from being

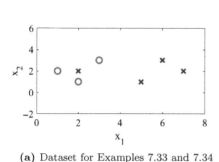 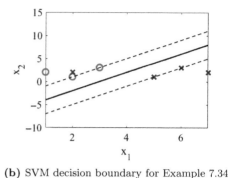

(a) Dataset for Examples 7.33 and 7.34   (b) SVM decision boundary for Example 7.34

FIGURE 7.27 Examples 7.33 and 7.34.

classified correctly, basically cut these points some slack. Mathematically, this is achieved by introducing slack variables for these points as below

$$\mathbf{w}^T\mathbf{x}^i + b \leq -1 + \epsilon_i \; \forall \, i \in S_{g1} \tag{7.29}$$

$$\mathbf{w}^T\mathbf{x}^i + b \geq 1 - \epsilon_i \; \forall \, i \in S_{g2} \tag{7.30}$$

When the slack variables take positive values, the first equation allows some points from class 1 to be in positive half of h_1 and the second equation allows some points from class 2 to be in the negative half of h_2. Addition of these constraints to the previous problem, after changing S_1 to $S_1 - S_{g1}$ and S_2 to $S_2 - S_{g2}$ will make the problem solvable. While this seems like a neat trick, one difficulty is that we would not know which points are in the gutter region, *a priori*.

However, this is not a difficult issue because one could include a slack variable for every one of the sample points (that is $S_{g1} = S_1$ and $S_{g2} = S_2$) and minimize the sum of all slack variables. Since for perfect classification all the slack variables need to take negative values, the optimizer will allow positive values for slack variables only when it is absolutely unavoidable. For linearly separable problems, all the slack variables will take negative values in the final solution. Since there are now two objectives, maximizing separation of hyperplanes and minimizing the sum of slack variables, a single objective function that combines both the objectives is optimized

$$\min_{\mathbf{w}, b, \epsilon_1 \ldots \epsilon_m} \quad \tfrac{1}{2}\mathbf{w}^T\mathbf{w} + C \sum_{i \in S} \epsilon_i$$
$$\mathbf{w}^T\mathbf{x}^i + b \leq -1 + \epsilon_i \; \forall \, i \in S_1$$
$$\mathbf{w}^T\mathbf{x}^i + b \geq 1 - \epsilon_i \; \forall \, i \in S_2 \tag{7.31}$$

This is also a quadratic programming problem. C is a positive constant that can be used to control overfitting.

Although the solution to the above formulation provides a decision boundary allowing for gutter points, the optimum is found such that all the slack variables are minimized. That is, the slack variables corresponding to perfectly classified points are reduced as well. Minimizing either an already negative slack variable or a positive slack variable by one unit has the same impact on the objective function. A better approach would be to minimize only the positive slack variables (corresponding to misclassified/gutter points). The corresponding optimization problem can be written as:

$$\min_{\mathbf{w},b,\epsilon_1\ldots\epsilon_m} \tfrac{1}{2}\mathbf{w}^T\mathbf{w} + C\sum_{i\in S}\max(0,\epsilon_i)$$
$$\mathbf{w}^T\mathbf{x}^i + b = -1 + \epsilon_i \ \forall\ i \in S_1$$
$$\mathbf{w}^T\mathbf{x}^i + b = 1 - \epsilon_i \ \forall\ i \in S_2 \tag{7.32}$$

The above optimization problem can also be written as:

$$\min_{\mathbf{w},b,\epsilon_1\ldots\epsilon_m} \tfrac{1}{2}\mathbf{w}^T\mathbf{w} + C\sum_{i\in S}\epsilon_i$$
$$\mathbf{w}^T\mathbf{x}^i + b \leq -1 + \epsilon_i \ \forall\ i \in S_1$$
$$\mathbf{w}^T\mathbf{x}^i + b \geq 1 - \epsilon_i \ \forall\ i \in S_2$$
$$\epsilon_i \geq 0 \ \forall\ i \in S \tag{7.33}$$

With the additional constraint $\epsilon_i \geq 0$, the feasible set is such that all the slack variables corresponding to perfectly classified points will be zero. Hence, the second term in the objective function is equivalent to the sum of positive slack variables. If $y_i = -1$ for Class-1 (samples in S_1) and $y_i = 1$ for class 2 (samples in S_2), equation 7.32 can also be rewritten as follows:

$$\min_{\mathbf{w},b} \frac{1}{2}\mathbf{w}^T\mathbf{w} + C\sum_{i\in S}\max(0, 1 - y_i(\mathbf{w}^T\mathbf{x}^i + b)) \tag{7.34}$$

Example 7.34. *Consider the data:*

Sample No.	1	2	3	4	5	6	7
x_1	1	2	3	5	6	7	2
x_2	2	1	3	1	3	2	2
x_3	0	0	0	1	1	1	1

Perform SVM considering possible gutter points and find the best hyperplane that separates the data. Predict the classes based on the hyperplane. Use $C = 10$.

Solution 7.34. *The optimization problem for SVM would be as follows:*

$$\min_{w_1,w_2,b,\epsilon_i} \quad 0.5(w_1^2 + w_2^2) + C\sum_{i\in[1,7]}\epsilon_i$$

$$s.t. \quad \mathbf{w}^T\mathbf{x}^1 + b = -1 + \epsilon_1$$
$$\mathbf{w}^T\mathbf{x}^2 + b \leq -1 + \epsilon_2$$
$$\mathbf{w}^T\mathbf{x}^3 + b \leq -1 + \epsilon_3$$

$$-\mathbf{w}^T\mathbf{x}^4 - b \le -1 + \epsilon_4$$
$$-\mathbf{w}^T\mathbf{x}^5 - b \le -1 + \epsilon_5$$
$$-\mathbf{w}^T\mathbf{x}^6 - b \le -1 + \epsilon_6$$
$$-\mathbf{w}^T\mathbf{x}^7 - b \le -1 + \epsilon_7$$
$$\epsilon_i \ge 0 \quad \forall\, i \in [1,7]$$

Solving the above optimization problem, we get

$$w_1 = 0.6667; \quad w_2 = -0.3333; \quad b = -2$$
$$\epsilon_7 = 2.333; \quad \epsilon_1 = \epsilon_2 = \epsilon_3 = \epsilon_4 = \epsilon_5 = \epsilon_6 = 0$$

Figure 7.27b shows the corresponding decision boundary. Note that ϵ_7 is positive implying that the seventh sample (2,2) is a gutter point. Except for seventh sample, the classifier has correctly identified the class. For the seventh sample, although the true class is 1, the developed SVM classifier identifies it as class 0. This can also be seen in Figure 7.27b.

Example 7.35. *For Example 7.35, comment on the effect of parameter C on decision boundary.*

Solution 7.35. *For low values of C, there is very little weightage given for minimization of positive slack variables. The priority is for minimization of weights. Moreover, unlike Example 7.33, there are no upper bounds for ϵ_i in this formulation. Hence, the solution obtained is such that the weights w_1 and w_2 are small while the constraints are satisfied by setting high positive values for the slack variables. For example, for $C = 0.001$, we get*

$$w_1 = 0.007; \quad w_2 = 0; \quad b = 0.0954$$

As we increase C, weightage for minimization of positive slack variables is increased. We get good classifiers for $C \ge 1$. The parameters of classifiers obtained for various values of C are provided in Table 7.8 and the corresponding decision boundaries are shown in Figure 7.28.

Example 7.36. *For the linearly separable data provided in Example 7.31, will the decision boundary change if the SVM formulation allowing gutter points as provided in equation 7.31 is used? Explain with reasons.*

Solution 7.36. *The decision boundary obtained for the formulation allowing gutter points would be different for different values of C. However, since the solution obtained in Example 7.31 satisfies all constraints, it is possible that the same solution as obtained in Example 7.31 is obtained for large enough values of C. For example, for $C = 10$, we get the same decision boundary.*

When a linear SVM doesn't provide satisfactory results, one could then look for nonlinear classifiers. The concept of lifting to higher dimensions and

TABLE 7.8

Classifiers for Various C Values (Example 7.35)

Decision Variable	$C = 0.001$	$C = 0.1$	$C = 1$	$C = 10$	$C = 20$	$C = 100$
w_1	0.007	0.412	0.667	0.667	0.667	0.667
w_2	0	−0.058	−0.33	−0.33	−0.33	−0.33
b	0.954	−1.296	−2	−2	−2	−2
ϵ_1	1.961	0	0	0	0	0
ϵ_2	1.968	0.469	0	0	0	0
ϵ_3	1.975	0.765	0	0	0	0
ϵ_4	0.011	0.296	0	0	0	0
ϵ_5	0.004	0	0	0	0	0
ϵ_6	0	0	0	0	0	0
ϵ_7	0.032	1.588	2.33	2.33	2.33	2.33

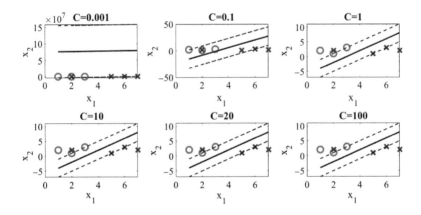

FIGURE 7.28 Classifiers for various C values (Example 7.35).

choice of kernels allows us to use the same SVM formulation for nonlinear problems. To illustrate this, let us consider the binary classification problem depicted in Figure 7.25. It is clear from the figure that an ellipsoidal boundary can solve the classification problem with the logic that points within the ellipse belong to Class-1 and outside the ellipse belong to Class-2. The equation of the ellipse is $\mathbf{x}^T Q \mathbf{x} + \mathbf{n}^T \mathbf{x} + k = 0$, where Q is a positive definite matrix. This equation can be written as a polynomial equation

$$ax_1^2 + bx_2^2 + cx_1 x_2 + dx_1 + ex_2 + k = 0 \tag{7.35}$$

Now the idea of lifting to higher dimensions is the following. If we are given labeled data matrix of size $m \times n$, we can add additional features to the data and increase the input dimension or the feature size to N, making the data matrix $m \times N$. The new dimensions or features are derived from the existing variables.

In the two-dimensional example above, given $\mathbf{x} = [x_1, x_2]$ as the original variables, a sample size of m will result in a data matrix of $m \times 2$. Now one could add extra features to generate $\mathbf{x}_{aug} = [x_1, x_2, x_1^2, x_2^2, x_1 x_2]$. Essentially, the $m \times 2$ data matrix can be converted to a $m \times 5$ data matrix. A linear SVM classifier in the original variables will be of the form $\mathbf{w}^T \mathbf{x} + b = 0$. A linear SVM classifier in the augmented variable space will be of the form $\mathbf{w}_{aug}^T \mathbf{x}_{aug} + b' = 0$. A quick comparison shows that this equation can represent equation 7.35, which is a nonlinear classifier in the original variable space. Essentially, a linear classifier in an augmented (lifted) variable space can model nonlinear classifiers in the original variable space. This simple yet powerful idea has a profound impact in ML algorithms. The parallels between this and linear-in-parameters models developed in Chapter 6 are hard to miss.

Example 7.37. *Consider the following nonlinearly separable data. Perform hard-margin SVM (no gutter points) to find the best decision boundary. Predict the classes based on the decision boundary.*

Sample No.	1	2	3	4	5	6
x_1	0	1	$-1/\sqrt{2}$	0	4	$-2\sqrt{2}$
x_2	1	0	$-1/\sqrt{2}$	4	0	$-2\sqrt{2}$
y	0	0	0	1	1	1

Solution 7.37. *The data given is not linearly separable as seen from Figure 7.29a. It is quite evident that an elliptical or circular boundary would be able to separate the classes. Thus, if we use features $x_3 = x_1^2$ and $x_4 = x_2^2$ instead of x_1 and x_2, the samples are linearly separable in x_3 and x_4 as can be seen in Figure 7.29b. Hence, a hard-margin SVM with x_3 and x_4 would be able to classify the data correctly. The SVM formulation is as follows:*

$$\min_{w_3, w_4, b, \epsilon_i} \quad 0.5(w_3^2 + w_4^2)$$

$$s.t. \quad w_3 x_3^1 + w_4 x_4^1 + b \leq -1$$
$$w_3 x_3^2 + w_4 x_4^2 + b \leq -1$$
$$w_3 x_3^3 + w_4 x_4^3 + b \leq -1$$
$$-w_3 x_3^4 - w_4 x_4^4 - b \leq -1$$
$$-w_3 x_3^5 - w_4 x_4^5 - b \leq -1$$
$$-w_3 x_3^6 - w_4 x_4^6 - b \leq -1$$

where x^i is the ith sample. Solving, we get, $w_3 = 0.1333$, $w_4 = 0.1333$, and $b = -1.1333$. Hence the boundary is $0.1333x_1^2 + 0.1333x_2^2 - 1.1333 = 0$ or in other words a circle at (0,0) with radius $r = \sqrt{1.1333/0.1333} = 2.916$. This circular boundary is shown in Figure 7.29c.

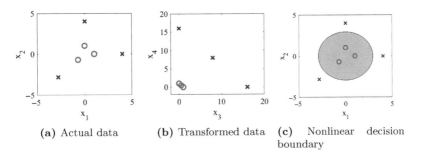

(a) Actual data   (b) Transformed data   (c)   Nonlinear   decision
                                                        boundary

FIGURE 7.29 Results for Example 7.37.

Example 7.38. *Consider the following data. Perform soft-margin SVM to find the best decision boundary. Predict the classes based on the decision boundary. Use $C = 10$.*

Sample No.	1	2	3	4	5	6	7
x_1	0	1	$-1/\sqrt{2}$	0	4	$-2\sqrt{2}$	0
x_2	1	0	$-1/\sqrt{2}$	4	0	$-2\sqrt{2}$	0
y	0	0	0	1	1	1	1

Solution 7.38. *The data given is non-separable as can be seen from Figure 7.30a. It is quite evident that an elliptical or circular boundary would be able to separate the classes if we allow some gutter points. Similar to the previous example, we use features $x_3 = x_1^2$ and $x_4 = x_2^2$ instead of x_1 and x_2 for formulating soft-margin SVM. Hence, a soft-margin SVM with x_3 and x_4 would be able to classify the data correctly. The SVM formulation is as follows:*

$$\min_{w_3, w_4, b, \epsilon_i} \quad 0.5(w_3^2 + w_4^2) + C \sum_{i \in [1,7]} \epsilon_i$$

$$s.t. \quad w_3 x_3^1 + w_4 x_4^1 + b \leq -1 + \epsilon_1$$

$$w_3 x_3^2 + w_4 x_4^2 + b \leq -1 + \epsilon_2$$

$$w_3 x_3^3 + w_4 x_4^3 + b \leq -1 + \epsilon_3$$

$$-w_3 x_3^4 - w_4 x_4^4 - b \leq -1 + \epsilon_4$$

$$-w_3 x_3^5 - w_4 x_4^5 - b \leq -1 + \epsilon_5$$

$$-w_3 x_3^6 - w_4 x_4^6 - b \leq -1 + \epsilon_6$$

$$-w_3 x_3^7 - w_4 x_4^7 - b \leq -1 + \epsilon_7$$

$$\epsilon_i \geq 0 \quad \forall \, i \in [1, 7]$$

Solving the above optimization problem for $C = 10$, we get $w_3 = 0.1333$, $w_4 = 0.1333$ and $b = -1.1333$. Also, the slack variables are all zero at optimum except for ϵ_7. The optimal value of ϵ_7 is 2.1333. Hence the boundary

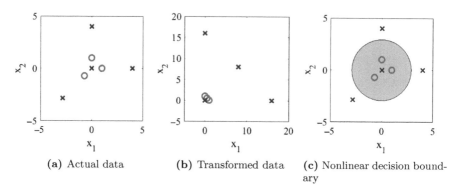

(a) Actual data   (b) Transformed data   (c) Nonlinear decision boundary

FIGURE 7.30 Results for Example 7.38.

is $0.1333x_1^2 + 0.1333x_2^2 - 1.1333 = 0$ or in other words a circle at (0,0) with radius $r = \sqrt{1.1333/0.1333} = 2.916$. The corresponding boundary is shown in Figure 7.30c. Note that the point (0,0) is mis-classified and the corresponding slack variable is the only positive value.

A final thought before we end this section on SVMs. For one to follow the description above, a user must decide the new features that should be included for efficient classification. In the specific example used in this book, it was clear that quadratic terms are required because of the knowledge that an ellipse will be able to perform the separation. For multidimensional problems, this approach must be generalized for unknown boundaries. There are several papers and algorithms that answer this question with rich theory behind them. A very simple way to think about this is through the Taylor's series expansion. Assume that we are trying to model an unknown nonlinear decision surface $h(\mathbf{x}) = 0$ for a binary classification problem. We could expand this nonlinear function as (for the sake of simplicity expanded around the origin)

$$h(\mathbf{x}) = h(\mathbf{x} = 0) + (\nabla h)^T|_{\mathbf{x}=0)}\mathbf{x} + \frac{1}{2}\mathbf{x}^T H|_{\mathbf{x}=0}\mathbf{x} + hot \qquad (7.36)$$

The Hessian matrix (H) terms will include the quadratic terms and one could add more higher order terms and so on. As a result, lifting to a reasonably higher dimension might provide satisfactory results. The number of terms to be included can be addressed using ideas very similar to the variable selection approaches used for function approximation. This idea goes under the name of polynomial kernels in the SVM approach. Several other kernels can also be used with different basis functions. There are many such non-polynomial basis functions (such as radial basis functions and so on).

Example 7.39. *For the data provided in Example 7.37, perform SVM using second-order polynomial kernel with and without Lasso reduction and compare the results.*

Solution 7.39. *For a second-order polynomial kernel, the input features are* $x_1, x_2, x_1^2, x_2^2,$ *and* $x_1 x_2$. *Define,*

$$x_3 = x_1^2; \quad x_4 = x_2^2; \quad x_5 = x_1 x_2;$$

Using the features $x_j \ \forall \ j \in [1,5]$, *soft-margin SVM formulation is as follows:*

$$\min_{w,b,\epsilon} \quad 0.5\mathbf{w}^T\mathbf{w} + C\sum \epsilon$$
$$s.t. \quad \mathbf{w}^T\mathbf{x}^i + b \leq -1 + \epsilon_j \quad \forall \ i \in [1,3]$$
$$-\mathbf{w}^T\mathbf{x}^i - b \leq -1 + \epsilon_i \quad \forall \ i \in [4,6]$$
$$\epsilon_i \geq 0 \quad \forall \ i \in [1,6]$$

where x^i *is* i^{th} *sample. Solving the above optimization problem for* $C = 10$, *we get* $w = \begin{bmatrix} 0.015 & 0.015 & 0.13 & 0.13 & 0.018 \end{bmatrix}$, $b = -1.145$ *and* $\epsilon_i = 0 \ \forall \ i \in [1,6]$.

With Lasso reduction, the SVM formulation is as follows:

$$\min_{w,b,\epsilon} \quad 0.5\mathbf{w}^T\mathbf{w} + C\sum \epsilon + L\sum |w|$$
$$s.t. \quad \mathbf{w}^T\mathbf{x}^i + b \leq -1 + \epsilon_j \quad \forall \ i \in [1,3]$$
$$-\mathbf{w}^T\mathbf{x}^i - b \leq -1 + \epsilon_i \quad \forall \ i \in [4,6]$$
$$\epsilon_i \geq 0 \quad \forall \ i \in [1,6]$$

Solving the above optimization problem for $C = 10$ *and* $L = 1$, *we get,* $\mathbf{w} = \begin{bmatrix} 0 & 0 & 0.133 & 0.133 & 0 \end{bmatrix}$, $b = -1.33$, *and* $\epsilon_i = 0 \ \forall \ i \in [1,6]$.

Classes predicted for each sample based on the developed decision boundaries are provided in table below. Note that the predicted classes for both the models are the same for all samples. However, the model without Lasso reduction uses all 5 features, whereas the model with Lasso reduction requires only 2 features (x_3 *and* x_4). *The weights of the remaining features were obtained as zeros.*

Sample No.	x_1	x_2	$x_3 = x_1^2$	$x_4 = x_2^2$	$x_5 = x_1 x_2$	y	$\hat{y}$ (without Lasso)	$\hat{y}$ (without Lasso)
1	0	1	0	1	0	0	0	0
2	1	0	1	0	0	0	0	0
3	$-1/\sqrt{2}$	$-1/\sqrt{2}$	1/2	1/2	1/2	0	0	0
4	0	4	0	16	0	1	1	1
5	4	0	16	0	0	1	1	1
6	$-2\sqrt{2}$	$-2\sqrt{2}$	8	8	8	1	1	1

7.3.4 Decision Trees and Random Forests

The basic concepts behind decision trees and random forests for classification problems are the same as the ones for the function approximation problems except that choices are made to serve classification. In function approximation problems, the choices of variables and split points are towards decreasing the error; however, in classification, different metrics such as Gini index are used. Consider a binary classification problem. As discussed in the function approximation chapter, each of the nodes in the tree can be thought of as holding a subset of data from the original dataset. We also described that to understand the tree development procedure, one needs to understand how one chooses a variable to branch on and the split point to use. These choices are made such that some objective function is maximized or minimized.

Let us describe the use of the Gini metric for these choices. Consider the jth node at the kth level of the tree. The data that is assigned to this node is represented by the set S_j^k. Let us assume that $n(j,k)$ datapoints are in this set. We now need to identify a variable and a split point so that two nodes at the next level characterized by sets S_l^{k+1} and S_m^{k+1} are identified. The objective that drives this identification is the Gini split index (GSI).

To calculate the Gini split index, we first define a Gini impurity for a node. Let us define $(f_i(j,k))$ as the fraction of datapoints in S_j^k that belong to Class-i. Now, a general definition of Gini impurity for a node in a multi-class (K classes) problem is the following

$$GI(j,k) = 1 - \sum_{i=1}^{K} f_i(j,k)^2 \tag{7.37}$$

Clearly, when a node has data from only one class then $GI = 0$ and the node is a pure node. Now for any choice of a variable to branch on and a split point, two children nodes (S_l^{k+1} and S_m^{k+1}) as shown in Figure 7.31 can be generated. A Gini split index GSI can be defined as below

$$GSI = GI(j,k) - \frac{n(l,k+1)}{n(l,k+1) + n(m,k+1)} GI(l,k+1)$$
$$- \frac{n(m,k+1)}{n(l,k+1) + n(m,k+1)} GI(m,k+1) \tag{7.38}$$

The variable and the split point that maximizes GSI is selected. From equation 7.38, it is clear that the GSI will always be less than GI. Since the GI for pure sets are zero, whenever the children nodes are purer, the terms that are subtracted in equation 7.38 will be small. Nodes that are pure are perfect from a classification viewpoint (because they categorically belong to a single class). Hence algorithms are designed to make choices that maximize the GSI when branching from a node. The process of growing the tree is continued until certain accuracy requirements are met and/or until the complexity of the tree becomes too much.

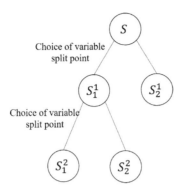

FIGURE 7.31 Decision tree for classification.

Similarly, entropy is another metric commonly used for selecting the best split point. Entropy for any node S_j^k is defined as $E(j,k) = -\sum_{i=1}^{K} p_i \log_2 p_i$ where $p_i = f_i(j,k)$. Let $E(l, k+1)$ and $E(m, k+1)$ be the entropies of the child nodes for a split point "s", then, weighted entropy of the split is defined as

$$E_w(s) = \frac{n_l}{n_l + n_m} E(l, k+1) + \frac{n_m}{n_l + n_m} E(m, k+1) \qquad (7.39)$$

where $n_l = n(l, k+1)$ and $n_m = n(m, k+1)$. Information gain (IG) is defined as the entropy gained due to the split, $IG(s) = E(j,k) - E_w(s)$. Higher the gain, better the split.

Random forests for classification problems follow the same notions of generating multiple datasets for multiple decision trees to be generated. When a new datapoint is presented to a random forest model, the decision trees that make up the model return their classification results. The aggregate random forest model prediction could be based on simple majority voting or through more sophisticated probability weighted voting mechanisms.

Example 7.40. *Consider the following data*

Sample No.	1	2	3	4	5	6	7	8
Hours studied, x	10.5	2.5	14	8.2	10.4	5	6.7	14.7
Pass (1)/Fail(0)	1	0	1	1	1	0	0	1

Find the optimal split point if a single node decision tree is used.

Solution 7.40. *Any split point "s" divides the data into two sets, S_1 and S_2. Let S_1 be the set of all points having $x \leq s$ and S_2 be the set of all points having $x > s$.*

a) Using Gini impurity:
For the given data, there are three samples of Class-0 and five samples of Class-1. Thus, the fraction corresponding to Class-0 is $f_0 = 3/8$ and fraction

corresponding to Class-1 is $f_1 = 5/8$. Gini impurity for the given data is $GI = 1 - (3/8)^2 - (5/8)^2 = 0.4688$.

Consider $x = 10.5$ as the split point. Then, S_1 contains samples 1, 2, 4, 5, 6, and 7, and the corresponding y values are 1, 0, 1, 1, 0, and 0. Since there are three samples of Class-0 and Class-1, the fraction of samples with Class-0 and Class-1 in S_1 are equal to 0.5.

$$f_{0,S_1} = 3/6 = 0.5; \quad f_{1,S_1} = 3/6 = 0.5$$

Hence, GI for S_1 is $GI_{S_1} = 1 - 0.5^2 - 0.5^2 = 0.5$.

Similarly, S_2 contains samples 3 and 8. Both samples in S_2 are from Class-1. Hence, S_2 is a pure node and $GI_{S_2} = 0$. Gini split index for 10.5 as the split point is given by,

$$GSI(10.5) = 0.4688 - (6/8) \times 0.5 - 0 = 0.0938$$

Similarly, GSI was calculated using other samples as split points. Results are tabulated in Table 7.9. Among all the cases, the case with 6.7 as split point has the highest GSI and thus is the best split point.

b) Using entropy:

For the given data set, there are three samples of Class-0 and five samples of Class-1. Thus, the probability corresponding to Class-0 is $p_0 = 3/8$ and probability corresponding to Class-1 is $p_1 = 5/8$. Entropy for the given data is $E = -(3/8)\log_2(3/8) - (5/8)\log_2(5/8) = 0.9544$.

Consider $x = 10.5$ as the split point. Then, S_1 contains samples 1, 2, 4, 5, 6, and 7, and the corresponding y values are 1, 0, 1, 1, 0, and 0. Since there are three samples of Class-0 and Class-1, the probability of samples with Class-0 and Class-1 in S_1 are equal to 0.5.

$$p_{0,S_1} = 3/6 = 0.5; \quad p_{1,S_1} = 3/6 = 0.5$$

Hence, entropy for S_1 is $E_{S_1} = -0.5\log_2(0.5) - 0.5\log_2(0.5) = 1$.

Similarly, S_2 contains samples 3 and 8. Both samples in S_2 are from Class-1. Hence, S_2 is a pure node and $E_{S_2} = 0$. Weighted entropy with 10.5 as the split point $E_w(10.5) = 6/8 \times 1 + 0 = 0.75$. The corresponding information gain (IG) is given by

$$IG(10.5) = 0.9544 - 0.75 = 0.2044$$

Similarly, IG was calculated using other samples as split points. Results are tabulated in Table 7.9. Among all the cases, the case with 6.7 as the split point has the highest IG with both children nodes being pure and thus, is the best split point. Based on split point 6.7, the prediction for S_1 would be "Fail" and that for S_2 would be "Pass" based on majority voting.

TABLE 7.9

Decision Trees for Classification – Calculations for Example 7.40

Split point (s)	Samples in S_1	Samples in S_2	Gini Impurity			Entropy		
			GI_{S_1}	GI_{S_2}	GSI	E_{S_1}	E_{S_2}	IG
10.5	1,2,4,5,6,7	3,8	0.5	0	0.094	1	0	0.2044
2.5	2	1,3,4,5,6,7,8	0	0.408	0.112	0	0.863	0.199
14	1,2,3,4,5,6,7	8	0.49	0	0.04	0.985	0	0.092
8.2	2,4,6,7	1,3,5,8	0.375	0	0.28	0.811	0	0.55
10.4	2,4,5,6,7	1,3,8	0.48	0	0.169	0.971	0	0.348
5	2,6	1,3,4,5,7,8	0	0.28	0.26	0	0.65	0.47
6.7	2,6,7	1,3,4,5,8	0	0	0.469	0	0	0.954
14.7	1,2,3,4,5,6,7,8	–	0.469	–	0	0.954	–	–

Example 7.41. *Consider the data given below containing the health status of 9 individuals categorised according to their age and the average distance travelled by them in cycle. Build a decision tree for determining the health status of an individual of age 40 and average cycling distance 15 km. Comment on the performance of the classification using both Gini impurity index and entropy measure.*

Age	22	38	26	35	35	54	9	27	14
km	7.25	71.2833	17.925	53.1	8.05	51.8625	21.075	11.1333	30.0708
Health status	0	1	1	1	0	0	0	1	1

Solution 7.41. *a) Gini Impurity: Decision tree classifier with Gini metric was built using inbuilt packages available. The optimal decision tree obtained is shown below:*

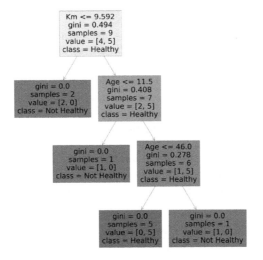

For the given individual's data with age $= 40$ and average cycling distance $= 15$ km, health status can be predicted using the above decision tree. In the first node, since $15 > 9.592$, the data is passed on to the right node where the age is compared with the splitting criteria $Age \leq 11.5$. Since age=40, the data is passed on to the next right node where the splitting criteria is satisfied as $40 < 46$ and the data sample is grouped into the left leaf node. Since the class corresponding to the left leaf node is Healthy, the individual is classified as belonging to Class-1 with Health Status 1.

b) Entropy
The optimal decision tree classifier based on entropy metric is given below:

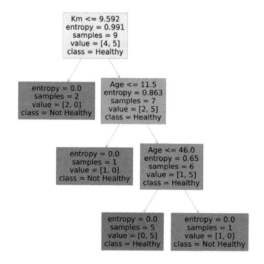

For the given individual's data with age $= 40$ and average cycling distance $= 15km$, using the above decision tree, health status is predicted as "Healthy".

We observe that both Gini impurity index and entropy measure yielded similar trees and outcomes for this example. However, in general, this may not be true. Further, it is to be noted that the outcome predicted by decision tree is sensitive to depth of the tree, the initial splitting criteria of the root node, tree structure and the input features used for prediction.

Example 7.42. *Consider the decision tree given in Figure 7.32 to determine the survival status of passengers in the fateful Titanic journey. Calculate accuracy of this decision tree model in predicting survival status of the two passengers below:*

Passenger class	Gender	Ticket Fare	Survival status
1	male	50.000	0
3	female	22.025	1
2	Male	20.192	0

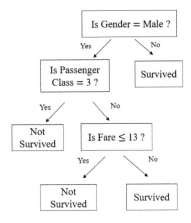

FIGURE 7.32 Decision tree for Example 7.43.

Solution 7.42. *For passenger 1, gender is male, passenger class is not 3 and the ticket fare is greater than 13. Hence, the passenger would belong to the 3rd leaf node (from left) and thus, the predicted survival status would be "Survived". Since the second passenger is female, the sample belongs to the last leaf node and hence, the predicted class would be "Survived". As the third sample is a male traveling in Class-2 with fare greater than 13, the predicted class is 0.*

$$Accuracy = \frac{No.\ of\ correct\ predictions}{No.\ of\ samples} = \frac{1}{3} = 0.33$$

Example 7.43. *Consider the data given below containing the details and survival status of 7 passenger in the fateful Titanic journey.*

Passenger Class	2	3	1	3	3	2	2
Gender	Male	Female	Male	Male	Male	Female	Male
Ticket Fare	39	15.5	26.55	15.5	7.9	13	13
Survived	1	1	1	0	0	1	0

Find the optimal decision tree based on Gini impurity. Test the performance of the optimal tree on the following test data.

Passenger class	Gender	Ticket Fare	Survival status
1	male	50.000	0
3	female	22.025	1
2	Male	20.192	0

Solution 7.43. *The optimal decision tree obtained using inbuilt packages using the given training data is shown in Figure 7.33. Note that the Gini impurity for all the leaf nodes are 0 indicating that all of them are pure nodes and thus, a good classification is obtained for the training data using this decision tree. The accuracy for the training set is 100%.*

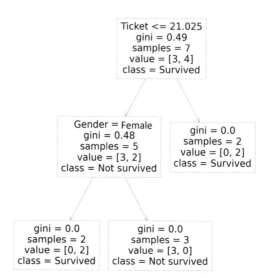

FIGURE 7.33 Optimal decision tree for Example 7.43.

For the first two test samples, the ticket prices are greater than 21.025 and hence these samples belong to the last leaf node and the predicted class is "Survived" for both samples. Since the third sample is a male traveling with ticket price less than 21, the predicted class is "Not Survived". The accuracy of the model is

$$Accuracy = \frac{No.\ of\ correct\ predictions}{No.\ of\ samples} = \frac{2}{3} = 0.67$$

Example 7.44. *Consider the medical test report used for diagnosis of severity level (Severity I (Class-0) and Severity II (Class-1) of a disease given below.*

Sl. No	Haemoglobin	Bilirubin	CPK	ALT	Severity
1.	13.17	18.66	85.98	534.6	Severity I
2.	13.49	22.3	86.91	561	Severity II
3.	10.95	21.35	71.9	371.1	Severity I
4.	13.64	16.34	87.21	571.8	Severity II
5.	18.65	17.6	123.7	1076	Severity I
6.	19.07	24.81	128.3	1104	Severity I
7.	8.196	16.84	51.71	201.9	Severity II
8.	19.21	18.57	125.5	1152	Severity I
9.	11.52	18.75	73.34	409	Severity II
10.	11.76	21.6	74.72	427.9	Severity II

Build a random forest classifier with feature selection (no data partition) for classifying the severity level of the disease. Use 5 decision trees with "Gini

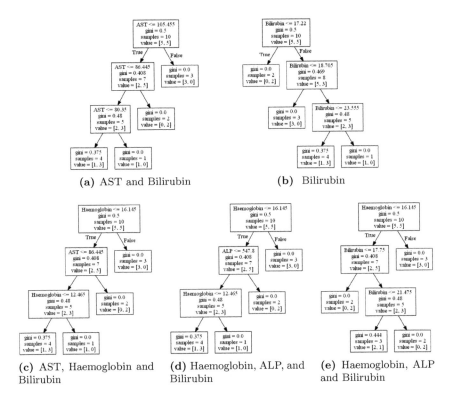

(a) AST and Bilirubin

(b) Bilirubin

(c) AST, Haemoglobin and Bilirubin

(d) Haemoglobin, ALP, and Bilirubin

(e) Haemoglobin, ALP and Bilirubin

FIGURE 7.34 Decision trees generated for Example 7.44. Input variables used for each decision tree are provided in the sub-captions.

impurity" as the criterion for split point. For each decision tree, set maximum depth to 3 and maximum leaf nodes to 6. Use the model built to comment on the severity level of the following patient:

Haemoglobin	Bilirubin	AST	ALP
11.94	18.24	75.71	437.6

Solution 7.44. *To implement random forest as an ensemble of 5 decision trees with randomized feature sub-sampling, each decision tree is provided with a subset of input features and is trained using the entire training data. Among the features used as inputs, all variables need not be used in the decision process. The 5 decision trees generated are shown in Figure 7.34.*

The true severity and severity predicted using the random forest model for all training samples are provided below:

Sample No	1	2	3	4	5	6	7	8	9	10
True Severity	I	II	I	II	I	I	II	I	II	II
Predicted Severity	I	II	II	II	I	I	II	I	II	II

Note that 9 out of 10 samples are correctly identified using the developed random forest. This data can be summarized using the following confusion matrix:

		Predicted Label	
		I	*II*
True Label	*I*	*4*	*1*
	II	*0*	*5*

For the test sample with "Haemoglobin" = 11.94, "Bilirubin" = 18.24, "AST = 75.71" and "ALP = 437.6", decision trees 1, 3 and 4 predicted "Severity II" and the remaining two trees predicted "Severity I". By majority voting, the diagnosis for the corresponding patient is "Severity II".

Example 7.45. *Consider the medical test report used for diagnosis of severity level (Severity I (Class-0) and Severity II (Class-1)) of a disease given in Example 7.44. Instead of using the entire training data, use a randomly selected subset of data samples for each decision tree. Use 5 decision trees with 'Gini impurity' as the criterion for split point. For each decision tree, set maximum depth to 3 and maximum leaf nodes to 6. Use the model built to comment on the severity level of the following patient:*

Haemoglobin	Bilirubin	AST	ALP
11.94	18.24	75.71	437.6

Solution 7.45. *To implement random forest as an ensemble of 5 decision trees with randomly selected subsets of training data samples, each decision tree is provided with a subset of training samples. Gini impurity is used as the metric for the selection of the best feature and its split point for growing the trees in the random forest. The 5 decision trees generated are shown in Figure 7.35.*

TABLE 7.10

Results for Example 7.45

Decision Tree	Samples Used	Predicted Class For Test Sample
Tree 1	7, 8, 5, 1, 3, 6, 10	Severity II
Tree 2	2, 7, 8, 5, 3, 6, 4	Severity II
Tree 3	6, 3, 5, 7, 8, 2, 4, 1, 9	Severity II
Tree 4	6, 7, 2, 3, 5	Severity II
Tree 5	7, 1, 4, 6, 5	Severity I

The sample numbers used for each decision tree are provided in Table 7.10. The true severity and severity predicted using the random forest model for all training samples are provided below:

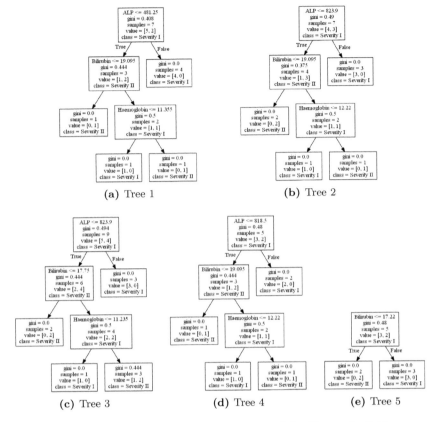

FIGURE 7.35 Decision trees generated for Example 7.45.

Sample No	1	2	3	4	5	6	7	8	9	10
True Severity	I	II	I	II	I	I	II	I	II	II
Predicted Severity	II	II	I	II	I	I	II	I	II	I

Note that 8 out of 10 samples are correctly identified using the developed random forest. This data can be summarized using the following confusion matrix:

		Predicted Label	
		I	II
True Label	I	4	1
	II	1	4

For the test sample with "Haemoglobin" = 11.94, "Bilirubin" = 18.24, "AST = 75.71" and "ALP = 437.6", the output predicted by the 5 trees are provided in Table 7.10. Since a majority of trees in the ensemble predicted the category as Severity II, the diagnosis for the patient is Severity II.

8 Conclusions and Future Directions

The fields of DS, ML, and AI are based on the three pillars of linear algebra, statistics, and optimization. After a general introduction to the terminology and important concepts in this field, three chapters described these topics from a DS/ML/AI perspective. The ML problems were categorized as two major problem types: function approximation and classification problems. One chapter was dedicated to each of the problem types, and algorithms that are used to solve these problems were described. Appreciation of these algorithms requires an understanding of the assumptions that underlie them. Hence, the key assumptions behind these algorithms were clearly articulated. Classification and function approximation algorithms can be categorized in multiple ways and these were described in detail.

The material covered in the book is vast, and the number of algorithms are growing almost on a daily basis. It is impossible to cover all the details of the algorithms in a single book. There are many other topics of importance from a data science perspective. These are: data imputation, handling class imbalance, bootstrapping, cross-validation, kernel trick/lifting to higher dimensions, boosting, analysis of complexity of models, over-parameterization, global optimization, and the notion of universal function approximation. These are topics of immediate interest to a reader who gains a reasonable understanding of the concepts covered in this book.

8.1 FUTURE DIRECTIONS

The fields covered in this book will continue to grow at a great pace in the next decade and beyond. Some topics of high topical interest currently are highlighted below.

8.1.1 Improvements in ML Techniques

Machine learning algorithms with certain desirable characteristics are likely to find industrial acceptance quickly. These are: the ability to model a wide variety of data, providing globally optimal estimates, graceful degradation, robustness, robotic ML, and computational tractability. The ability to model a wide variety of data comes from representational capability. For example, neural networks with one hidden layer have been shown to be universal function approximators. Similarly, the representational ability of any new network architecture needs to be established. While representational ability is a necessary condition, this doesn't imply that a training algorithm will deliver a

DOI: 10.1201/b23276-8

model that will represent the data to its best possible accuracy. This is because identifying the best possible representation is predicated on parameters in the model being identified to global optimality. This is very difficult to guarantee in most cases; many training algorithms provide convergence guaranties only to locally optimal solutions. Convex programming approaches and stochastic algorithms are being investigated to mitigate some of these issues.

Graceful degradation is the property that the ML solutions do not fail dramatically around edge cases, rather, any eventual failure be gradual. Robustness is the property of not having high variability in the trained solution due to minor variations in data (quality and quantity) or tuning parameters. There is also considerable interest in automating the process of problem formulation and algorithm selection. This is sometimes referred to as robotic ML. Automatic algorithm selection as a trial-and-error process is much easier to automate than problem formulation. Finally, improving computational tractability of the algorithms while guaranteeing a certain level of performance is a topic of interest.

8.1.2 Deep Learning

Deep networks have made a substantial impact in ML/AI. Deep networks try to address two traditional problems with backpropagation and fully-connected neural networks discussed in this book. The first problem is that the network structure is opaque and not amenable to further analysis or generation of insights. The second problem is that the backpropagation network may get mired in local optimum, leading to performance loss.

Deep networks have a large number of layers when compared to standard neural networks. When there are several layers, interpretations regarding how knowledge is hierarchically organized are possible. Furthermore, deep networks are not fully connected networks. There are layers for convolution, max pooling, and so on, which allow for more structured representation and interpretation. This addresses the first problem with standard neural networks to some extent. The second disadvantage is addressed by a combination of several ideas in the structure and training of these networks. In the structural part, other than the deep architecture, convolution and pooling layers described above, significant improvements in convergence results when ReLU activation functions are used. Further, layer-by-layer training through auto-encoding also helps mitigate local minima problems. Finally, as described in Chapter 4, the use of stochastic optimization approaches while training is another approach to combat local minima problems.

8.1.3 Reinforcement Learning (RL)

In all the ML approaches that were described, the key aim was to learn a model, whether it is for function approximation or classification. Models are generally used in an optimization framework to identify actions that attain certain goals. RL is an approach that allows problems to be solved gradually through an action-reward cycle. As powerful unsupervised learning algorithms are combined with efficient dynamic programming approaches in RL, a wide range of difficult problems can be solved. The biggest success of RL can be seen in the dominance of chess playing algorithms that use RL over the traditional computational chess programs. The action-reward idea is easily explained using a child and a flame example. In supervised learning, a parent might tell the child not to touch the flame because it will burn the child's fingers. However, the child will still learn not to touch the flame without any supervised learning. The child will learn by touching the flame (action) and receiving a negative reward or penalty. This action-reward cycle is at the heart of RL algorithms.

Another example is someone learning to ride a bicycle on their own. In this case, through multiple attempts, a person will learn to ride the cycle at a desired speed with appropriate tilts, speed, handle bar positions, force on the pedal, and application of breaks. This unsupervised learning happens through the same reward/penalty idea where positive actions (no fall) are reinforced, and negative actions (fall) are discouraged. Any model required for this is implicitly learnt through the action-reward cycle using actions that explore and exploit.

8.1.4   Integrating Domain Knowledge in ML/AI

We used to live in a "data-poor" and "computation-poor" world. This was because data storage systems were clunky, expensive and computational capabilities were limited a few decades back. In such a situation of limited data and computational power, mental and conceptual models were critical for extrapolation of knowledge and decision-making. The focus was on identifying simple overarching principles that govern the physical systems and developing mathematical models that capture the essence of these physical principles. These models constituted domain knowledge that could be used in extrapolating to situations where prior data does not exist. Improvements in semiconductor technologies have largely shattered the storage and computational challenges. This new data and computation tsunami has washed away the importance of first principles modeling and domain knowledge in the last decade. The thought process seems to be that everything that needs to be understood can be achieved through data alone - leading to popular catch-phrases like "data is the new oil" and so on. However, the best approaches are likely to be the ones that integrate the data-based ML approaches with domain and first principles knowledge. This is an area for further fertile research.

References

1. Ethem Alpaydin. *Introduction to machine learning*. MIT press, 2020.
2. Naomi S Altman. An introduction to kernel and nearest-neighbor nonparametric regression. *The American Statistician*, 46(3):175–185, 1992.
3. Howard Anton and Chris Rorres. *Elementary linear algebra: applications version*. John Wiley & Sons, 2013.
4. Alexander Barvinok. *A course in convexity*, volume 54. American Mathematical Soc., 2002.
5. Richard Bellman. On the theory of dynamic programming. *Proceedings of the National Academy of Sciences of the United States of America*, 38(8):716, 1952.
6. Dimitri P Bertsekas. *Convex optimization theory*. Athena Scientific Belmont, 2009.
7. George EP Box, William H Hunter, Stuart Hunter, et al. *Statistics for experimenters*, volume 664. John Wiley and sons New York, 1978.
8. Stephen Boyd, Stephen P Boyd, and Lieven Vandenberghe. *Convex optimization*. Cambridge university press, 2004.
9. Auguste Bravais. *Analyse mathematique sur les probabilites des erreurs de situation d'un point*. Impr. Royale, 1844.
10. Wayne W Daniel. The spearman rank correlation coefficient. *Biostatistics: A Foundation for Analysis in the Health Sciences*, 1987.
11. Anthony V Fiacco and Garth P McCormick. *Nonlinear programming: sequential unconstrained minimization techniques*. SIAM, 1990.
12. Roger Fletcher. *Practical methods of optimization*. John Wiley & Sons, 2013.
13. John E Freund, Irwin Miller, and Marylees Miller. *John E. Freund's Mathematical Statistics: With Applications*. Pearson Education India, 2004.
14. Michel Gendreau, Jean-Yves Potvin, et al. *Handbook of metaheuristics*, volume 2. Springer, 2010.
15. John J Grefenstette. Genetic algorithms and machine learning. In *Proceedings of the sixth annual conference on Computational learning theory*, pages 3–4, 1993.
16. SC Gupta et al. *Fundamentals of statistics*. Himalaya publishing house New Delhi, India, 2011.
17. D.O. Hebb. *The Organization of Behavior*. New York: Wiley & Sons, 1949.
18. Reiner Horst, Panos M Pardalos, and Nguyen Van Thoai. *Introduction to global optimization*. Springer Science & Business Media, 2000.
19. Richard A Johnson, Irwin Miller, and John E Freund. *Probability and statistics for engineers*, volume 2000. Pearson Education London, 2000.
20. William Karush. Minima of functions of several variables with inequalities as side constraints. *M. Sc. Dissertation. Dept. of Mathematics, Univ. of Chicago*, 1939.
21. Maurice G Kendall. A new measure of rank correlation. *Biometrika*, 30(1/2): 81–93, 1938.

22. Serkan Kiranyaz, Turker Ince, and Moncef Gabbouj. *Multidimensional particle swarm optimization for machine learning and pattern recognition.* Springer, 2014.

23. Mykel J Kochenderfer and Tim A Wheeler. *Algorithms for optimization.* Mit Press, 2019.

24. HW Kuhn and AW Tucker. Nonlinear programming, 1951.

25. Serge Lang. *Introduction to linear algebra.* Springer Science & Business Media, 2012.

26. David G Luenberger. *Optimization by vector space methods.* John Wiley & Sons, 1997.

27. J MacQueen. Classification and analysis of multivariate observations. In *5th Berkeley Symp. Math. Statist. Probability,* pages 281–297, 1967.

28. Olvi L Mangasarian. *Nonlinear programming.* SIAM, 1994.

29. Warren S McCulloch and Walter Pitts. A logical calculus of the ideas immanent in nervous activity. *The bulletin of mathematical biophysics,* 5(4):115–133, 1943.

30. Carl D Meyer. *Matrix analysis and applied linear algebra,* volume 71. Siam, 2000.

31. Mehryar Mohri, Afshin Rostamizadeh, and Ameet Talwalkar. *Foundations of machine learning.* MIT press, 2018.

32. Douglas C Montgomery, George C Runger, and Norma F Hubele. *Engineering statistics.* John Wiley & Sons, 2009.

33. William Cyrus Navidi. *Statistics for engineers and scientists.* McGraw-Hill Higher Education New York, NY, USA:, 2008.

34. Karl Pearson. Liii. on lines and planes of closest fit to systems of points in space. *The London, Edinburgh, and Dublin philosophical magazine and journal of science,* 2(11):559–572, 1901.

35. Michael A Poole and Patrick N O'Farrell. The assumptions of the linear regression model. *Transactions of the Institute of British Geographers,* pages 145–158, 1971.

36. Singiresu S Rao. *Engineering optimization: theory and practice.* John Wiley & Sons, 2019.

37. Stuart J Russell. *Artificial intelligence a modern approach.* Pearson Education, Inc., 2010.

38. John Schiller, R Srinivasan, and Murray Spiegel. *Probability and statistics.* McGraw-Hill Education, 2012.

39. Alex Schwarzenberg-Czerny. On matrix factorization and efficient least squares solution. *Astronomy and Astrophysics Supplement Series,* 110:405, 1995.

40. Charles Spearman. The proof and measurement of association between two things. 1961.

41. David Spiegelhalter. *The art of statistics: Learning from data.* Penguin UK, 2019.

42. Gilbert Strang. *Introduction to linear algebra,* volume 3. Wellesley-Cambridge Press Wellesley, MA, 1993.

43. Roman G Strongin and Yaroslav D Sergeyev. *Global optimization with non-convex constraints: Sequential and parallel algorithms,* volume 45. Springer Science & Business Media, 2013.

44. Gabor J Szekely, Maria L Rizzo, et al. Hierarchical clustering via joint between-within distances: Extending ward's minimum variance method. *Journal of classification*, 22(2):151–184, 2005.
45. Stephen Wright, Jorge Nocedal, et al. Numerical optimization. *Springer Science*, 35(67-68):7, 1999.
46. Xin-She Yang. *Optimization techniques and applications with examples*. John Wiley & Sons, 2018.
47. Jacek Zurada. *Introduction to artificial neural systems*. West Publishing Co., 1992.

Index

For Product Safety Concerns and Information please contact our EU representative GPSR@taylorandfrancis.com Taylor & Francis Verlag GmbH, Kaufingerstraße 24, 80331 München, Germany